INTERNATIONAL UNION OF CRYSTALLOGRAPHY
CRYSTALLOGRAPHIC SYMPOSIA

INTERNATIONAL UNION OF CRYSTALLOGRAPHY
CRYSTALLOGRAPHIC SYMPOSIA

Crystallographic Computing 4

TECHNIQUES AND NEW TECHNOLOGIES

Papers presented at the
International School on Crystallographic Computing
held at
The Flinders University of South Australia
Adelaide, Australia
August 22–29, 1987

Edited by

N. W. Isaacs

St. Vincent's Institute of Medical Research,
Melbourne

and

M. R. Taylor

School of Physical Sciences,
The Flinders University of South Australia
Adelaide

INTERNATIONAL UNION OF CRYSTALLOGRAPHY

OXFORD UNIVERSITY PRESS

1988

Oxford University Press, Walton Street, Oxford OX2 6DP

Oxford New York Toronto
Delhi Bombay Calcutta Madras Karachi
Petaling Jaya Singapore Hong Kong Tokyo
Nairobi Dar es Salaam Cape Town
Melbourne Auckland

and associated companies in
Beirut Berlin Ibadan Nicosia

Oxford is a trade mark of Oxford University Press

Published in the United States
by Oxford University Press, New York

© The contributors listed on pp. xi–xiii, 1988

British Library Cataloguing in Publication Data
International School on Crystallographic
Computing (1987: Adelaide)
Crystallographic computing 4: technique
and new technologies: papers presented
at the International School on Crystallographic
Computing held at the Flinders University
of South Australia, Adelaide,
Australia August 22–29, 1987.
1. Crystallography, Applications of
computer systems
I. Title   II. Isaacs, N. W.   III. Taylor,
M. R.   IV. International Union of
Crystallography   V. Series
548'028'54
ISBN 0-19-855282-3

Library of Congress Cataloging in Publication Data
Data available

Printed in Great Britain
at the University Printing House, Oxford
by David Stanford
Printer to the University

Preface

The International School on Crystallographic Computing held at The Flinders University of South Australia, Adelaide from August 22–29 1987 was the eleventh such school organized since 1960 under the auspices of the IUCr Commission on Crystallographic Computing. The school, a satellite meeting to the XIVth Congress in Perth, was the first held in the Southern Hemisphere (making it a winter school) and was attended by 108 participants from 24 countries.

In selecting the programme for the school, the organizing committee attempted to satisfy two objectives. It was felt that the school should provide instruction on techniques for younger crystallographers or for crystallographers who have not, because of geographical factors, had the opportunity to attend a previous school. At the same time, the rapid advances in computer technology demanded that these developments should be addressed. The title of this book reflects these two themes in the school.

The organization of the school was the work of two committees; a programme committee of N. W. Isaacs (Chair), V. I. Adrianov, H. D. Flack, I. E. Grey, S. R. Hall, W. A. Hendrickson, K. Huml, A. D. Rae, J. S. Rollett, H. Schenk, M. R. Taylor, D. Viterbo, and T. R. Welberry, and a local committee of M. R. Taylor (Chair), D. J. M. Bevan, J. Mohyla, A. Pring, M. Rossi, P. Self, E. Summerville, E. Tiekink, L. Vilkins, and J. Westphalen.

The school was sponsored by the IUCr Commission on Crystallographic Computing and the Society of Crystallographers in Australia. The organizers gratefully acknowledge the generous financial support provided by the International Union of Crystallography; South Australian Government; Society of Crystallographers in Australia; Digital Equipment Corporation; The Flinders University of South Australia; Australian Institute of Physics; Specialist Group for Geochemistry, Mineralogy, and Petrology; and Australian Airlines. In addition the following are thanked for the loan of computing equipment etc. Digital Equipment Corporation; Techway Ltd.; Silicon Graphics; T.C.G. Systems Automation Pty. Ltd.;

Tektronix Aust. Pty. Ltd.; and ANZ Bank.

The logo for the Computing School was inspired by the association of the Braggs with Adelaide. W. L. Bragg was born in Adelaide in 1890 and received his education at St. Peter's College, and The University of Adelaide where W. H. Bragg was Professor of Mathematics and Physics (1886–1908). It also serves to remind us of the discovery of Bragg's Law seventy-five years ago and of the beginnings of crystallographic computation.

Melbourne and   N.W.I.

Adelaide   M.R.T

October 1987

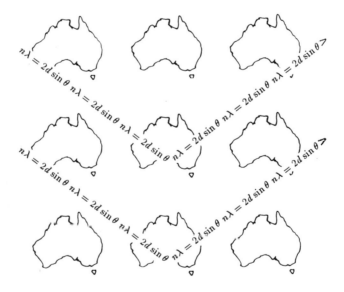

Contents

Contributors

Frank H. Allen
Crystallographic Data Centre,
University Chemical Laboratory,
Lensfield Road,
Cambridge CB2 1EW
England

Norman Badler
The University of Pennsylvania,
Philadelphia,
Pennsylvania U.S.A

G. Bergerhoff
Anorganisch–chemisches Institut
der Universität Bonn,
D–5300 Bonn
West Germany

Helen M. Berman
The Institute for Cancer Research,
The Fox Chase Cancer Center,
Philadelphia,
Pennsylvania U.S.A

Philip E. Bourne
Howard Hughes Medical Institute,
Department of Biochemistry &
Molecular Biophysics,
Columbia University,
630 W. 168th Street,
New York, N.Y. 10032 U.S.A.

Gérard Bricogne
L.U.R.E.,
Bâtiment 209D,
91405 ORSAY Cedex
France

Axel T. Brünger
The Howard Hughes Medical Insti-
tute and Department of Molecular Bio-
physics and Biochemistry,
Yale University,
P.O. Box 6666,
New Haven, CT 06511
USA

H.L. Carrell
The Institute for Cancer Research,
The Fox Chase Cancer Center,
Philadelphia,
Pennsylvania U.S.A

B.M Craven
Department of Crystallography,
University of Pittsburgh,
Pittsburgh,
PA 15260 U.S.A

John E. Davies
Crystallographic Data Centre,
University Chemical Laboratory,
Lensfield Road,
Cambridge CB2 1EW
England

E. J. Dodson
Department of Chemistry,
University of York,
Heslington, York YO1 5DD
England

H. D. Flack
Laboratoire de Cristallographie,
University of Geneva,
24 quai Ernest-Ansermet,
CH - 1211 Genève 4
Switzerland

Suzanne Fortier
Department of Chemistry,
Queen's University,
Kingston, Canada K7L 3N6

Eric Gabe
National Research Council
Canada,
Division of Chemistry,
Ottawa, Canada K1A OR6

Sydney Hall
Crystallography Centre,
The University of Western Australia,
Nedlands 6009
Australia

Wayne A. Hendrickson
Howard Hughes Medical Institute,
Department of Biochemistry &
Molecular Biophysics,
Columbia University,
New York, N.Y. 10032 U.S.A.

Karel Huml
Institute of Macromolecular Chem-
istry,
Czechoslovak Academy of Sciences,
162 06 Prague 6,
Czechoslovakia

Wolfgang Hummel
Universität Bern,
Laboratorium für chemische
und mineralogioche Kristallographie,
CH – 3012 Bern,
Switzerland

Frank J. Mannion
The Institute for Cancer Research,
The Fox Chase Cancer Center,
Philadelphia,
Pennsylvania U.S.A

R.P. Millane
Whistler Center for Carbohydrate Re-
search,
Smith Hall,
Purdue University,
West Lafayette, Indiana 47907
U.S.A.

Ward T. Robinson
Department of Chemistry,
University of Canterbury,
Christchurch 1
New Zealand

J.S. Rollett
Oxford University Computing Laboratory,
8–11 Keble Road,
Oxford OX3 9JW
England

H. Schenk
Laboratory for Crystallography,
University of Amsterdam,
Nieuwe Achtergracht 166,
1018 WV Amsterdam
The Netherlands

P.G. Self
CSIRO Division of Soils,
Private Bag No. 2,
Glen Osmond,
South Australia 5062
Australia

George M. Sheldrick
Institut für Anorganische Chemie
der Universität Göttingen,
Tammanstr. 4,
D 3400 Göttingen
W. Germany

Uri Shmueli
School of Chemistry,
Tel Aviv University,
Ramat Aviv, 69 978 Tel Aviv
Israel

Brian Skelton
Crystallography Centre,
The University of Western Australia,
Nedlands 6009
Australia

James Stewart
Chemistry Department ,
University of Maryland,
College Park,
MD 20742 U.S.A.

Robert K. Stodola
The Institute for Cancer Research,
The Fox Chase Cancer Center,
Philadelphia,
Pennsylvania U.S.A

D. Viterbo
Department of Chemistry,
University of Calabria,
87030 Arcavacata di Rende (CS)
Italy

David Watkin
Chemical Crystallography Laboratory,
University of Oxford,
9 Parks Road,
Oxford OX1 3PD
England

William P. Wood, Jr.
Smith Kline & French,
Swedeland,
Pennsylvania U.S.A

Participants in the International School on Crystallographic Computing held at
The Flinders University of South Australia, Adelaide. 22–29 August 1987 See p.xvi for the key

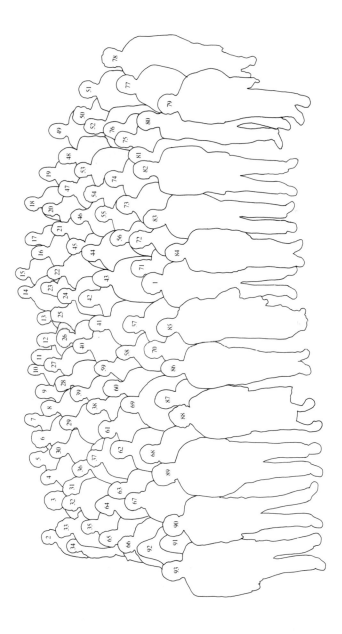

Key to photograph on p. xiv.

1. Dr Max Taylor
2. Prof. James Stewart
3. Dr Kay Fair
4. Dr John Davies
5. Dr H. L. Carrell
6. Dr Dale Tronrud
7. Dr T. R. Welberry
8. Mr Michael Taylor
9. Dr Anthony Baker
10. Dr Frank Allen
11. Prof. Bev Robertson
12. Prof. Dennis Engel
13. Prof. Bryan Craven
14. Prof. Henk Schenk
15. Dr Alpana Seal
16. Prof. Gert Kruger
17. Dr Roeli Olthof
18. Dr Rudolf Wartchow
19. Dr John Erickson
20. Dr Rolf Kruesemann
21. Dr Ward T. Robinson
22. Dr Chris Gilmore
23. Dr Suzanne Fortier
24. Dr Brian Skelton
25. Miss Alison Edwards
26. Dr Gillian Norris
27. Dr John Rollett
28. Mr Ben Bax
29. Mr Harren Jhoti
30. Dr Zygmunt Derewenda
31. Mrs Eleanor Dobson
32. Dr J. Wilson Quail
33. Dr Anthony Willis
34. Dr Lata Prasad
35. Dr Staffan Sundell
36. Mr Joseph Noonan
37. Dr A. D. Rae
38. Dr Howard Flack
39. Dr David Waller
40. Mr Edgar Weckert
41. Dr Richard Ortega
42. Mr Wolfgang Hummel
43. Prof. Rick Millane
44. Dr Peter Self
45. Miss Marie Fraser
46. Mr Ray Stevens
47. Dr Axel Brunger
48. Dr Gerard Bricogne
49. Dr Kunio Miki
50. Miyako Miki
51. Prof. Xinkan Yao
52. Dr Eric Gabe
53. Dr Kenneth Haller
54. Dr Jorge Garin
55. Dr Alan Hazell
56. Miss Nancy L. Morris
57. Dr John Orbell
58. Miss Sandra Cowan
59. Ms Norma Duke
60. Mr Zihe Rao
61. Mr Luke Guddat
62. Mr William Tulip
63. Miss Susan Hambling
64. Mrs Rita Hazell
65. Prof. Gunter Bergerhoff
66. Dr Daniel Vercauteren
67. Mr K. Sivakumar
68. Dr Philip Bourne
69. Dr Pedro Alzari
70. Dr Hendrikus Van
    Koningsveld
71. Dr Kiwako Sakabe
72. Prof. Noriyoshi Sakabe
73. Mr Robert Stodola
74. Dr Manfred Moeller
75. Dr J. K. Dattagupta
76. Ms Louise Vilkins
77. Dr Z. Baran
78. Dr Peter Seidel
79. Dr Sundar Gopal Biswas
80. Mr Erdong Wu
81. Dr Jane Burridge
82. Dr David Watkin
83. Dr Phathana Phavanantha
84. Dr Chaveng Pakawatchai
85. Dr Sukla Paul
86. Dr Virginia Pett
87. Miss Gabriele Lampert
88. Dr Miriam Rossi
89. Dr Jury Mohyla
90. Prof. Uri Shmueli
91. Ms Diana Tomchick
92. Ms Natalie Strynadka
93. Prof. Frank Herbstein

Data collection and structure solution

1

DIFFRACTOMETER DATA COLLECTION

Ward T. Robinson

The intensities of X-ray beams diffracted by single crystals have been measured using film (Friedrich, Knipping, & Laue, 1912) and ionization chambers (Bragg, 1913) with many different camera, diffractometer and goniometer head designs (Arndt and Willis, 1966). Twenty-five years ago most investigators had to measure each reflection intensity sequentially, using either a manually driven diffractometer or photometer, or by eye-ball comparison with a calibrated intensity set. Manually driven diffractometers had been used to measure neutron beam intensities since 1945. Evolution proceeded through various ingenious mechanical and electrome-chanical devices so that there were many punched card and paper tape controlled machines operating by the early 1960's. There were alo a very few facilities where diffractometers were directly on-line with the world's most powerful computers. The full potential of computer control could not be realised in off-peak hours available on these machines, however, and it is only now that dedicated personal computers with virtually un-limited processing power and memory can be attached to diffractometers, at negligible cost, that we are beginning to see anything like optimal real-isation of that potential. One consequence of the small storage that went with the early small computers, such as the 4K DEC PDP-8, was that initial programming efforts were most obviously successful in automating the simple measurements of intensity and background which are repeated thousands of times for each crystal structure analysis. Automatic intervention to check crystal orientation, if the intensities of standard reflections altered significantly, was soon seen to be particularly important in protein data collections where crystals, bathed in mother liquor, may move during diffractometry.

Several laboratories have developed original and imaginative algorithms, to facilitate other important diffractometer operations. Some of these will be outlined during this description of some data collection procedures which could be implemented on any 4-circle or equivalent diffractometer.

In our laboratory, a Nicolet R3m diffractometer, driven by a Nova 4/c computer, operates routinely to produce data for three (20-200 atom) accurate structure analyses per week. This average can only be maintained by using low temperature (-150°C) continuously and control programs which reduce essential human interruption to a minimum.

1   Preliminary Matters

Tube alignment routines should be available and used to minimise the sphere of confusion around the ideal intersection point of all the rotation axes on any diffractometer. Procedures specific to various models and monochromator configurations are supplied by manufacturers. This task can only be accomplished, competently, by experienced operators with a thorough understanding of diffraction geometry.   A binocular microscope, with a good lens to sample working distance, should be used to select the most suitable single crystal for diffractometry. Whatever orientations are achieved during data collection the chosen crystal should always lie entirely inside that cross section of the beam which has homogeneous intensity. A maximum dimension of $0.75mm$ can be handled on most commercial X-ray diffractometers but smaller dimensions may be more appropriate for heavily absorbing materials and much smaller ($< 0.1mm$) crystals can yield adequate structure solutions specially if high power X-ray sources are available.

X-radiation characteristic of a molybdenum anode is suitable for almost all small unit cell crystal structure analyses. Copper radiation can be preferable for unit cells with axes longer than 40Å, and is essential for proteins with very large cells. It is usually less satisfactory for organometallic and inorganic work.

Since diffractometer axis rotations achieve any desired orientations extra arcs on goniometer heads are unnecessary sources of mechanical instability. A single crystal is normally mounted on a non-diffracting glass fibre fixed to a metal pin held tightly in the goniometer head by a grub screw. The slow curing of adhesives used at these attachment points can cause gradual movement of the crystal, particularly during room temperature data collection experiments. This problem is rarely encountered at low temperatures where oils, liquid at room temperature, freeze with a vice-like grip on the crystal. Pins used in low temperature gas streams need to be streamlined to minimize turbulence and consequent icing at the crystal. Diffractometer geometry should be arranged so the goniometer head and mounting pin are subject to constant temperature conditions, otherwise relative intensity data will suffer from movements caused by thermal expansion and contraction of metal parts. Low temperature dry nitrogen streams provide perfectly adequate protection from moisture and oxygen attack thus largely obviating the need for capilliary mountings of air sensitive compounds.

In order to minimize Renninger effects it is best if no prominent reciprocal lattice row coincides precisely with the rotation axis of a goniometer head. The relative magnitude of absorption corrections may be reduced if long crystal dimensions do coincide approximately with this axis.

2   Optical Centering

Very great care should be taken to ensure the centroid of the crystal is at the true centre of rotation of all the diffractometer circles. Microscope cross hairs only provide frames of reference for these operations and should not be trusted as defining the true centre without careful calibration. Control programs usually rotate the crystal between settings $180°$ apart and appropriate for separate adjustment of each of three, orthogonal, goniometer head, translational movements. Adjustments are usually determined from crystal extremity positions and well developed faces, viewed edge on, can be particularly helpful during this important task.

3   Bragg Reflection Geometry

Intelligent and efficient use of diffractometers is bound to be enhanced in proportion to a user's familiarity with reciprocal lattices and their corresponding direct lattices. Understanding unit cell geometry, and the Laue symmetry characteristic of diffraction patterns, will prove essential to anybody embarking on a number of crystal structure analyses.

Some of the strategies described below can only be understood if the concepts of Bragg reflection, bisecting mode geometry, ω-scans and $\omega - 2\theta$ scans are also fully appreciated.

Location of any X-ray reflection defines one reciprocal lattice vector. An $\omega/2\theta$ scan along this vector must encounter other orders of Bragg reflections if these have detectable intensity. Location of any three noncoplanar reciprocal lattice vectors must define one cell which can enable judicious use of $\omega/2\theta$ scans to completely, accurately, and rapidly determine the true unit cell constants and orientation.

4   Locating and Centering Reflections

Reciprocal lattice points can be detected by systematic or random rotations about the instrument circle axes with the X-ray beam on. Significant signal to noise discrimination, and consequent adjustments of circle settings to optimize this ratio, are operations rapidly and accurately controlled by computer. There are algorithms to make best use of different machine features such as horiontal and vertical, or diagonal, slits. Though slits can speed up the centering process appreciably, they do not necessarily produce a more accurate result and, if not properly aligned, an

introduce systematic errors into unit cell dimension and orientation determinations.

Initially each reflection centred should be subjected to an ω-scan
and its profile plotted, on a screen or line printer. The crystal quality, indicated by its mosaic spread, may be unacceptable. Any decision to change
crystals needs to be made as quickly as possible so that time is not wasted.
Crystals with shocking profiles can give good structure analyses but the
chances of wasting difractometer, computer and scientist time should be
weighed carefully against the chemist's time required for a recrystallisation
attempt.

Immediately three non-coplanar reflections have been centred a
(3 x 3) orientation matrix can be calculated. $\omega/2\theta$ searches can then
be initiated down all independent reciprocal lattice vectors for Miller indices between -1 and +1 then for other vectors with at least one index
+ or -2. There are 49 independent vectors in this scheme and maximum
and minimum 2θ values can be chosen to match the expected diffracting power of the crystal. Starting at high Bragg angle ensures optimal
accuracy in ultimate cell constant and orientation determinations. The
continuous scan down many vectors greatly reduces the chance of finding
a unit cell which fits all the centred reflections but still has an incorrect
axis multiplicity. The high angle reflections should be centred as they are
detected and, when there are enough of them, automatically used to find
the unit cell consistent with all their centred angle values. Diffractometer
programs can carry out further tests and report other lattice symmetries
acceptable within prescribed unit cell constant tolerances. It is prudent to
check a set of reflections which should have equal intensities, on account
of Laue symmetry, before using high symmetry to limit that portion of
the reciprocal lattice covered in data collection. Symmetry can lead to
imposition of appropriate constraints on unit cell constants used in the
subsequent analysis.

Computer control can accomplish all of the tasks described so far
without interruption. A fast data collection of the $h00, 0k0$ and $00l$ re-

flections will reveal suitable reflections for use as checks on physical and chemical stability during data collection. Similarly fast scans at half wavelength, of all reflections with $+1 \leq h, k, l \leq +3$ can reveal axis multiplicity errors. If significant intensity is detected at any reciprocal lattice point with any Miller index odd the corresponding dffractometer axis angles should be included in new cell determination calculations.

5   Data Collection Strategies

Prudent investigators plot strong low and high angle reflection profiles, in three roughly orthogonal directions and using the proposed data collection scan mode, to determine a safe, minimum, adequate scan range for every intensity measurement. The modes most often used are ω-scans, $\omega/2\theta$ scans and peak top scans. Diffractometer control software and hardware permits varying degrees of on-line data processing including variable scan range measurements, full profile fitting and regular monitoring of the intensities of check reflections.

The scan speed selected for data collection should be entirely appropriate to the goal of each structure analysis. Thus routine conformational analysis, where very accurate molecular dimensions are not required, can be accomplished at low temperature using very high speeds ($59°/m$) (Hope and Nichols, 1981) whereas electron density determinations may require very slow speeds ($< 1°/m$). Aside from scientific criteria scan speeds should always be slow enough that diffractometers are never idle at night.

Most control programs proceed between adjacent reflections on reciprocal lattice vectors parallel to the axial vectors. This approach can be far from optimal with respect to minimizing wear and tear on the drive mechanisms and often results in unproductive use of machine time. If it is possible to compute a significantly better sequence for the Miller indices, and store it in the controlling computer hardware, real benefits can accrue. There is no simple solution to this huge three-dimensional travelling salesman probem. It is best tackled on another computer, off-line from the diffractometer, during the time (1-2 hours) usually taken to centre high angle reflections. Some time can also be saved by omitting systematically absent reflections, where there is no chance of space group ambiguity. Other strategies such as omitting reflections showing weak pre-scans are useful in some situations.

6   Absorption Corrections

The dimensions of a crystal should be measured using an accurately calibrated optical micrometer or graticule as early as possible in an analysis, in case a crystal is lost. Once the unit cell and orientation are established but not necessarily refined, the crystal's faces can be indexed. There are several systematic ways of doing this but the simplest is to rotate the crystal so that one of its real lattice axes coincides with the optic axis of the diffractometer telescope. The controlling computer can then print out a reciprocal lattice grid, properly scaled and oriented, from which Miller indices can be read for any well developed face which is edge-on in the field of view. If this is done for two out of the three real lattice axes most faces will be indexed. Those which are not should be reoriented so that their reciprocal lattice vectors are in a standard orientation, different for different instruments, for which the controlling computer can report back appropriate Miller indices.

7   Comment

It has only been possible to describe a few of the ways in which routine tasks, of central importance to most crystal structure analyses, can be facilitated by computers. There are many other useful applications both in the literature (Clegg, 1981*a, b*) and, unfortunately, unpublished.

A four-circle diffractometer is a very precise, complex and expensive instrument. It should be used efficiently to produce the best possible data in the time allocated to each structure analysis. Computers can help achieve this goal but can never compensate for inadequate understanding of the experiment.

REFERENCES

Arndt, U. W. and Willis, B. T. M., (1966). *Single Crystal Diffractometry*, Cambridge University Press.

Bragg, W. H., (1913). *Nature*, 91.

Clegg, W., (1981*a*). *Acta Cryst.*, **A37**, 22–28.

Clegg, W., (1981*b*). *Acta Cryst.*, **A37**, 437.

Friedrich, W., Knipping, P. and von Laue, M. (1912). *Proc. Bavarian Acad. Sci.*, 303.

Hope, H. and Nichols, B. G., (1981). *Acta Cryst.*, **B37**, 158-161.

2

DIRECT METHODS PROCEDURES FOR THE SOLUTION OF SMALL AND MEDIUM SIZED MOLECULES

D. Viterbo

1 Introduction

The development of practical direct methods has seen a rapid growth in the past 15 years. Several procedures have been set up as computer programs capable of solving crystal structures of increasing complexity.

The most popular program is certainly MULTAN (Main *et al.*, 1980), which is a highly optimized way of using triple-phase invariants (triplets or Σ_2 relations)

$$T = \phi_{-\vec{h}} + \phi_{\vec{k}} + \phi_{\vec{h}-\vec{k}} \simeq 0 \tag{1}$$

Following Cochran (1955), their values are estimated to be centered around zero with a variance decreasing with increasing values of the concentration parameter

$$C_{\vec{h}\vec{k}} = \frac{1}{\sqrt{N}} |E_{\vec{h}} E_{\vec{k}} E_{\vec{h}-\vec{k}}| \tag{2}$$

where N is the number of atoms in the unit cell.

The following steps are common to most phasing procedures:

1. Calculation of the normalized structure factors |E| from the observed magnitudes;

2. Setting up of triplets relating reflexions with large | E| (Σ_2 listing);

3. Selection of a limited number of reflexions, the phases of which are assigned *a priori* (<u>starting set</u> including origin and enantiomorph fixing reflexions);

4. Phase extension and refinement;

5. Selection of 'best' phase set(s) and calculation of E-map(s).

When the *Symbolic Addition* process is used in step 3, symbolic values are assigned to the unknown starting set phases, while in the *Multisolution* approach these phases are assigned permuted numerical values.

Despite the enormous successes of direct methods, it is common experience that a certain percentage of structures still resist all attempts at being solved. The most important factor is certainly the number of atoms in the unit cell which appears in (2): as N increases C decreases and the variance of the zero estimate for the triplets increases. But complexity is not the only factor and difficulties are often encountered when the structure has a very regular geometrical pattern and/or shows superstructure effects. In fact in these cases, the assumption that the atomic position vectors form a set of random variables breaks down and the probability formulae loose their validity, unless these factors are explicitly accounted for.

In the last years efforts have been made to increase the power of direct methods, following two main directions:

(i) An improvement of the techniques mainly based on triple-phase relationships.

(ii) Use also of other structure invariant (s.i.) and seminvariant (s.s.) phase relationships and development of new probability formulae for an improved estimate of their values.

Both directions are pursued in parallel and very often they are combined to give important contributions to the strengthening of direct methods.

In this lecture these developments and the computational procedures which have been set up, will be analyzed with particular emphasis on multisolution techniques. The scope of this analysis is to give an idea of the different procedures, so that one may select those which are more appropriate for the structural problem at hand. Often troublesome structures may be solved by some programs and not by others. It is therefore a recommended strategy to try different procedures in an intelligent way.

2 Developments of techniques based on triplets

Use of magic integers — It is obvious that the larger is the number of phases in the starting set, the greater are the chances of avoiding 'weak links' in the phase expansion process and therefore of solving the structure. Unfortunately, in a multisolution procedure, any increase of the starting set results in a large increase of the number of permutations to be expanded and the computing time soon exceeds any practical value. In order to reduce this effect White and Woolfson (1975) indicated the possibility of representing a limited number n of phases by means of a single variable x, through relations of the type where the m_i's are the 'magic integers'. Given a sequence of n integers and the appropriate value of x, (3) will give approximate values of the n phases and the error with which the phases are represented is related to the values chosen for the integers m_i.

Later Main (1977,1978) gave the mathematical basis of this idea and showed how to define the best sequences of magic integers both in terms of minimum mean phase-error and of even distribution of the errros among the represented phases. He also indicated a practical way of using magic integers to reduce considerably the number of permutations in the multisolution process (the different permutations are obtained by exploring x in the range 0–2π with steps Δx of the same order of the r.m.s. deviation of the magic integer representation). The procedure was soon incorporated in MULTAN.

The use of fairly large sets of phases (up to 50–60 phases), represented in terms of magic integers is at the basis of two multisolution procedures: MAGIC developed by Declercq, Germain and Woolfson (1975) and MAGEX developed by Hull, Viterbo, Woolfson and Zhang (1981).

Approximately 10 phases (*primary* phases, P) are expressed in terms of magic integers and then all strong triplets relating two primary phases with a new phase (*secondary* phases, S) are set up. Their form will be

$$\phi_S \simeq \phi_{P1} + \phi_{P2} + b \simeq (m_1 + m_2)x + b \tag{4}$$

where b is a constant angle arising from translational symmetry; thus S will also be expressed in terms of magic integers. In MAGEX a procedure has been developed to select the P reflexions in such a way that the error accumulation in the definition of the S-phases by (4) is minimized. The best values of x are found as maxima in the ψ-map: this is a Fourier series obtained as the weighted (using the C values as weights) sum of the cosines of all triplets relating P and S reflexions, i.e.

$$\psi(x) = \sum_t C_t \cos \left(M_t + B_t \right) \tag{5}$$

where M_t is a combination of magic integers and B_t a constant angle. Since the cosines of triplets with large C are expected to be close to $+1$, we will select the values of x which maximize $\psi(x)$. In a typical run, approximatly hundred values of x, corresponding to the highest peaks in the ψ-map, are selected giving rise to an equal number of phase sets to be expanded and refined by the tangent formula.

Use of random phases — In testing a more stable and robust alternative to the tangent formula, Baggio, Woolfson, Declercq and Germain (1978) proposed a least-squares solution of the overdetermined system of linear equations formed by the set of available triplets. In order to use this approach it is first necessary to overcome the 2π ambiguity. In fact triplets must be given as $T = 2\pi n$ without reducing the right-hand term modulo 2π and an initial determination of the values of n requires an initial estimate of the phases (the old 'phase problem' coming in in a disguised way). While testing the radius of convergence of the method, starting from phases with increasing mean error, it was found that it is so large that the process can converge to the correct solution even when assigning random values to the initial phases.

This was the beginning of the very fruitful *random approach* to the phase problem. YZARC was the first computer program based on this idea as described in the same paper by Baggio *et al.* (1978).

Later Yao Jia-Xing (1981) proved that a weighted tangent formula

$$\tan \phi_{\bar{h}} = \frac{\sum_{\bar{k}} w_{\bar{k}} w_{\bar{h}-\bar{k}} C_{\bar{h}\bar{k}} \sin(\phi_{\bar{k}} + \phi_{\bar{h}-\bar{k}})}{\sum_{\bar{k}} w_{\bar{k}} w_{\bar{h}-\bar{k}} C_{\bar{h}\bar{k}} \cos(\phi_{\bar{k}} + \phi_{\bar{h}-\bar{k}})} = \frac{T_{\bar{h}}}{B_{\bar{h}}} \tag{6}$$

is also capable of converging to the correct solution when starting from random phases and proposed the RANTAN procedure. Convergence depends on the weights assigned to the phases during the refinement and the following scheme was found to be optimum: an initial $w = 1.0$ is given to the origin and enantiomorph fixing reflexions, while $w = 0.25$ is assigned to the reflexions with random phase values; at each cycle a new weight for each reflexion is computed as a function of

$$\alpha_{\bar{h}} = (T_{\bar{h}}^2 + B_{\bar{h}}^2)^{\frac{1}{2}} \tag{7}$$

as

$$w_{\bar{h}} = \min (0.2\alpha_{\bar{h}}, 1.0); \tag{8}$$

only when the new weight is greater than 0.25 the phase is updated.

Both with YZARC and RANTAN the number of trial random starting sets of phases to be refined in order to hit the correct solution is in general quite limited (50–150).

More recently new procedures have been developed such as XMY proposed by Debaerdemaker and Woolfson (1983) or SAYTAN proposed by Debaerdemaker, Tate and Woolfson (1985). In the first the initial random phases are refined by a parameter-shift process to yield maximum values of the function

$$F = \sum_{\bar{h}} [X_{\bar{h}} - Y_{\bar{h}}] \tag{9}$$

where

$$X_{\bar{h}} = \sum_{\bar{k}} |E_{\bar{h}} E_{\bar{k}} E_{\bar{h}-\bar{k}}| \cos(\phi_{-\bar{h}} + \phi_{\bar{k}} + \phi_{\bar{h}-\bar{k}})$$

and

$$Y_{\bar{h}} = \sum_{\bar{k}} |E_{\bar{h}} E_{\bar{k}} E_{\bar{h}-\bar{k}}| \sin(\phi_{-\bar{h}} + \phi_{\bar{k}} + \phi_{\bar{h}-\bar{k}}) \qquad (10)$$

In SAYTAN random phases are refined by means of a modified tangent formula derived from Sayre's equation.

Use of structural information — Main (1975) analyzed in detail the problem of exploiting all available types of structural information in a direct method procedure. He considered the following cases:

(a) known stereochemistry of a molecule or of a molecular fragment, but unknown position and orientation;

(b) also the orientation is known;

(c) both orientation and position of a molecular fragment are known (partial structure).

In all cases appropriate normalized structure factors are computed and the estimates of the triplets are also modified. This idea was soon incorporated in MULTAN.

A very efficient procedure to handle partial structure information was developed by Beurskens *et al.* (1985): DIRDIF operates on difference structure factors phased by the partial structure, which may either be a molecular fragment or consist of one or more heavy atoms (in general or in special or even in pseudo-special positions). The input phases and amplitudes are refined by a weighted tangent procedure. If the known atoms do not define the structure uniquely, symbolic addition is used to solve the ambiguities. Connected to DIRDIF are two programs for the orientational and translational vector search of the Patterson function, which may be used to define the initial molecular fragment.

An alternative way for completing a partial structure, based on a probabilistic approach proposed by Giacovazzo (1983), has been developed by Camalli, Giacovazzo and Spagna (1985). The structure factors for the known fragment are calculated and appropriate pseudo-normalized structure factors are obtained. These are then used in the new probability formulae for the estimation of s.i.'s, derived both for centro and for non-centrosymmetric structures. On the basis of the estimated α's computed from the above formulae a *convergence/divergence* procedure will define the best starting set of phases, which will include both permuted phases and phases calculated from the partial structure. Phase expansion is performed by means of a new weighted tangent formula and the correct solution is seeked using two appropriate figures of merit (*fom 's*) $R(p)$ and $\psi_0(p)$ which are related to the traditional R_α and ψ_0 *fom 's*, and are computed using the pseudo-normalized structure factors obtained from the partial structure. The method has been successfully applied to several cases in which recovering of the crystal structure is not trivial.

An integrated direct methods and Patterson procedure (PATSEE) for the search of fragments of known geometry has been set up by Egert and Sheldrick (1985) and can be used in connection with the SHELXS system (Sheldrick, 1986).

3 Use of probabilistic estimates of different s.i.'s and s.s.'s

As we have seen the Σ_2 listing may be regarded as an overdetermined set of equations linking the unknown phases and the use of other relations than triplets is a way of increasing the power of direct methods by further increasing the number of equations. At the same time the use of more accurate probability formulae will improve the quality of the equations.

A *basic principle*, which has constituted the grounds for probabilistic direct methods since their beginning, but only in the mid 70' was formulated in a more explicit and general way, is at the basis of these developments:

> *It is possible to obtain a good estimate of s.i.'s and s.s.'s, given 'appropriate' sets of structure factor moduli, which are statistically the most effective in determining the values of the given s.i.'s or s.s.'s.*

Cochran's (1955) formula for triplets is a trivial example of this principle, as it states that the estimate $T \simeq 0$ depends on the three magnitudes $|E_{\bar{h}}|$, $|E_{\bar{k}}|$, $|E_{\bar{h}-\bar{k}}|$.

Early attempts at using quartet invariants were unsuccessful, because, as shown by Simerska (1954)

$$Q = \phi_{\bar{h}} + \phi_{\bar{k}} + \phi_{\bar{l}} + \phi_{-\bar{h}-\bar{k}-\bar{l}} \simeq 0 \tag{11}$$

with a variance depending on

$$B = \frac{1}{N}|E_{\bar{h}}E_{\bar{k}}E_{\bar{l}}E_{\bar{h}+\bar{k}+\bar{l}}| \tag{12}$$

because of the $\frac{1}{N}$ factor B can rarely assume a sufficiently large value to yield reliable indications. Later Schenk (1973 a,b) showed by an empirical procedure, that the values of Q are determined not only by the four basis magnitudes in (12), but also by the cross magnitudes $|E_{\bar{h}+\bar{k}}|$, $|E_{\bar{h}+\bar{l}}|$, $|E_{\bar{k}+\bar{l}}|$. When the cross magnitudes are all large the indication $Q \simeq 0$ is strengthened, but, rather unexpectedly, it turned out that when all three cross magnitudes are small $Q \simeq \pi$ (*negative quartets*).

Hauptman (1974) then derived the probability distribution of a quartet, given the seven basis and cross magnitudes, confirming Schenk's predictions. Subsequently Hauptman (1975,1976) explicitly formulated the above principle, indicated as the *neighborhood principle*, and gave heuristic rules for identifying the 'appropriate' magnitudes (*neighborhoods*) contributing to some s.i.'s and s.s.'s.

Giacovazzo (1977,1980) then formulated the *representation theory*, which is a systematization of the above ideas, since:

(a) It allows one to define in a very general way those structure factor moduli which are statistically the most effective in estimating s.i.'s and s.s.'s. They are the *phasing magnitudes* and are ranked in order of increasing effectiveness to form the so called *phasing shells*.

(b) It allows a full use of the space group symmetry.

Given the set of magnitudes, $\{|E|\}_1$, $\{|E|\}_2$, etc. in the first, second, etc. phasing shell of a given s.i. or s.s., Φ, such that $\{|E|\}_1 \supset \{|E|\}_2 \supset ....$, it is then possible to derive the corresponding conditional probability distributions $P(\Phi|\{|E|\}_n)$. Their accuracy will be greater the higher is the order n of the phasing shell considered. Of course, on the other hand, as the number of magnitudes increases the formulae become more and more complex.

High order s.i.'s found almost immediate application, as in the case of the use of negative quartets to compute a very powerful figure of merit (NQUEST by De Titta, Edmonds, Langs and Hauptman, 1975). As illustrated in Professor Schenk's lecture, quartets and quintets were quite soon introduced in the SIMPEL program, but only in the past few years were theoretical developments on the estimation of s.i.'s and s.s.'s incorporated in practical direct method procedures in a more systematic and efficient way.

Let us first mention MITHRIL, an integrated direct method program set up by Gilmore (1984), with the aim of collecting together most of the existing procedures and at the same time allowing the use of high order s.i.'s estimated with a variety of formulae.

However the first attempt at a systematic use of these new techniques is represented by the SIR program (Burla *et al.*, 1987). SIR stands for SemInvariant Representation, and was set up with the aim of allowing the practical use of the representation theory for estimating s.i.'s and s.s.'s in all space groups in a completely general way.

A description of the most important features of the SIR program will be given in the rest of this lecture.

4   Description of the SIR program

The program can be conveniently described in terms of the following steps:

1 - *Normalization* — The basic calculations for the evaluation of normalized structure factors are performed in a standard way.

The program includes the subroutine SYM of Burzlaff and Hountas (1982) to derive, directly from the space group symbol, all symmetry operators and from them also all the information necessary to identify the structure seminvariants (used by the following steps).

It is also possible to perform a statistical analysis of the intensities, capable of identifying, in most cases, the presence and type of pseudotranslational symmetry (Cascarano, Giacovazzo and Luić, 1985) and suitably renormalize the intensities.

Finally the atomic coordinates of a structural fragment can be input to start the completing procedure (Camalli *et al.*, 1985) previously described.

2 - *One and two-phase seminvariants* — One-phase s.s.'s are estimated by means of their second representation as described by Cascarano and Giacovazzo (1983), Cascarano *et al.* (1984a), Burla *et al.* (1980) and Giacovazzo (1978).

Two-phase s.s.'s are estimated by means of their first representation as proposed by Giacovazzo *et al.* (1979) and Cascarano *et al.* (1982).

The optimization procedure for one-phase s.s.'s (Burla *et al.*, 1981), which uses multipoles of special two-phase s.s.'s (formed by two one-phase s.s.'s) can also be employed.

All the estimated s.s.'s are stored; those evaluated with highest reliability will be used actively, while the others will be employed to compute two *fom* 's.

3 - *Triplets and quartets* — All triplets relating reflexions with $|E|$ greater than a given treshold are generated and those with the highest value of C are stored.

At the same time also triplets (ψ_0 triplets), relating two reflexions with large $|E|$ and one with $|E|$ close to zero, are generated and will be used to compute several *fom* 's.

Negative quartets are generated by combining the ψ_0 triplets in pairs, and those with cross-magnitudes smaller than a given treshold are estimated by means of their first representation, as described by Busetta *et al.* (1980). These quartets will be essentially used to compute a *fom* .

It is also possible to improve the estimates of triplets by using their second representation ($P10$ formula) according to Cascarano *et al.* (1984b). The concentration parameter of the new von Mises (i.e. of the same form of Cochran's) distribution is given by

$$G = C(1 + Q) \qquad (13)$$

where Q is a function of all the magnitudes in the second representation of the triplet. The G values are rescaled on the C values and the triplets are ranked in decreasing order of G. The top relationships represent a better selection of triplets, with value close to zero, than obtained when ranking according to C. These triplets will be actively used in the phase determination process. Triplets estimated with a negative G represent a sufficiently good selection of relationships close to π to be used for the calculation of a powerful *fom* . Triplets with G close to zero are expected to have values around $\pm\left(\frac{\pi}{2}\right)$ and are used to compute an enantiomorph sensitive *fom* . A similar *fom* is also computed using quartets estimated with a very small concentration parameter.

When a pseudotranslation has been detected by the normalization program, then the parameter C of the Cochran distribution is suitably modified taking into account the indeces of reflexions $\vec{h}, \vec{k}$ and $\vec{h} - \vec{k}$ and the type of pseudotranslation (Cascarano, Giacovazzo, and Luić 1986).

4 - *Convergence/divergence procedure* — The convergence method (Germain, Main and Woolfson, 1970) is a very convenient way of defining an optimum starting set of phases to be expanded by the tangent formula or by any other algorithm.

In the SIR program the most reliable one-phase s.s.'s are treated as known phases. Not only triplets are used, but also a few tens of two-phase s.s.'s estimated with highest accuracy. An option is open to use actively a limited number of negative quartets, but this option should be selected with great caution. Each relationship is used with its proper weight: concentration parameter of the first representation for two-phase s.s.'s and C or G for triplets.

Special care is taken in eliminating redundant relationships such as triplets formed by the combination of one and two-phase s.s.'s

Once the starting set has been defined by the convergence procedure, the best pathway for phase expansion is determined by a divergence procedure. In the divergence map, starting from the reflexions in the starting set, each new reflexion is linked to the preceding ones with the highest value of

$$< \alpha_{\vec{h}} > = \sum G \frac{I_1(G)}{I_0(G)} \tag{14}$$

where $I_0(G)$ and $I_1(G)$ are modified Bessel functions of zero and first order. The summation in (14) is over all relationships of weight G defining reflexion $\vec{h}$.

A development of this step uses a new weighting procedure (Burla *et al.*, 1987). Each reflexion is assigned a weight which also takes into account the accuracy with which the phases contributing to it have been previously determined.

5 - *Phase extension and refinement* — The process is carried out by means of the tangent formula using both triplets and two-phase s.s.'s. The starting set defined by the preceding step is usually formed by the origin (and enantiomorph) fixing reflexions, a few one-phase s.s.'s and a number of other phases to be permuted. Among these, the general phases are assigned magic integer values.

After expansion and refinement of each set of phases several *fom* 's are computed using all invariants and seminvariants estimated by means of the representation method. Their form and meaning is described in the paper by Cascarano, Giacovazzo and Viterbo (1986) together with an optimized way of combining all the computed functions to give a highly selective combined *fom* .

6 - *Calculation and interpretation of the E-maps* — The sets of phases generated by the tangent routine can be passed to the EXFFT and SEARCH programs of the MULTAN system for the calculation and interpretation of the *E*-maps.

A new feature of SIR is represented by a *random procedure* proposed by Burla, Giacovazzo and Polidori (1987). Triplets which are involved in the initial stages of phase determination, are assigned a random value distributed according to von Mises distributions with G 's as concentration parameters. The random value is then added as a phase-shift to each triplet and the phase expansion and refinement is performed by the tangent formula. Tests carried out on several crystal structures show that the procedure is very effective.

5   Examples of the use of the SIR program

The application of the SIR program to the solution of three crystal structures will now be illustrated in order to demonstrate its features and its versatility.

RIFOLE — It is an antibiotic, 21-acetoxy-11-(R)- rifamycinol S (Brufani *et al.*, 1986), the structure of which is shown in figure 1. It belongs to the space group $P2_1$ with two molecules, $C_{39}H_{49}NO_{13} \cdot CH_3OH \cdot H_2O$, in the unit cell. This MULTAN resistant structure was first solved by an early version of the program and is now the test structure distributed with SIR. The first solution revealed a 22 atom fragment (the planar part plus a few atoms linked to it) and was completed by Karle (1968) recycling. Later the structure was used as a test of the partial structure procedure; starting from the 8 atom fragment (11% of the total number of electrons) surrounded by a dotted line in Figure 1, the procedure is capable of yielding the complete structure.

BZDIA — The second structure is 7-chloro-2- (diethylamino-ethylthio)-4-3H,1,5-benzodiazepine, $C_{21}H_{24}N_3SCl$, space group $Pca2_1$, $Z = 8$ (Giordano, 1987) and it is shown in Figure 2. The key to the solution was the use of the $P10$ formula to obtain more reliable estimates of triplets.

FEGAS — As a third example we shall consider the structure of the polytype *2H* of the $Fe_2Ga_2S_5$ family (Cascarano, Dogguy-Smiri and Nguyen-Huy-Dung, 1987), space group $P6_3/mmc$ with $Z = 2$. This type of very regular structure is known to be troublesome for direct methods, and this particular one, shown in Figure 3, had resisted several attempts at solving it also by Patterson methods. The SIR program identified the presence of a pseudotranslational symmetry $(u = \frac{c}{3})$ and the E-map computed with the best set of phases revealed the positions of the Fe and S atoms; the structure was then easily completed by Fourier methods.

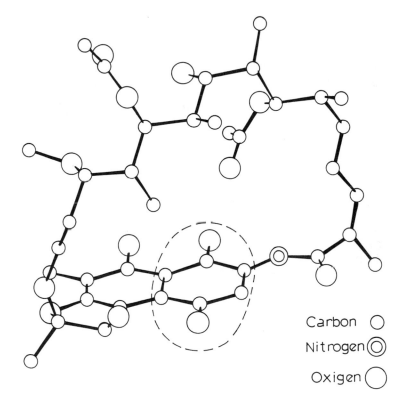

FIGURE 1 — The structure of RIFOLE solved by the SIR program. The fragment outlined by the dotted line can yield the complete molecule when input as a partial structure.

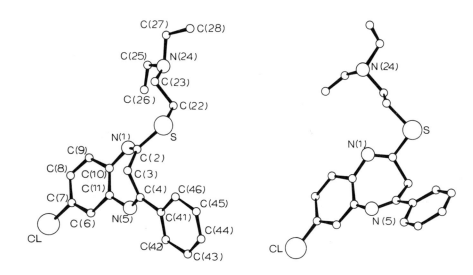

FIGURE 2 — The structure of BZDIA solved by the SIR program using triplets estimated by the *P*10 formula.

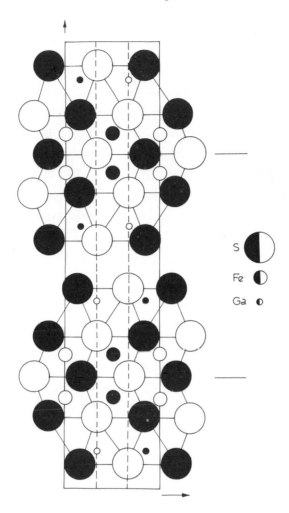

FIGURE 3 — $Fe_2Ga_2S_5, 2H$ polytype: an example of structure with pseudotranslational symmetry solved by SIR. Projection on the (110) plane

REFERENCES

Baggio R., Woolfson M. M., Declercq J. P., and Germain G. (1978). *Acta Cryst.*, **A34**, 883.

Beurskens P. T., Bosman W. P., Doesburg H. M., Gould R. O., Van den Hark Th. E. M., Prick P. A. J., Noordik J. H., Beurskens G., Parthasarathi V., Bruins Slott H. J., and Haltiwanger R. C. (1985). *Program*

System DIRDIF, Tech. Rep. 1985, Crystallography Laboratory, Univ. of Nijmegen.

Burla M. C., Camalli M., Cascarano G., Giacovazzo C., Nunzi A., Polidori G., Spagna R., and Viterbo D. (1987). *The SIR Program for the Direct Solution of Crystal Structures Using the Seminvariant Representation Method,* Univs. of Bari, Perugia, Torino and CNR Lab., Rome.

Burla M. C., Cascarano G., Giacovazzo C., Nunzi A., and Polidori G. (1987). *Acta Cryst.,* in press.

Burla M. C., Giacovazzo C., and Polidori G. (1987). *Acta Cryst.,* in press.

Burla M. C., Nunzi A., Giacovazzo C., and Polidori G. (1981). *Acta Cryst.,* **A37**, 677.

Burla M. C., Nunzi A., Polidori G., Busetta B., and Giacovazzo C. (1980). *Acta Cryst.,* **A36**, 573.

Burzlaff H., and Hountas A. (1982). *J. Appl. Cryst.,* **15**, 464.

Busetta B., Giacovazzo C., Burla M. C., Nunzi A., Polidori G., and Viterbo D. (1980). *Acta Cryst.,* **A36**, 68.

Camalli M., Giacovazzo C., and Spagna R. (1985). *Acta Cryst.,* **A41**, 605.

Cascarano G., Dogguy-Smiri, and Nguyen-Huy-Dung (1987). Submitted.

Cascarano G., and Giacovazzo C. (1983). *Z. Kristallogr.,* **165**, 169.

Cascarano G., Giacovazzo C., Calabrese G., Burla M.C., Nunzi A., Polidori G., and Viterbo D. (1984). *Z. Kristallogr.,* **167**, 34.

Cascarano G., Giacovazzo C., Camalli M., Spagna R., Burla M. C., Nunzi A., and Polidori G. (1984). *Acta Cryst.,* **A40**, 278.

Cascarano G., Giacovazzo C., and Luić M. (1985). *Acta Cryst.,* **A41**, 544.

Cascarano G., Giacovazzo C., and Luić M. (1987). *Acta Cryst.,* **A43**, 14.

Cascarano G., Giacovazzo C., Polidori G., Spagna R., and Viterbo D. (1982). *Acta Cryst.,* **A38**, 663.

Cascarano G., Giacovazzo C., and Viterbo D. (1987). *Acta Cryst.,* **A43**, 22.

Cochran W. (1955). *Acta Cryst.,* **8**, 473.

Debaerdemaeker T., and Woolfson M. M. (1983). *Acta Cryst.,* **A39**, 193.

Debaerdemaeker T., Tate C., and Woolfson M. M. (1985). *Acta Cryst.*, **A41**, 286.

Declercq J. P., Germain G., and Woolfson M. M. (1975). *Acta Cryst.*, **A31**, 367.

De Titta G. T., Edmonds J. W., Langs D. A., and Hauptman H. (1975). *Acta Cryst.*, **A31**, 472.

Egert E., and Sheldrick G. M. (1985). *Acta Cryst.*, **A41**, 262.

Germain G., Main P., and Woolfson M. M. (1970). *Acta Cryst.*, **B26**, 274.

Giacovazzo C. (1977). *Acta Cryst.*, **A33**, 933.

Giacovazzo C. (1978). *Acta Cryst.*, **A34**, 562.

Giacovazzo C. (1980). *Acta Cryst.*, **A36**, 362.

Giacovazzo C. (1983). *Acta Cryst.*, **A39**, 685.

Giacovazzo C., Spagna R., Vicković I., and Viterbo D. (1979). *Acta Cryst.*, **A35**, 401.

Gilmore C. J. (1984). *J. Appl. Cryst.*, **17**, 42.

Giordano F. (1987) personal communication.

Hauptman H. (1974). *Acta Cryst.*, **A30**, 472.

Hauptman H. (1975). *Acta Cryst.*, **A31**, 680.

Hauptman H. (1976). *Acta Cryst.*, **A32**, 934.

Hull S. E., Viterbo D., Woolfson M. M., and Zhang Shao-Hui (1981) *Acta Cryst.*, **A37**, 566.

Karle J. (1968). *Acta Cryst.*, **B24**, 182.

Main P. (1976). *Crystallographic Computing Techniques*, Munksgaard, Copenhagen, p. 97.

Main P. (1977). *Acta Cryst.*, **A33**, 750.

Main P. (1978). *Acta Cryst.*, **A34**, 31.

Main P., Fiske S. J., Hull S. E., Lessinger L., Germain G., Declercq J. P., and Woolfson M. M. (1980). - *MULTAN80, a System of Computer Programs for the Automatic Solution of Crystal Structures from X-Ray Diffraction Data* , Univ. of York, England.

Sheldrick G. M. (1986). SHELXS-86 in *Crystallographic Computing* , Oxford University Press, p. 175.

Schenk H. (1973a). *Acta Cryst.*, **A29**, 77.

Schenk H. (1973b). *Acta Cryst.*, **A29**, 480.

Simerska M. (1956). *Czech. J. Phys.*, **6**, 1.

White P. S., and Woolfson M. M. (1975). *Acta Cryst.*, **A31** , 53.

Yao Jia-Xing (1981). *Acta Cryst.*, **A37**, 642.

3

THE PRACTICAL USE OF HIGHER INVARIANTS

H. Schenk

1   Introduction.

In present direct-method procedures higher invariants play a more and more significant role. They serve in nearly all stages of the phase determination from the starting set procedures to the determination of the correct solution by means of figures of merit, and also in extension and refinement processes.

Higher invariants are known from the beginning of direct methods. One of the Sigma-relations of Hauptman and Karle (1953) is at present known as the quartet relation

$$\phi_4 = \phi_H + \phi_K + \phi_L + \phi_{-H-K-L} \tag{1}$$

Because the indices of the four main reflections H, K, L, and -H-K-L sum to zero ϕ_4 is a structure invariant, and for the reason that it involves more than three reflections it is called a higher invariant.

However, to estimate ϕ_4, Hauptman and Karle considered only the $|E|$'s of the reflections H, K, L and -H-K-L to be important, which makes this relation inferior compared with the triplet relation.

The first useful approach to higher invariants was invented some 20 years later when the $|E|$ magnitudes of the seven reflections H, K, L, -H-K-L, H+K, H+L and K+L were used to estimate the quartet phase sum ϕ_4 (Schenk, 1973). It was shown that, when the cross terms H+K,

H+L, and K+L are used in addition to the main terms, more reliable estimates of ϕ_4 are obtained. Later, these empirical observations were theoretically founded and also other higher invariants (such as quintets) and seminvariants were investigated. The latter are generally estimated via higher invariants.

It is the purpose of this paper to present some applications of higher invariants. Higher invariants are in particular useful when the normal direct methods fail due to the symmetry of the space group.

2   Symmetry and Induced Problems.

In principle a Direct-Method application should lead directly to the image of the structure. In practice, however, quite a few structures remain un-solved. This may have several reasons, such as space group symmetry, size of the structure and special features in the atomic positions. The effect of symmetry is the most serious reason and will be dealt with here in some more detail. It can easily be seen that the trivial set of phases with all $\phi_H = 0$ fulfills all triplet relations

$$\phi_H + \phi_K + \phi_{-H-K} = 0 \tag{2}$$

This implies that in unfavourable cases no solution of the phase prob-lem will be found. Fortunately, in many cases the space group symmetry adds constraints to the phases. e.g. in space group $P2_1/c$ the phase con-straints $\phi_{hkl} = 0$ or π, $\phi_{hkl} = \phi_{\overline{hkl}} = \phi_{\overline{h}k\overline{l}}$ for $k+l = 2n$ and $\phi_{hkl} = \phi_{\overline{hkl}} = \pi + \phi_{\overline{h}k\overline{l}}$ for $k+l = 2n+1$ follow from the symmetry and notably the latter condition implies that the set of all phases $\phi_H = 0$ is no longer a solution of the relations (2). In practice, in this space group the most consistent solution under (2) gives nearly always a correct image of the structure.

However, also in space group $P\overline{1}$, where no symmetry constraints are present, the triplet relations (2) are reliable enough to carry out a phasing procedure and thus, by using a good starting set of origin- defining and symbolic phases, a phase set can be built up containing among other solutions both the trivial and the correct one. Since the triplet relations (2) tend to indicate the trivial solution as the correct one, a FOM based on these relations will not find the correct solution, and this implies that other FOM's must be used for this purpose.

In this chapter it is not possible to deal with the symmetry aspects of all space groups in detail. Nevertheless, one other problem must be mentioned, namely that of the refinement of phases in polar non- centrosymmetric space groups. As an example space group $P2_1$ is chosen, in which the origin may lie on any point of the two-fold screw axis. This implies that there is an infinite number of correct phase sets instead of just one, and they are related by the expression

$$\phi_{hkl} = a \times k + \phi_{hkl} \tag{3}$$

Now, a ϕ_H derived from a triplet (2) carrying an error, will in general be in error as well. This wrong phase ϕ_H, however, can be interpreted as being correct with respect to another origin or to the enantiomorph of the structure. As a result, solutions of the phase using (2) often consist of several partial structures related to different origins of both enantiomorphs. Because of this, one finds that in space groups such as $P2_1$ application of (2) often leads to centric phases, whereas similar refinements in $P2_12_12_1$, where the origin is fixed, show stable non-centric phases (e.g. Schenk, 1982). It needs no argument that this behaviour of (1) in space group $P2_1$ induces extra problems in phase-determination processes.

In summary some final rules (Schenk, 1972) are given. There are two categories of centrosymmetric space groups: a) the symmorphic space groups (no translational symmetry, e.g. $P\bar{1}$ for which the most consistent solution under (2) is the trivial one $\phi_H = 0$ and b) the other space groups (e.g. $P2_1/c$) for which the most consistent solution usually is the correct one. For the non-centrosymmetric space groups there are three categories: c) the symmorphic space groups (e.g. $P1$), for which the relations (2) produce again the $\phi_H = 0$ solution, but also less consistent solutions of (2) mostly having centric phases; d) the polar space groups with an infinite number of possible origins (e.g. $P2_1$), for which the phasing process usually ends with centric phases and e) the other space groups (e.g. $P2_12_12_1$) with an origin restricted to a finite number of positions, and for which the relations produce the correct phases.

To overcome the above difficulties, higher invariants play an important role.

3   Probability Distributions for Phase Relationships.

Once the reflections controlling the value of the phase sum of a (sem)-invariant are identified, a conditional joint-probability distribution can be derived, describing the probability that the phase sum has a certain value, given the E's of all reflections involved. Throughout the years many scientists have been contributing to this part of direct methods, e.g. Hauptman, Karle, Cochran, Kitaigorodsky, Klug, Bertaut, Naya, Nitta, Oda, Giacovazzo and Heinerman, and, more recently, Bricogne and Peschar.

Starting point of the derivation of a joint-probability distribution is a process to mimic structures. Then, given such a structure, the formalism of probability theory is used to derive a joint-probability distribution. Finally, the conditional joint-probability distribution is found by specifying some parameters of the joint-probability distribution.

For the process of mimicking structures in general randomly chosen co-ordinates are given to all atoms. Thus, at every position in the unit cell, there is an equal chance to find an atom. Of course, for real structures this is an incorrect assumption, because atoms are situated at sites which satisfy chemical and physical rules such as chemical bonding and Van der Waals forces. Inclusion of these rules in the process of mimicing structures complicates the procedure to arrive at a conditional joint-probability distribution enormously.

In the second step probability theory is used. In general, however, approximations are invoked to make the derivation feasible. Different authors use different approximations and may thus arrive at different joint-probability distributions for the same phase relationship. This paper is not intending to give a full account of theoretical direct methods. In stead, just some recent results in the field of quartet relations are compared. After the first empirical observations concerning 7-magnitude quartets (Schenk, 1973a) theories were developed by Hauptman (1975) and Giacovazzo (1976) along the lines described above. The conditional probability distribution of Giacovazzo (1976) is given by

$$P(\phi_4, R_1, R_2, R_3, R_4, R_5, R_6, R_7) =$$
$$C^{-1} exp[-2N^{-1} R_1 R_2 R_3 R_4 (R_5^2 - 1)(R_6^2 - 1)(R_7^2 - 1)] \qquad (4)$$

and the one of Hauptman (1975) by

$$P(\phi_4, R_1, R_2, R_3, R_4, R_5, R_6, R_7) =$$
$$C^{-1} exp[-2N^{-1} R_1 R_2 R_3 R_4 cos\phi_4] \times$$
$$I_0(2N^{-1/2} R_5 Z_5) I_0(2N^{-1/2} R_6 Z_6) I_0(2N^{-1/2} R_7 Z_7) \qquad (5)$$

with

$$Z_5 = [R_1^2 R_2^2 + R_3^2 R_4^2 + 2R_1 R_2 R_3 R_4 cos\phi_4]^{1/2}$$

and similar expressions for Z_6 and Z_7. In all expressions R_1 is the random variable associated with $|E_H|$, R_2 with $|E_K|$, R_3 with $|E_L|$, R_4 with $|E_{-H-K-L}|$, R_5 with $|E_{H+K}|$, R_6 with $|E_{H+L}|$, and R_7 with $|E_{K+L}|$.

The form of expression (4) shows that the maximum of P is either found at $\phi_4 = 0$ or at $\phi_4 = \pi$, according to the values of R_5, R_6 and R_7. For large values of these cross terms $\phi_4 = 0$ and for small values the sign of the exponential part changes and consequently $\phi_4 = \pi$. The formula of Hauptman, however, is capable to predict all values in the range $0 \leq \phi_4 \leq \pi$, and has therefore a wider range of possible application. Nevertheless, also this expression has its shortcomings. In particular, the fact that in real structures the predicted ϕ_4-values show systematic errors for both the modes and the means of the conditional probability distributions is a limiting factor for successful application. The Giacovazzo- and

Hauptman- expressions have been derived neglecting higher-order terms. In order to judge whether this leads to systematic errors of the ϕ_4 estimates, in my laboratory Peschar(1986) followed a computerized approach to the joint-probability distribution. This approach makes it possible to use higher-order terms to any specified order of N. Therefore, the characteristic function C of the joint- probability distribution (jpd) is brought into a special form, and then expanded in a Taylor series. The result can be Fourier transformed term by term. In the final form of the jpd, products of the Hermite and Laguerre polynomials play an important role. In order to make this process feasible a computer program has been written which does all the hard work from both the mathematical and administrative point of view. The program derives the expression for the conditional jpd to any specified order of N and delivers the result as a series of terms on a disk file. Typically, the conditional jpd correct to the order of N^{-3} contains 1010 terms, the one correct to N^{-4} contains 9133 terms. These numbers show the impossibility to derive the expressions by hand. The fact that the pd is available as a computer disk file makes it easy to test it, and this has been done extensively.

A first test shows that convergence is normally reached using the N^{-3} expression. Only when the reflections involved are very strong the N^{-4} expression has to be used in order to ensure convergence. Another test shows that the estimates of ϕ_4 do not contain any systematic error with respect to the true phase sums, when the mean values of the N^{-3} cjpd are used, whereas the systematic errors in the case of expression (5) are as big as 15 degrees on average. Moreover, there is no systematic error variation over the whole range of $0 \le \phi_4 \le \pi$, whereas the differences in the case of (5) vary from 20 to 10 degrees for different averages. A last advantage of the new formula is its lower error level: the absolute errors of the estimated ϕ_4 with respect to the true ϕ_4 are reduced by approximately 15%. In conclusion, it has been shown from this investigation that the incorporation of higher-order terms gives rise to superior jpd's for the quartet phase relationship.

Along the same route, taking advantage of similar computer programs, also other jpd's are derived with similar results which will be published in due course.

4 Application of Higher Invariants in Practical Direct Methods.

In practical direct methods applications of estimates of phase relations
based on higher invariants vary from well-established widely used ones to
applications which have not been developed yet beyond the research stage.

Among the well-established applications are the use of quartets in the
starting-set determination and the use of figures of merit based on quartets, quintets and seminvariants. These applications are implemented in
several Direct Method programs, including our own SIMPEL program. A
number of other applications makes use of the fact that for many invariants and seminvariants absolute values of sums of phases can be estimated
which differ from 0 and π. These estimates are therefore enantiomorph
sensitive, and are used for enantiomorph-specific procedures. However,
these procedures are still in a research stage and as such only useful at
the places where they were invented.

The applications will be dealt with hereafter in more detail.

5 The Selection of a Starting Set.

The selection of a good starting set, or rather the best starting set, is
essential for a successful phase determination. Quartets can be used fruit-
fully to solve this problem because many of the strongest and reliably
estimated quartets relate the phases of the 30 to 50 strongest E-values
(Schenk, 1973), whereas the strongest triplet relations are usually found
within the group of 200 to 300 large E-values. In the program system
SIMPEL a procedure based on the mixed use of quartets and triplets of
the 30 to 60 strongest reflections is adopted and this leads in general to

a good starting set. However, a subsequent accessibility test, to check whether all phases can be reached via the triplet relations (the divergence procedure), indicates in quite a few cases that additional unknowns have to be included.

6   Figures of Merit.

As was shown before, in space group $P\bar{1}$ the most consistent solution under (2) corresponds to the set of trivial phases. Nevertheless, among the other solutions the set of correct phases may wel be present, but, of course, this set will only be found by using appropriate FOM's. FOM's, which in recent years proved to be very valuable, are those based on phase relations equal to π, e.g. the negative quartets

$$\phi_H + \phi_K + \phi_L + \phi_{-H-K-L} \approx \pi \tag{6}$$

and the special negative quartets

$$2\phi_H + \phi_{-H+K} + \phi_{-H-K} \approx \pi \tag{7}$$

When applied in space group $P\bar{1}$, these FOM's make the number of successful phase determinations as large as that reached by using the triplet consistency in spacegroup $P2_1/c$. The first FOM of this kind was HKC (Schenk and De Jong, 1973), based on the special quartets, soon followed by the so-called Negative Quartet Criterion (NQC Schenk, 1974). Later other FOM's were defined on the basis of the same relations, which give rise to similar and not better results. To get an idea of the strength of these FOM's, results of 12 structures are collected in Table 1 taken from Schenk, 1983). It can easily be seen that NQC is the best FOM of all, followed by ϕ_0 and HKC. The triplet consistency Q is not very reliable at all. With respect to the ϕ_0 -test (Cochran and Douglas, 1957) it may be remarked that it is related to the NQC FOM. In this test there are, however, some adjustable parameters. As a result, its application in routine phasing is more troublesome, in particular in symbolic addition programs. Recently we developed a FOM (Kiers and Schenk, 1984) based on the positive quartet relation:

$$\phi_H + \phi_K + \phi_L + \phi_{-H-K-L} = 0 \qquad (8)$$

Of course, in symmorphic space groups the trivial solution is indicated by this FOM as the most probable one, like the Σ_2 consistency. Still, the positive quartet criterion appears to be more reliable than the Σ_2-consistency FOM.

TABLE 1

Figures of Merit for 12 structures in space group P1.

Structural code	Number of atoms in unit cell	Number of solutions of (1)	Rank of the correct solution after FOM				
			Q	HKC	NQC	ψ_0	Σ_1
ADEN	54	32	>10	2	1	2	1
LLOYD3	108	2048	1	13	1	4	>20
FUKS	76	128	2	1	1	5	8
BAR2	40	16	13	7	1	1	8
FABRE4	104	512	1	>20	1	17	5
LAB26	46	16	8	3	1	1	10
LAB25	58	64	3	1	2	5	2
MARIN	30	64	3	1	1	2	16
SEVIL	28	16	6	11	1	1	16
VITA	44	8	7	1	1	4	7
QUINOL	80	128	3	3	1	1	>25
BORAAN	58	64	13	7	1	1	1

In the FOM columns the rank of the correct solution after the FOM in question has been given; in this connection it may be noted that in general only a few highest-ranked solutions are considered for the calculation of a Fourier summation. This means that a structure with a rank of 5 or 6 will not be solved using that particular FOM. In total 5 different FOM's are given: 1) the triplet consistency Q, 2) the special negative quartet criterion HKC, 3) the negative quartet criterion NQC, 4) the ϕ_0 check, and 5) the Σ_1 criterion.

7   Enantiomorph-Specific Procedures.

In polar space groups (e.g. $P2_1$) the triplet consistency tends to produce centric phases and in these space groups it is therefore not a very reliable FOM either. Nevertheless, the symbolic phases may be correct but, in order to find the structure, it is necessary to use an enantiomorph-specific FOM. It was shown that reliable estimates of $\pi/2$ can be obtained for quintet phase sums (Van der Putten and Schenk, 1977), and the same applies for quartet phase sums (Hauptman, 1975). On the basis of these estimates two enantiomorph-specific FOM's were developed (Van der Putten and Schenk, 1979):

$$ENQUAC = \sum_i [W_i |\phi_{H_i}\phi_{K_i}\phi_{L_i} + \phi_{-H_i-K_i-L_i} + S_i|P_i|] \qquad (9)$$

and

$$ENQUIC = \sum_i [W_i |\phi_{H_i}\phi_{K_i}\phi_{L_i} + \phi_{M_i} + \phi_{-H_i-K_i-L_i-M_i} + S_i|q_i|] \, (10)$$

in which P_i and q_i are the estimated values of quartet and quintet phase sums, respectively. To calculate an ENQUAC value all phases are substituted on the right- hand side and the signs S_i are chosen such that for each term $\phi_{H_i} + \phi_{K_i} + \phi_{L_i} + \phi_{-H_i-K_i-L_i} + S_i|P_i|$ is closest to zero. ENQUAC is zero for the correct phases and the correct $|p|$-values. In practice, as long as the errors in $|p|$ are random, the FOM is enantiomorph specific, and it can be expected that the smallest value of ENQUAC corresponds to the set of approximately correct phases. Any set of centric phases will have much larger ENQUAC values. Similar reasoning and results are valid for the FOM ENQUIC.

Also enantiomorph-conserving refinement procedures have been developed employing estimates $|p|$ and $|q|$ for quartet and quintet phase sums. e.g. for quartets a modified tangent formula (Van der Putten and Schenk, 1979) is given by:

$$tan\phi_H = \frac{\sum_K \sum_L E_4 sin(\phi_k + \phi_L + \phi_{H-K-L} + s|p|)}{\sum_K \sum_L E_4 cos(\phi_k + \phi_L + \phi_{H-K-L} + s|p|)} \qquad (11)$$

Here s is chosen such that $-\phi_H + \phi_K + \phi_L + \phi_{H-K-L} + s|p|$ is closest to 0. This refinement will maintain the enantiomorph and introduces random errors in the phases ϕ_H only. The strongest practical procedure combines the use of triplets, quartets and quintets in a way similar to (11). (Van der Putten *et al.*, 1979). This procedure has also been adopted in Mithril by Gilmore.

8   Seminvariants and Higher Invariants.

Other applications of higher order phase relations are their use in the estimation of lower-order phase relations. This is a very broad field in theoretical direct methods with up to the present only a few applications for routine phase determination. In particular, in the SIR system one- and two-phase seminvariants are used extensively. (See these proceedings). Also in SIMPEL a FOM is based on one-phase seminvariants: the Σ_1 relation. This most simple phase relation correlates in spacegroup $P\bar{1}$ the phase of 2H with $|E_H|$ and $|E_{2H}|$: if both $|E|$'s are large the phase ϕ_{2H} is most probably zero; this is one of the so-called Σ_1 relations. Σ_1 relations in space groups of higher symmetry (Hauptman and Karle, 1953) can be very reliable, although they are always few in number. Some authors advise the use of Σ_1 relations in the starting-set procedure, but we have better experiences with their use as a FOM (Overbeek and Schenk, 1976). As an example, in Table 1 the rank of the correct solution with the Σ_1 criterion has also been given. It is the weakest FOM in that table; in monoclinic and orthorhombic space groups, however, its reliability approximates the strength of the negative-quartet criterion.

9 Estimating E's of non-Observed Reflections.

A completely different application of quartets is the estimation of structure-factor amplitudes of non-observed reflections. This approach is based on the observation that if $|E_H|, |E_K|, |E_L|, |E_{-H-K-L}|, |E_{H+K}|,$ and $|E_{H+L}|$ are large, as a rule $|E_{H+K}|$ is also large, and if the first four are large and $|E_{H+K}|$ and $|E_{H+L}|$ are small, then often $|E_{K+L}|$ is also small. In case a non-observed reflection acts as K+L in many quartets, with the other six reflections in each quartet observed, in general a rather firm statement can be made regarding the $|E|$ of that reflection. Using these observations as a starting point, a theory was developed, upon which a practical procedure could be based. The theory leads to rather good estimates of a number of $|E|$-values of non-observed reflections (Van der Putten, Schenk and Tsoucaris, 1982). This successful procedure does not imply that the resolution of the structure determination is improved; this can only be realized by measuring more data. The importance of the procedure lies in the fact that a re-ordering of the information is realized which sometimes facilitates the phase determination enormously.

10 The Program System SIMPEL.

In the program system SIMPEL, written by my group, most of the above options are realised. SIMPEL is available as standalone program for CDC-cyber computers and also in the program systems SDP from ENRAF-NONIUS and XTAL. The SDP-version of SIMPEL is at the time of writing an exact copy of the Cyber version. The XTAL-version uses where possible GENTAN-routines and shows therefore a number of other feature. The stand-alone version is very CDC-specific and will not be ported easily to other computers. The SDP version can be obtained through ENRAF-NONIUS and the XTAL-version through the University of Western Australia (see these proceedings).

11   Conclusion

In conclusion, Direct Methods procedures have profited substantially from the incorporation of higher-invariant features. Moreover, since many potential applications are not yet available in the form of user programs, and since research in this field is still proceeding, future Direct Methods procedures will become even more powerful by further exploiting higher invariants.

REFERENCES

Giacovazzo, C. (1976). *Acta Cryst.*, **A32**, 91.

Giacovazzo, C. (1977). *Acta Cryst.*, **A33**, 933.

Harker, D. and Kasper, J. S. (1948). *Acta Cryst.*, **1**, 70.

Hauptman, H. (1975). *Acta Cryst.*, **A31**, 680.

Hauptman, H. and Karle, J. (1953). *A.C.A. Monograph No. 3*. Pittsburgh, Polycrystal.

Karle, J. and Hauptman, H. (1950). *Acta Cryst.*, **3**, 181.

Kiers, C. T. and Schenk, H. (1984). unpublished results.

Kitaigorodsky, A. I. (1954). *Dokl. Acad. Nauk. SSSR*, **94**, 225.

Peschar, R. and Schenk, H. (1986). *Acta Cryst.*, **A42**,309–317.

Peschar, R. and Schenk, H. (1987a). *Acta Cryst.*, **A43**,84–92.

Peschar, R. and Schenk, H. (1987b). *Acta Cryst.*, in press.

Sayre, D. (1952). *Acta Cryst.*, **5**, 60.

Schenk, H. (1972), *Acta Cryst.*, **A28**, 412.

Schenk, H. (1973). *Acta Cryst.*, **A29**, 77.

Schenk, H. and de Jong, J. G. H. (1973). *Acta Cryst.*, **A29**, 31.

Schenk, H. (1973b). *Acta Cryst.*, **A29**, 480.

Schenk, H. (1982) in *Computational Crystallography*, (ed. D. Sayre) p. 231. Oxford.

Van der Putten, N. and Schenk, H. (1977). *Acta Cryst.*, **A33**, 378.

Van der Putten, N. and Schenk, H. (1979). *Acta Cryst.*, **A35**, 381.

Van der Putten, N., Schenk, H. and Tsoucaris, G. (1982). *Acta Cryst.*, **A38**, 98.

4

DIRECT METHODS APPLICATIONS TO MACROMOLECULES

Suzanne Fortier

1 Introduction

The term 'direct methods' normally refers to crystal structure determination methods that take advantage of relationships among the structure factors to extract missing phase information from the observed intensities. An initial model of the structure can thus be calculated with little or no *a priori* chemical information. While in general the chemical content of the unit cell is known, even this information is not essential for the successful application of direct methods. Indeed there are numerous examples of crystal structures solved with unknown or presumably known, but erroneous, chemical formulae.

Because the values of individual phases depend not only on the structure but on the choice of origin as well, it is the origin independent linear combinations of phases, the structure invariants, that are estimated. Among those, the most widely used one is the three-phase invariant,

$$\omega = \phi_H + \phi_K + \phi_L \tag{1}$$

where

$$H + K + L = 0 \tag{2}$$

From the method of joint probability distribution, the value of ω can be estimated by:

$$P(\Omega \mid |E_H E_K E_L|) = 1/K \ exp(A \ cos\Omega) \tag{3}$$

where

$$|E| = \text{normalized structure factor magnitude}$$
$$K = \text{normalization constant}$$
$$A = 2\sigma_3/\sigma_2^{3/2} |E_H E_K E_L|$$

with

$$\sigma_n = \sum_{j=1}^{N} Z_j^n$$

(Cochran, 1955; Hauptman, 1976).

Equation (3) shows that the distribution has a unique maximum, in the range 0 to $360°$, at $\Omega = 0°$. The larger the A value, the more reliable this estimate is. However since the coefficient $\sigma_3/\sigma_2^{3/2}$ (is approximately equal to $1/N^{1/2}$, the larger the number of atoms in the unit cell, the smaller the number of reliably estimated invariants. For that reason, the standard machinery of direct methods, as it is applied so successfully to small molecules, cannot be applied directly to macromolecular problems. What will be discussed, therefore, are phasing tools that have evolved from the techniques of direct methods but have been adapted to macromolecular problems. The adjective 'direct' is thus to be taken loosely in this context.

There have been considerable efforts to make use of direct methods in three areas of macromolecular phasing: a) the determination of the heavy atom/anomalous scatterer substructure, b) the refinement and extension of phases, and c) the primary phasing of single isomorphous replacement (SIR) and single wavelength anomalous scattering (SAS) data. The first two topics have been discussed and reviewed fairly recently (Karle, 1984; Giacovazzo, 1980; Schenk, 1982) and so will not be covered in the present article.

2   Theory and applications of direct methods to SIR

Most of the macromolecular structures determined to date have relied on the successful application of the multiple isomorphous replacement (MIR) method. Because of the obvious benefit, there has always been, however, much interest in devising techniques that could be used with data from a native and a single derivative (SIR).

In the SIR case, from the knowledge of the heavy atom substructure and therefore of its structure factors, amplitudes and phases, and the measured amplitudes of the native and derivative structure factors, two possible values are calculated for the native or derivative phases. These two solutions are enantiomorphic with respect to the heavy atom substructure vector. In the absence of any other information, the two solutions are equally probable.

Blow and Rossman (1961) proposed to compute a Fourier map using the average of the two possible phases and multiplying the structure factor amplitudes by suitable weights. Although they showed that the method could be used successfully, it has severe limitations, largely due to the high level of background noise. Most of the efforts to make use of SIR data have thus been directed at breaking down the two-fold ambiguity. Among the methods proposed are the use of anomalous scattering (Bijvoet, 1954), non-crystallographic symmetry (Bricogne, 1976), noise filtering techniques (Wang, 1985) and direct methods.

The first direct methods attempts to resolve the SIR phase ambiguity made use of the tangent formula (Coulter, 1965). The results, however, were not promising and the approach has not been pursued. In recent years a great deal of effort has been directed, instead, at combining the techniques of direct methods and isomorphous replacement at the theoretical level, with formulae specifically derived for the SIR case. Of the methods proposed, the probabilistic approach of Hauptman (1982) and algebraic approach of Karle (1983) go from intensities to phases by the most 'direct' route with a minimum requirement of *a priori* chemical information. Other approaches (Fan Hai-fu *et al.*, 1984; Fortier, Moore and Fraser, 1985; Langs, 1986) make use of heavy atom substructure information which is integrated in the formulae. To our knowledge none of the methods has yet been used successfully to determine an unknown macromolecular structure.

The initial testings on error free data have, however, been very promising. For example, the first calculations published by Hauptman, Potter and Weeks (1982), on the protein cytochrome C(550), molecular weight of 14 500, and its $PtCl_4^{2-}$ derivative, showed that several tens of thousands of three-phase invariants could be reliably estimated as having values close to 0 or 180°. In a set of 25 000, generated from 1 000 native and 1 000 derivative phases, an error of 30° is calculated. It was further shown (Fortier, Moore and Fraser, 1986) that the error could be decreased by a factor of two after incorporating heavy atom substructure information into the formulae, which allowed cosine estimates in the full range of -1 to +1. Finally, it was demonstrated by Langs (1986), using pig insulin data, that phases could be determined and refined to an average error of 6° if a full invariant set with perfect modular estimates of -1 or +1 was available. Similarly, initial test calculations by Karle (1983) and Fan Hai-fu *et al.* (1984) have yielded very encouraging results.

The theoretical formulae are often presented in a form suitable for computation but impermeable to straightforward interpretation in terms of the experimental parameters. Such an interpretation is useful in determining the scope and nature of direct methods in the solution of the phase problem in the SIR case (Fortier, Weeks and Hauptman, 1984). In that respect, the formulae obtained by Hauptman (1982) are particularly interesting because they were derived with a minimum of *a priori* physical information. In fact the only assumption made is that diffraction data are available for a pair of isomorphous structures for which the unit cell content is known.

For each reciprocal-lattice vector $\mathbf{H}$, there exist two normalized structure factors E_H and G_H. For a triplet of reciprocal lattice vectors $\mathbf{H}$, $\mathbf{K}$, $\mathbf{L}$ satisfying $\mathbf{H} + \mathbf{K} + \mathbf{L} = 0$, there exist eight structure invariants

$$\omega_1 = \phi_H + \phi_K + \phi_L,$$
$$\omega_2 = \phi_H + \phi_K + \psi_L,$$
$$\omega_3 = \phi_H + \psi_K + \phi_L,$$
$$\omega_4 = \psi_H + \phi_K + \phi_L,$$
$$\omega_5 = \phi_H + \psi_K + \psi_L,$$
$$\omega_6 = \psi_H + \phi_K + \psi_L,$$
$$\omega_7 = \psi_H + \psi_K + \phi_L,$$
$$\omega_8 = \psi_H + \psi_K + \psi_L,$$

$$(4)$$

where the ϕ's and the ψ's are the phases associated with the isomorphous pair of structures.

Let

$$|E_H| = R_1, \quad |E_K| = R_2, \quad |E_L| = R_3;$$
$$|G_H| = S_1, \quad |G_K| = S_2, \quad |G_L| = S_3,$$

$$(5)$$

The conditional probability distributions of the three-phase structure invariants ω_i given the six magnitudes $|E_H|, |E_K|, |E_L|, |G_H|, |G_K|, |G_L|$ in their first neighborhood are given by

$$P_i((\Omega_i \mid R_1, R_2, R_3, S_1, S_2, S_3) = (1/K_i) \, exp(A_i cos\Omega_i),$$
$$i = 1, 2, \ldots 8,$$

$$(6)$$

where

$$K_i = 2\pi I_0(A_i) \tag{7}$$

and I_0 is the modified Bessel function (Hauptman, 1982). The A_i values are given by

$$A_i = 2\{\beta_1 \tau_1 R_1 R_2 R_3$$
$$+ \beta_2 [\tau_{21} R_1 R_2 S_3 + \tau_{22} R_1 S_2 R_3 + \tau_{23} S_1 R_2 R_3]$$
$$+ \beta_3 [\tau_{31} R_1 S_2 S_3 + \tau_{32} S_1 R_2 S_3 + \tau_{33} S_1 S_2 R_3]$$
$$+ \beta_4 \tau_4 S_1 S_2 S_3\}$$

$$(8)$$

where the β's are functions of the atomic scattering factors, and $\tau = C_1 C_2 C_3$ is obtained by comparing the ith structure factor associated with the coefficient of τ with the ith structure factor associated with the invariant. If they are of the same type, i.e., both R or both S, then $C_i = 1.0$, $i = 1,2,3$. If one is of type R and the other of type S, then

$$C_i = I_1(2\gamma R_i S_i) / I_0(2\gamma R_i S_i), \qquad i = 1, 2, 3, \tag{9}$$

where I_1 and I_0 are the modified Bessel functions and τ is a function of the atomic scattering factors. For the special case of a native protein and an heavy atom isomorphous derivative, if we assume that the atomic content of the derivative equals the atomic content of the native protein (P) plus the heavy-atom content (H) then

$$\gamma = \alpha_{20}^{1/2} \alpha_{02}^{1/2} / (\alpha_{02} - \alpha_{20}), \tag{10}$$

$$\beta_1 = \frac{-(\alpha_{03} - \alpha_{30}) \alpha_{20}^{3/2}}{(\alpha_{02} - \alpha_{20})^3}, \tag{11}$$

$$\beta_2 = \frac{(\alpha_{03} - \alpha_{30}) \alpha_{20} \alpha_{02}^{1/2}}{(\alpha_{02} - \alpha_{20})^3}, \tag{12}$$

$$\beta_3 = \frac{-(\alpha_{03} - \alpha_{30}) \alpha_{20}^{1/2} \alpha_{02}}{(\alpha_{02} - \alpha_{20})^3}, \tag{13}$$

$$\beta_4 = \frac{(\alpha_{03} - \alpha_{30}) \alpha_{02}^{3/2}}{(\alpha_{02} - \alpha_{20})^3}, \tag{14}$$

$$\alpha_{mn} = \sum_{j=1}^{N} f_j^m g_j^n, \tag{15}$$

where f_j and g_j denote atomic structure factors for a corresponding pair of isomorphous structures.

From equations (10) to (14) it is seen that the distribution does not depend, as in the case of the traditional three-phase invariant, on the total number of atoms per unit cell but rather on the scattering difference between the native protein and the derivative — that is, on the scattering of the heavy atoms in the derivative. For the special case in which the heavy atoms in the derivative are of equal weight and equal occupancy, the distribution depends on the number of heavy atoms in the derivative. Since this number is usually small, it becomes clear that the distribution is capable of yielding extremely reliable estimates, as demonstrated by the calculations of Hauptman, Potter and Weeks (1982). When the τ functions approach 1.0, i.e. when the $2\gamma R_i S_i$'s are large,

$$
\begin{aligned}
A_i \;=\; & \sum Z_H^{3/4} (\sum Z_P^2)^{3/2} [R_1 R_2 R_3 \\
& + 2\gamma (R_1 R_2 \Delta_3 + R_1 \Delta_2 R_3 + \Delta_1 R_2 R_3) \\
& + 4\gamma^2 (R_1 \Delta_2 \Delta_3 + \Delta_1 R_2 \Delta_3 + \Delta_1 \Delta_2 R_3) \\
& + 8\gamma^3 \Delta_1 \Delta_2 \Delta_3]
\end{aligned}
\tag{16}
$$

$$
\sum_{j \varepsilon P} Z_j^n \;=\; \sum Z_P^n,
$$

$$
\sum_{k \varepsilon H} Z_k^n \;=\; \sum Z_H^n.
\tag{17}
$$

$$
\Delta_i \;=\; S_i - R_i
\tag{18}
$$

$$
2\gamma \;=\; (1 + 2 \sum Z_P^2 / \sum Z_H^2).
\tag{19}
$$

The coefficient 2γ is related to the diffraction ratio. This ratio, which is a measure of the average change in intensity due to the addition of heavy atoms, is estimated at low resolution as

$$
(2\Sigma Z_H^2 / \Sigma Z_P^2)^{1/2},
\tag{20}
$$

(Crick and Magdoff, 1956).

Hence, $2\gamma = [1 + 4/ \,(\text{diffraction ratio})^2]$. The term 2γ is usually large compared to unity while the term $\Sigma Z_H^{3/4} (\Sigma Z_P^2)^{3/2}$ is usually very small. Thus the eight magnitude product terms of the distribution, $R_1 R_2 R_3$, $R_1 R_2 \Delta_3$, $R_1 \Delta_2 R_3$, $\Delta_1 R_2 R_3$, $R_1 \Delta_2 \Delta_3$, $\Delta_1 R_2 \Delta_3$, $\Delta_1 \Delta_2 R_3$, $\Delta_1 \Delta_2 \Delta_3$, do not contribute equally to the A_i term. The predominant term is the $\Delta\Delta\Delta$ term and a result comparable to Karle's simple rule (Karle, 1983) is obtained. If the

$\triangle\triangle\triangle$ product is positive, than the invariant is estimated as having a cosine value close to $+1$ while if it is negative the cosine value is estimated as being close to -1.

From the initial applications, it is clear that direct methods have the potential to determine very accurately the value of SIR phase invariants. What is still unclear, though, is whether or not the individual phases, needed to compute an interpretable density map, can be obtained from the set of reliably estimated invariants.

The τ functions used in estimating the A values are equal to either 1.0 or the Bessel-function ratio as defined in (9). This Bessel-function ratio is the expected value of the cosine of the phase difference $(\psi_i - \phi_i)$ associated with the two magnitudes R_i and S_i (Sim, 1960). A τ function is thus equal to one when the phase difference between the native and derivative is expected to be 0^0. Therefore the form of the distribution shown in equation (16), and for which there is a correspondence with Karle's simple rule, corresponds to the special case in which the phases of all three reflections participating in the invariants have approximately the same value for the native and derivative.

In this special case,

$$
\begin{aligned}
\phi_P &= \phi_{PH} = \phi_H & if \ \triangle &> 0 \\
\phi_P &= \phi_{PH} = \phi_H + 180^0 & if \ \triangle &< 0
\end{aligned}
\tag{21}
$$

Since the heavy atom substructure normally consists of a small number of atoms, it has associated phases that form three-phase invariants with values very close to $0°$. The corresponding native and derivative invariants will thus have estimates of 0 or $180°$ depending on whether the product $\triangle\triangle\triangle$ is larger or smaller than zero. It is possible to obtain very accurate estimates of this special class of invariants. It should be noted however that the phases that participate in these invariants are precisely those for which the two-fold ambiguity vanishes.

To look at the cases for which the τ functions differ from 1.0, we can substitute in the distribution the equivalent cosine functions. Let

$$\psi_i - \phi_i = \pm\alpha_i \qquad (22)$$

then, for example, the conditional probability distribution of the ω_1 invariant can be written as

$$P_1(\Omega_1) = \frac{1}{K_1}exp(A_1cos\Omega_1), \qquad (23)$$

where

$$
\begin{aligned}
A_1cos\Omega_1 \;=\; & 2cos\Omega_1\{\beta_1 R_1 R_2 R_3 \\
& + \beta_2[R_1 R_2 S_3 cos\alpha_3 + R_1 S_2 R_3 cos\alpha_2 + S_1 R_2 R_3 cos\alpha_1] \\
& + \beta_3[R_1 S_2 S_3 cos\alpha_2 cos\alpha_3 + S_1 R_2 S_3 cos\alpha_1 cos\alpha_3 \\
& + S_1 S_2 R_3 cos\alpha_1 cos\alpha_2] + \beta_4 S_1 S_2 S_3 cos\alpha_1 cos\alpha_2 cos\alpha_3\}. (24)
\end{aligned}
$$

For the sake of simplicity, let us assume that $\alpha_2 = \alpha_3 = 0$ and $|\alpha_1| \neq 0$. Equation (24) then becomes

$$
\begin{aligned}
A_1cos\Omega_1 \;=\; & 2cos\Omega_1\{\beta_1 R_1 R_2 R_3 \\
& + \beta_2[R_1 R_2 S_3 + R_1 S_2 R_3] + \beta_3 R_1 S_2 S_3\} \\
& \{cos(\Omega_1 + \alpha_1) + cos(\Omega_1 - \alpha_1\} \; \times \\
& \{\beta_2 S_1 R_2 R_3 + \beta_3[S_1 R_2 S_3 + S_1 S_2 R_3] + \beta_4 S_1 S_2 S_3\}. \;(25)
\end{aligned}
$$

While the expected value of the phase difference $\psi_1 - \phi_1$ can be estimated, its sign is not known. In the form of the distribution shown in equation (25), both signs are considered equally probable and their contributions are averaged, as is done in the standard SIR technique (Blow and Rossmann, 1961).

If instead, the $A_1cos\Omega_1$ expression is calculated twice, assuming first the one sign and then the other, i.e.,

$$
\begin{aligned}
A_1cos\Omega_1 \;=\; & 2cos\Omega_1\{\beta_1 R_1 R_2 R_3 \\
& + \beta_2[R_1 R_2 S_3 + R_1 S_2 R_3] + \beta_3 R_1 S_2 S_3\} \\
& + 2cos(\Omega_1 + \alpha_1)\{\beta_2 S_1 R_2 R_3 \\
& + \beta_3[S_1 R_2 S_3 + S_1 S_2 R_3] + \beta_4 S_1 S_2 S_3\}. \qquad (26)
\end{aligned}
$$

and

$$A_1 cos\Omega_1 \ = \ 2cos\Omega_1 \{\beta_1 R_1 R_2 R_3$$
$$+ \ \beta_2 [R_1 R_2 S_3 + R_1 S_2 R_3] + \beta_3 R_1 S_2 S_3\}$$
$$+ \ 2cos(\Omega_1 - \alpha_1)\{\beta_2 S_1 R_2 R_3$$
$$+ \ \beta_3 [S_1 R_2 S_3 + S_1 S_2 R_3] + \beta_4 S_1 S_2 S_3\}. \qquad (27)$$

then estimates of Ω_1 ranging over the full 0 to 360° interval are obtained. Furthermore, it is clear that the two sign possibilities yield enantiomorphic estimates of Ω_1 , and thus determine uniquely the cosine of the invariant.

In the general case where $|\alpha_1|$, $|\alpha_2|$, and $|\alpha_3|$ are all non-zero, again the magnitudes of the phase differences can be calculated while the signs are unknown. There exist, therefore, eight possible sign combinations. Calculation of the distribution for each of the eight sign combinations yields four enantiomorphic pairs of Ω_1 estimates or four cosine-invariant estimates.

In general, not all three α's are significantly different from zero. Therefore, although four possible cosine invariants are estimated, they do not in many cases differ much from one another. By using a weighted average, very accurate cosine invariant estimates can be obtained, particularly if the cos α's are calculated once the heavy atom substructure has been determined (Fortier, Moore and Fraser, 1985).

Several important points emerge from a careful analysis of the formulae obtained through the joint probability distribution approach. Firstly, it is seen that even without any heavy atom substructure information, the method is able to identify very accurately those reflections for which the native and derivative have the same phase angle, and — furthermore — to determine whether this value is the same or 180° away from the heavy atom substructure phase. An estimate of the absolute value of the derivative-native phase difference — and, consequently, of the heavy atoms substructure vector — falls out of the derivation. These results are obtained simply from the knowledge of the unit cell content of the isomorphous pair of structures. Secondly, it is seen that the 'inherent' SIR two-fold ambiguity does not disappear magically. Rather it is integrated in the formulae and, through a weighting scheme similar to that used by Blow and Rossman (1961), unique invariant estimates are obtained. It is encouraging to see that many of the standard SIR practices, which have emerged from careful experimental observations, fall out naturally in the mathematical derivation. On the other hand, this may indicate that the reliability of unique invariant estimates will decrease as the protein and derivative phases diverge from one another and thus as they diverge from their standard SIR values. As a result, it remains unclear whether or not the method is capable of yielding resolved native or derivative phases, as has been observed in the applications by Xu*et al.* (1984).

3   Theory and Applications of Direct Methods to SAS

It has been known for some thirty years that structure amplitude differences due to anomalous scattering can be used to obtain phase information. Reviews on the various proposed phasing techniques based on this approach can be found in several publications (Ramaseshan and Abrahams, 1975; Sayre, 1982; Ramachandran, 1964). As in the SIR case, the SAS experiment yields estimates of phases bearing a two-fold ambiguity. Unlike the SIR case, however, the two estimates are not equally probable. This is because the two phase estimates are enantiomorphic with respect to the *imaginary* part of the anomalous scatterer substructure vector. The phase ambiguity can thus be resolved by selecting the phase closest to the anomalous scatterer substructure phase. However, as the ratio of heavy atom to light atom scattering decreases, the probability that this choice will be the correct one decreases. This method has been used sucessfully in the solution of the lysine hydrochloride structure (Raman, 1959) and more recently, in a probabilistic fashion, in the solution of the protein crambin (Hendrickson and Teeter, 1981).

The first results on the use of the integrated direct methods –

anomalous dispersion technique for the estimation of the three-phase structure invariant were presented by Kroon, Spek and Krabbendam (1977). By extending the method proposed by Peederman and Bijvoet (1956), Ramachandran and Raman (1956) and Okaya and Pepinsky (1956) to the three-phase structure invariant, they showed that three-phase sine invariants could be estimated from the observed intensities.

Using the method of joint probability distribution, Hauptman (1982) and, subsequently, Giacovazzo (1983) obtained formulae which give unique estimates of the two-phase and three-phase structure invariants and thus unique estimates of the phases themselves. Unique estimates have also been obtained by Karle (1984 and 1985) from the use of formulae derived through an algebraic approach. As in the probabilistic approach of Hauptman (1982) and Giacovazzo (1983), this is done without any information about the anomalous scatterer substructure. Finally, Fan Hai-fu *et al.* (1984) and Langs (1986) have proposed direct methods approaches that make use of anomalous scatterer substructure information to resolve the SAS phase ambiguity. Results on the application of the latter approach have not been published.

In the initial applications made by Hauptman (1982), Giacovazzo (1983), Karle (1984) and Qian *et al.* (1985) on macromolecular data, it was seen, however, that although unique estimates were obtained substantial errors persisted, even when the calculations were done using error-free data. In the probabilistic formulae, it was observed that the magnitude of the error was not well predicted, as the variance of the distributions was systematically underestimated. In order to locate the source of these systematic errors, the distributions were reexamined (Fortier, Fraser and Moore, 1986). The analysis has identified certain approximations appearing in the formulae which are responsible for the ability to obtain the unique estimates but are also largely responsible for the errors observed. More specifically, a probabilistic estimate of the two-phase invariant $\phi_{\mathbf{H}} + \phi_{\overline{\mathbf{H}}}$ which appears in the distribution gives the sum of $\phi_{\mathbf{H}} + \phi_{\overline{\mathbf{H}}}$ a positive value. This corresponds to the solution for which the phase angles are closest to the anomalous scatterer substructure phase angles. Thus the SAS ambiguity is resolved essentially in the usual manner. It is important to note, though, that this is done without any anomalous scatterer substructure information. Large errors are observed in invariants that contain a phase angle whose value is not closest to the anomalous scatterer substructure phase angle.

To our knowledge, the fused direct methods-SAS phasing techniques have not been used yet to determine an unknown macromolecular structure. The probabilistic formulae of Hauptman (1982) were used however by Furey *et al.* (1986) to locate a 4-Cd cluster in the crystal structure determination of Cd, Zn Metallothionein.

4   Concluding Remarks

A substantial theoretical base exists for the integration of direct methods into the single isomorphous replacement (SIR) and single wavelength anomalous scattering (SAS) phasing techniques. Naturally, the role that these phasing tools will play in the determination of macromolecular structures cannot be predicted until the problems associated with errors in the data and errors resulting from the normalization of the structure factors have been assessed. While the test calculations appear promising, it is important to keep in mind that the ability to estimate accurately a large set of invariants is not necessarily a guarantee of success. It is seen that often unique and accurate estimates are obtained precisely in cases when the SIR or SAS ambiguity vanishes. Success can only be judged therefore by an analysis of the complete set of phases needed to compute an interpretable map.

In small molecule applications, much of the direct methods power resides in their ability to rank reliably a redundant set of invariant estimates. In the extensive calculations done to date on macromolecular data, this ability is also observed. An optimistic but cautious view of the present state of the direct methods - macromolecular phasing tools is that the results obtained to date are not qualitatively much different from the standard SIR and SAS practices. An important gain though may be in the ability to rank the redundant information reliably.

Acknowledgements: Financial assistance from the Natural Sciences and Engineering Research Council of Canada is gratefully acknowledged.

REFERENCES

Bijvoet, J. M. (1954). *Nature (London)*, **173**, 888-891.

Blow, D. N. and Rossmann, M. G. (1961). *Acta Crystallographica*, **14**, 1195-1202.

Bricogne, G. (1976). *Acta Crystallographica*, **A32**, 832-847.

Cochran, W. (1955). *Acta Crystallographica*, **8**, 473-478.

Coulter, C. L. (1965). *Journal of Molecular Biology*, **12**, 292-295.

Crick, F. H. C. and Magdoff, B. (1956). *Acta Crystallographica*, **9**, 901-908.

Fan Hai-Fu, Han Fu-son, Qian Jin-zi and Yao Jia-xing (1984). *Acta Crystallographica*, **A40**, 489-495.

Fortier, S., Fraser, M. E. and Moore, N. J. (1986). *Acta Crystallographica*, **A42**, 149-156.

Fortier, S., Moore, N. J. and Fraser M. E. (1985). *Acta Crystallographica*, **A41**, 571-577.

Fortier, S., Weeks, C. M. and Hauptman, H. (1984). *Acta Crystallographica*, **A40**, 544-548.

Furey, W. F., Robbins, A. H., Clancy, L. L., Winge, D. R., Wang, B. C. and Stout, C.,D. (1986). *Science*, **231**, 704-710.

Giacovazzo, C. (1980). *Direct Methods in Crystallography*, pp. 373-378. Academic Press, London.

Giacovazzo, C. (1983). *Acta Crystallographica*, **A39**, 585-592.

Hendrickson, W. A. and Teeter, N. M. (1981). *Nature (London)*, **290**, 107-113.

Hauptman, H. (1976). *Acta Crystallographica*, **A32**, 877-882.

Hauptman, H. (1982). *Acta Crystallographica*, **A38**, 289-294.

Hauptman, H. (1982). *Acta Crystallographica*, **A38**, 632-641.

Hauptman, H., Potter, S. and Weeks, C. M. (1982). *Acta Crystallographica*, **A38**, 294-300.

Karle, J. (1983). *Acta Crystallographica*, **A39**, 800-805.

Karle, J. (1984). *Acta Crystallographica*, **A40**, 4-11.

Karle, J. (1984). In *Methods and Applications in Crystallographic Computing*, (ed. S. R. Hall and T. Ashida) pp. 120–140. Clarendon Press, Oxford.

Karle, J. (1985). *Acta Crystallographica*, **A41**, 387-394.

Kroon, J., Spek, A. L. and Krabbendam, H. (1977). *Acta Crystallographica*, **A33**, 382-385.

Langs, D.A. (1986). *Acta Crystallographica*, **A42**, 362-368.

Okaya, Y. and Pepinsky, R. (1956). *Physics Review*, **103**, 1645-1647.

Peederman, A. F. and Bijvoet, J. M. (1956). *Proceedings of the Koninklijke Nederlandse Akademie van Wetenschaffen*, **B59**, 312-313.

Qian, Jin-zi, Fan, Hai-fu and Gu, Yuan-xin (1985). *Acta Crystallographica*, **A41**, 476-478.

Ramachandran, G. N. (1964). Editor. *Advanced Methods of Crystallography*, Academic Press, London and New York.

Ramachandran, G. N. and Raman, S. (1956). *Current Science*, **25**, 348-351.

Raman, S. (1959). *Zeitschrift für Kristallographie*, **111**, 301-317.

Ramaseshan, S. and Abrahams, S. C. (1975). Editors. *Anomalous Scattering*, Munksgaard, Copenhagen.

Sayre, D. (1982). Editor. *Computational Crystallography*. Clarendon Press, Oxford.

Schenk, H. (1982). In *Computational Crystallography*, (ed. D. Sayre) pp. 231-241. Clarendon Press, Oxford.

Sim, G. A. (1960). *Acta Crystallographica*, **13**, 511-512.

Wang, B. C. (1985). In *Methods in Enzymology*, Vol. 115 (ed. H. W. Wyckoff, C. H. W. Hirs and S. N. Timasheff) pp. 90-112. Academic Press, Inc., London.

Xu, Z. B., Yang, D. S. C., Furey, W. Jr., Sax, M., Rose, J. and Wang, B. C. (1984). *Proceedings of the American Crystallographic Association Meeting*, 20-25 May 1984, Lexington, Kentucky, U.S.A. Abstract PC2.

5

MAXIMUM ENTROPY METHODS IN THE X-RAY PHASE PROBLEM

Gérard Bricogne

Introduction

The purpose of this contribution is to serve as a guide for readers who wish to become familiar with the contents of a previous paper on maximum entropy methods (Bricogne 1984), hereafter referred to as MEFDM) and of a series of forthcoming articles which generalize and extend the scope of this new approach. It presents a survey of a universal statistical formulation of all the computational processes by which phase information is generated in the course of a structure determination. The notion of a maximum-entropy prior distribution of atoms plays a fundamental role in the derivation of joint probability distributions of structure factors from a model of randomly placed atoms, which is the basis for this entire approach. In its barest form, this computational process gives rise to an extension of direct methods capable of deriving strong phase relations between the structure factors of proteins; in the last part of this paper, it is adapted to a variety of situations of increasing complexity, culminating in a 'multichannel' formalism encompassing all sources of phase information.

The style of this contribution is as informal and as 'formula-free' as is compatible with giving meaningful examples, the full details and proofs being available in the articles themselves, be they already published or to appear.

1 The basic mechanism: approximating joint distributions of structure factors and conditional distributions of phases

Direct phase determination by statistical methods, first proposed and investigated by Hauptman and Karle (1953), was shown by the later work of Bertaut (1955 *a,b,c*) and of Klug (1958) to be based on well-known techniques of analytical probability theory, usually associated with investigations of the Central Limit Theorem.

The general argument goes as follows. If we assume that a crystal structure is made up of N atoms thrown randomly and — for the sake of tractability — independently of each other into the asymmetric unit of the crystal, then we may call upon standard mathematical devices (characteristic or moment-generating functions, cumulant expansions, Gram-Charlier and Edgeworth series) to calculate, to various degrees of approximation, the *joint probability distribution* (jpd), denoted $P(\mathbf{F})$, of any collection $\mathbf{F} = (F_{\mathbf{h}_1}, F_{\mathbf{h}_2}, \ldots, F_{\mathbf{h}_n})$ of structure factors we like. Thus we can get indications that not all combinations of structure factor values will occur with equal probability: some combinations are more probable than others.

The application to the Phase Problem then seems obvious: since we know (by measurement) the moduli of the structure factors, let us simply *substitute* the values of these moduli into the expression for the jpd $P(\mathbf{F})$ of the structure factors: then, only the phases will remain as free variables, and we will obtain the *conditional distribution* of these phases. It is upon this conditional distribution that we will then rely to infer that, given the values of the moduli measured for the structure at hand, certain combinations of phases are more probable than others.

2   Living with asymptotic expansions: the need for recentering

The basic idea just formulated is so simple that one would expect its implementation to be straightforward. Unfortunately, there are several hidden pitfalls, which were examined in detail in a previous paper (MEFDM, § 2).

The first is what one might nickname 'Humpty Dumpty's fall'. Here the great fall has been to fragment the jpd $P(\mathbf{F})$ into small marginal distributions (e.g. Cochran distributions) involving only a few phases in the form of invariant or seminvariant combinations. This seems to be a legacy of history: in the early days, only these small-base distributions could be studied by hand calculation, and the computer was later used to try and piece them back into some ersatz of joint distribution. But, just as in the rhyme, the pieces cannot be put together again: the resulting approximate jpd is not, and cannot be, as good as if we had instead used the computer to construct a large-base jpd from the outset, so as to keep Humpty Dumpty in one piece. The latter task has very recently been addressed by Peschar and Schenk (1987).

The second pitfall is that Humpty Dumpty (i.e. $P(\mathbf{F})$), even if prevented from falling, has always been sitting in the wrong place anyway! Indeed, the methods traditionally used to construct approximations to $P(\mathbf{F})$ (the Gram-Charlier or the Edgeworth series) always assume a *uniform* distribution of atoms, and thus yield approximations to $P(\mathbf{F})$ which are good only near $\mathbf{F} = \mathbf{0}$, i.e. for *small* moduli; yet the expressions obtained, as well as other results from determinant theory for example, show that the conditional distributions of the phases will be essentially featureless unless the moduli are *large*. One is thus faced with the rather frustrating dilemma that the approximate conditional distributions of phases are most accurate where least informative, and least accurate where potentially most informative, as illustrated in Figure 1. This is what was meant by Humpty Dumpty sitting in the wrong place: the origin $\mathbf{F} = \mathbf{0}$ is equally far from all places of interest. The difficulty boils down to the fact that a functional expression suitable only for *locally* approximating $P(\mathbf{F})$ near $\mathbf{F} = \mathbf{0}$ has been mistakenly assumed to be the *global* functional form for that function 'in the large'. In fact, *there is no tractable unique expression for* $P(\mathbf{F})$ into which we could substitute the measured values of the large moduli to get the conditional distribution of the corresponding phases, as we naively assumed at the beginning. Matters are made worse by the fact that the functional expressions in question are in fact asymptotic expansions, which are much more difficult to handle than convergent power series.

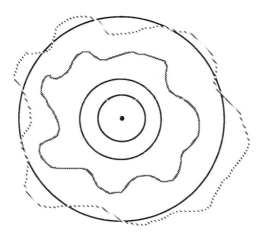

Figure 1. This figure is in structure factor space. The large circle represents the locus M defined by the values of the structure factor moduli. The asymptotic expansion of $P(\mathbf{F})$ given by the standard Edgeworth series is symbolized by its contour surfaces. The approximation is good near the origin, but probability indications on M, which are used to calculate conditional distributions of phases, are given only by the 'tail' of the expansion. Thus the conditional distributions obtained are highly unreliable.

These two observations show that the traditional form of direct methods is a rather rudimentary implementation of the ideas of probability
theory on which it is based. Fortunately, better analytical devices can be
found which circumvent these difficulties; the reader is again referred to
MEFDM, $\int 2$ for a full exposition, which will be freely summarized here.
The main idea is that, if the locus M (a high-dimensional torus) defined by the large moduli is too large for a single asymptotic expansion of
$P(\mathbf{F})$ to be acccurate everywhere on it, then we should break M up into
subregions, and construct *a collection of local approximations* to $P(\mathbf{F})$, a
different one for each subregion (see Figure 2). This is analogous to the
process of analytic continuation of a power-series in complex function theory; in fact, this turns out to be more than an analogy, as will be seen
later in connection with the saddlepoint approximation. Each subregion
will consist of a 'patch' of M, surrounding a point $\mathbf{F}^* \neq \mathbf{0}$ located on M.
Such a point $\mathbf{F}^*$ is obtained by assigning 'trial' phase values to the known
moduli, but these trial values do not have to be viewed as assumptions
concerning the true values of the phases: rather, they should be thought
of as 'search lights', pointing to a patch of M and to a specialised asymptotic expansion of $P(\mathbf{F})$ which will be custom-built (or Taylor-made) so
as to be the most accurate approximation possible to $P(\mathbf{F})$ on that patch.
With a sufficiently rich collection of such constructs, we have an 'atlas'
of M by means of which $P(\mathbf{F})$ can be accurately calculated anywhere on
M.

We are thus led inevitably to the notion of *recentering*. Recentering
the usual Gram-Charlier or Edgeworth asymptotic expansion for $P(\mathbf{F})$
away from $\mathbf{F} = \mathbf{0}$, by making trial phase assignments which define a point
$\mathbf{F}^*$ on M, is equivalent to using a non-uniform prior distribution of atoms
$q(\mathbf{x}) \not\equiv 1/V$, reproducing among its Fourier coefficients the coordinates of
$\mathbf{F}^*$.

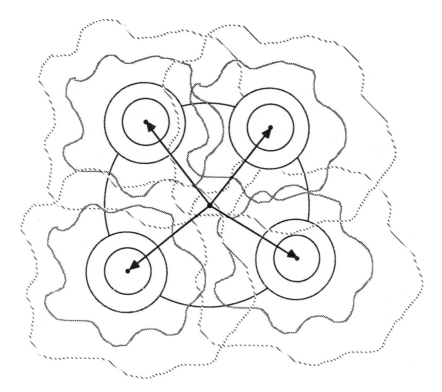

Figure 2. By using recentered asymptotic expansions, it is possible to better approximate $P(\mathbf{F})$ on local patches of M. The advantage is that $P(\mathbf{F})$ is well approximated where it is large. Where it is small, it is less well represented; but since the asymptotic expansion is also small, and since the difference between two small positive numbers is smaller than either of them, the approximation cannot be too bad. Such recentered expansions will therefore give a good approximation to the integral of $P(\mathbf{F})$ over M, which is used to calculate likelihoods.

3   The Saddlepoint Approximation: maximum-entropy non-uniform prior distributions of atoms

Let us look more closely at this recentering operation, i.e., in real-space terms, at the operation of switching over to a suitable non-uniform prior distribution of the atoms. There is something disturbing about this distribution: it is only constrained to reproduce a small number of Fourier coefficients (those specified by $\mathbf{F}^*$), and hence it is highly *indeterminate*.

It turns out, however, that there is a uniquely defined 'best' choice for it: it is that distribution which has *maximum entropy* under these constraints, high entropy being a measure of closeness to uniformity.

This result can be justified in two ways (MEFDM, $\oint 3$ and $\oint 5$).

(1) The first way invokes Shannon's theory of information (Shannon and Weaver 1949), according to which any decrease in the entropy of the distribution of atoms $q(\mathbf{x})$ tends to diminish the size of the statistical ensemble of structures which can be generated using $q(\mathbf{x})$; then it invokes Jaynes's heuristic principle (Jaynes 1957) that one should rule out as few structures as possible on the basis of the limited information available in $\mathbf{F}^*$, to conclude that $q(\mathbf{x})$ should have maximum entropy (call it $q^{\mathrm{ME}}(\mathbf{x})$).

(2) The second way relies on a purely analytical procedure, leading to an improvement over the traditional Edgeworth series called the 'saddlepoint approximation'; the remarkable thing is that, without ever mentioning entropy nor invoking any semi-philosophical arguments, it yields an identical expression for the best choice of a non-uniform prior $q(\mathbf{x})$. The saddlepoint approximation $P^{\mathrm{SP}}(\mathbf{F})$ to $P(\mathbf{F})$ is then the Gaussian approximation built from $q^{\mathrm{ME}}(\mathbf{x})$.

This equivalence result is of fundamental importance to a proper understanding of the role of the process of entropy maximisation in phase determination. The maximum entropy criterion intervenes only in the construction of $q^{\mathrm{ME}}(\mathbf{x})$ under constraint values $\mathbf{F}^*$. The distribution $q^{\mathrm{ME}}(\mathbf{x})$ does not have the status of a 'preferred map': it is *not* a map, but a prior distribution of randomly placed atoms (this difference will become essential in § 6) which is merely a computational intermediate in obtaining an optimal approximation $P^{SP}(\mathbf{F})$ to $P(\mathbf{F})$ by a multivariate Gaussian on a patch of M surrounding $\mathbf{F}^*$. Thus the really useful result of entropy maximisation is not $q^{ME}(\mathbf{x})$, but the jpd $P^{SP}(\mathbf{F})$, which keeps us within the framework of direct methods. In the sequel, we will use unitary structure factors $\mathbf{U}$, which for point atoms of unit weight are simply $\mathbf{U} = (1/N)\mathbf{F}$.

A remarkable feature of the approximate jpd $P^{SP}(\mathbf{U})$ is that it extrapolates non-zero structure factor values beyond the 'basis' reflexions for which explicit phase choices were made initially and were used as constraints in the entropy maximisation. In other words, $q^{ME}(\mathbf{x})$ has non-zero coefficients even where no constraint values were specified via $\mathbf{U}^*$. $P^{SP}(\mathbf{U})$ is a multivariate Gaussian centered around the vector $\mathbf{U}^{ME}$ corresponding to q^{ME}, and its covariance matrix $\mathbf{Q}$ is essentially the Toeplitz (or Karle-Hauptman) matrix associated to q^{ME} (MEFDM, § 4.2), which will be denoted by writing $\mathbf{Q} - T(q^{ME})$. The conditional distribution of the phases is then obtained as the restriction of this multivariate Gaussian to the manifold M specified by the moduli of non-basis reflexions, the corresponding phases being used as coordinates on M.

A crude but useful approximation to $P^{SP}(\mathbf{U})$, which may be called the 'diagonal approximation', consists in making the covariance matrix equal to the identity matrix, i.e. to the Karle-Hauptman matrix of a uniform distribution $q(\mathbf{x}) \equiv 1/V$. Thus we retain the first-moment information (U^{ME}) generated by the non-uniformity of q^{ME}, but discard the off-diagonal second-moment information $(\mathbf{Q})$. This decouples the various $U_{\mathbf{h}}$'s so that the resulting $P^{SP}_{diag}(\mathbf{U})$ may be written as:

$$\prod_{\mathbf{H}} P^{SP}_{diag}(U_{\mathbf{H}})$$

each factor $P^{SP}_{diag}(U_{\mathbf{H}})$ being a 1D or 2D Gaussian distribution offset from the origin by $U^{ME}_{\mathbf{H}}$. Thus, for $\mathbf{H}$ acentric,

$$P^{SP}_{diag}(U_{\mathbf{H}}) = 2N|U_{\mathbf{H}}| \exp\{-N[|U_{\mathbf{H}}|^2 + |U^{ME}_{\mathbf{H}}|^2]\} \exp[2N|U_{\mathbf{H}}||U^{ME}_{\mathbf{H}}|$$
$$\cos(\phi_{\mathbf{H}} - \phi^{ME}_{\mathbf{H}})] \tag{1}$$

The conditional distribution of each phase ϕ_H is then, for $\mathbf{H}$ acentric:

$$P(\phi_\mathbf{H}) = \exp[X\cos(\phi_\mathbf{H} - \phi_\mathbf{H}^{ME})]/2\pi I_0(X) \qquad (2)$$

where $X = 2N|U_\mathbf{H}||U_\mathbf{H}^{ME}|$. For $\mathbf{H}$ centric, the normalisation factor $2\pi I_0(X)$ becomes $2\cosh X$, and the cosine is then a mere sign.

This crude approximation can already yield useful results. Thus if q^{ME} is constructed from specified values for $U_\mathbf{h}$ and $U_\mathbf{k}$, then it can be shown (MEFDM, eq. 3.23) that

$$U_{\mathbf{h+k}}^{ME} = U_\mathbf{h}U_\mathbf{k} \qquad (3)$$

which is equivalent to

$$\begin{cases} |U_{\mathbf{h+k}}^{ME}| &= |U_\mathbf{h}||U_\mathbf{k}| \\ \phi_{\mathbf{h+k}}^{ME} &= \phi_\mathbf{h} + \phi_\mathbf{k} \end{cases}$$

so that the conditional distribution of $\phi_{\mathbf{h+k}}$ according to (2) is

$$P(\phi_{\mathbf{h+k}}) = \exp[X\cos(\phi_{\mathbf{h+k}} - \phi_\mathbf{h} - \phi_\mathbf{k})]$$

where $X = 2N|U_\mathbf{h}||U_\mathbf{k}||U_{\mathbf{h+k}}|$. This is the classical Cochran distribution (Cochran 1955). In the centric case, a similar argument leads immediately to the tanh formula of Cochran and Woolfson (1955).

4   Likelihood functions and phase refinement: a generalisation of quartets

The calculations carried out so far convey the (correct) impression that entropy maximisation has a useful built-in ability to extrapolate phase information. Indeed, even in the case of the small protein Crambin, very significant phase extension was shown to take place in conditions where direct methods are powerless (MEFDM, § 7). But in this test case, the starting phases to 3Å resolution were exact, and the question immediately arose of knowing whether, in a real situation, errors in the starting phases would not be amplified so as to make the extrapolated phases worthless. In other words maximum-entropy phase extension looks like an inherently divergent process, capable of some extrapolation from good starting phases, but incapable of *refining* bad starting phases prior to their extrapolation. This section will be devoted to showing that this is not the case: phase refinement is possible through the use of *likelihood functions* introduced in MEFDM($\oint 4.2.2$).

We need to first introduce the notion of likelihood in this context. The approximate jpd P^{SP} depends on a certain number of parameters, which are essentially the phases chosen to make up the vector $\mathbf{U}^*$ of constraints used to build up $q^{\mathrm{ME}}(\mathbf{x})$. For given values for these parameters, yielding vector $\mathbf{U}^*$, we may look at the conditional distribution $P^{\mathrm{SP}}(\mathbf{U}_{\perp} \mid \mathbf{U}^*)$ of any set $\mathbf{U}_{\perp}$ of structure factors for which no phases have yet been chosen, and integrate it with respect to the phases of the $\mathbf{U}_{\perp}$ to get the *conditional marginal distribution of moduli* $P^{\mathrm{SP}}(|\mathbf{U}_{\perp}| \mid \mathbf{U}^*)$. The latter distribution will differ from the Wilson distribution (corresponding to $q^{\mathrm{ME}} \equiv 1/V$) in two respects:

- P^{SP} is centered around $\mathbf{U}_{\perp}^{\mathrm{ME}}$, not the origin;
- P^{SP} has covariance matrix $\mathbf{Q} = T(q^{\mathrm{ME}})$, not the identity matrix.

Thus, through the use of the saddlepoint approximation, phase choices for the basis reflexions $\mathbf{U}^*$ induce a *deformation* of the conditional marginal distribution of the moduli $|\mathbf{U}_{\perp}|$ away from their usual (Wilson) distribution. Hypotheses about phase values in $\mathbf{U}^*$ may then be tested as hypotheses about the distribution of the moduli $|\mathbf{U}_{\perp}|$. For this purpose, we define in the usual way the *likelihood* $\wedge$ of the parameter values in $\mathbf{U}^*$ as the conditional marginal probability of the observed moduli:

$$\wedge (\mathbf{U}^*) = P^{\mathrm{SP}}(|\mathbf{U}_{\perp}|_{\mathrm{obs}} \mid \mathbf{U}^*) \tag{4}$$

We may take this likelihood as a 'figure of merit' for the phase choices contained in $\mathbf{U}^*$, since it measures the degree to which the deformation of P^{SP} (away from the Wilson distribution) induced by these choices has enabled it to better predict (hence give a higher marginal probability to) the observed distribution of the other moduli. In practice, it is convenient to normalise this quantity with respect to the null hypothesis (H_o) that the distribution of atoms is uniform, (i.e. that $\mathbf{U}^* = \mathbf{0}$), and to use the likelihood ratio

$$\frac{\wedge(\mathbf{U}^*)}{\wedge(\mathbf{O})} = \frac{P^{SP}(|\mathbf{U}_\perp|_{obs} \mid \mathbf{U}^*)}{P^{SP}(|\mathbf{U}_\perp|_{obs} \mid \mathbf{O})}$$

or its logarithm $L(\mathbf{U}^*) - L(\mathbf{O})$, where $L = \log \wedge$.

A full expression for this likelihood has been obtained for a general multivariate Gaussian P^{SP}, and will be reported elsewhere. If however we make a further approximation and use P^{SP}_{diag} instead of P^{SP}, explicit expressions for likelihoods may be obtained (MEFDM, $\int 4.2.2$). For an acentric reflexion $\mathbf{H}$ belonging to $\mathbf{U}_\perp$, integration of (1) with respect to $\phi_\mathbf{H}$ gives for the conditional marginal distribution of the modulus $|U_\mathbf{H}|$ a Rice distribution:

$$P^{SP}_{\text{diag}}(|U_\mathbf{H}| \mid \mathbf{U}^*) = 2N|U_\mathbf{H}|\exp\{-N[|U_\mathbf{H}|^2 + |U_\mathbf{H}^{ME}|^2]\}I_0(2N|U_\mathbf{H}||U_\mathbf{H}^{ME}|) \tag{5}$$

which reduces to the Wilson distribution when $|U_\mathbf{H}^{ME}| = 0$. The likelihood ratio of the phase choices $\mathbf{U}^*$, in view of the measured modulus $|U_\mathbf{H}|_{obs}$, is then:

$$\frac{\wedge_\mathbf{H}(\mathbf{U}^*)}{\wedge_\mathbf{H}(\mathbf{O})} = \exp\{-N|U_\mathbf{H}^{ME}|^2\}I_0(2N|U_\mathbf{H}||U_\mathbf{H}^{ME}|)$$

and the global log-likelihood criterion from all moduli in $\mathbf{U}_\perp$ is then:

$$L(\mathbf{U}^*) = \sum_{\mathbf{H}\in\mathbf{U}_\perp} [\log I_0(2N|U_\mathbf{H}|_{obs}|U_\mathbf{H}^{ME}|) - N|U_\mathbf{H}^{ME}|^2] \tag{6}$$

The simplest example of the use of likelihood occurs in the case of quartets. Let the space-group be P1, let $h + k + l + m = 0$, and let $\Phi = \phi_\mathbf{h} + \phi_\mathbf{k} + \phi_\mathbf{l} + \phi_\mathbf{m}$. Then, if $|U_\mathbf{h}|, |U_\mathbf{k}|, |U_\mathbf{l}|,$ and $|U_\mathbf{m}|$ are known, the method at the end of $\int 3$ gives for Φ a conditional distribution:

$$p_0(\Phi) = \frac{\exp(X \cos \Phi)}{2\pi I_0(X)}$$

with $X = |U_\mathbf{h}||U_\mathbf{k}||U_\mathbf{l}||U_\mathbf{m}|$. But phase assignments to the four basis reflexions $\mathbf{h}, \mathbf{k}, \mathbf{l}, \mathbf{m}$ will induce ME-extrapolated structure factors at $\mathbf{h} + \mathbf{k}$,

$\mathbf{h}+\mathbf{l}$ and $\mathbf{h}+\mathbf{m}$ according to (3); because, for instance, $\mathbf{h}+\mathbf{k} = -(\mathbf{l}+\mathbf{m})$, $U_{\mathbf{h}+\mathbf{k}}^{\mathrm{ME}}$ will be made of two contributions of the form (3), so that (see Fig 3):

$$U_{\mathbf{h}+\mathbf{k}}^{\mathrm{ME}} = U_{\mathbf{h}}U_{\mathbf{k}} + U_{-\mathbf{l}}U_{-\mathbf{m}}$$

i.e.

$$U_{\mathbf{h}+\mathbf{k}}^{\mathrm{ME}} = |U_{\mathbf{h}}||U_{\mathbf{k}}| \exp[i(\phi_{\mathbf{h}} + \phi_{\mathbf{k}})] + |U_{\mathbf{l}}||U_{\mathbf{m}}| \exp[-i(\phi_{\mathbf{l}} + \phi_{\mathbf{m}})]$$

and hence

$$|U_{\mathbf{h}+\mathbf{k}}^{\mathrm{ME}}| = \{|U_{\mathbf{h}}|^2|U_{\mathbf{k}}|^2 + |U_{\mathbf{l}}|^2|U_{\mathbf{m}}|^2 + 2|U_{\mathbf{h}}||U_{\mathbf{k}}||U_{\mathbf{l}}||U_{\mathbf{m}}| \cos \Phi\}^{\frac{1}{2}}$$

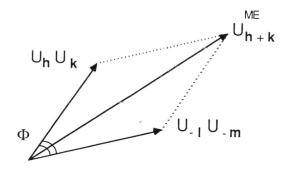

Figure 3. The extrapolated structure factor $U_{\mathbf{h}+\mathbf{k}}^{\mathrm{ME}}$ is made up of two contributions, and the quartet invariant is the angle between them.

The knowledge of the observed value of $|U_{\mathbf{h}+\mathbf{k}}|$ will then give rise to a likelihood whose Φ-dependent part may be written

$$\exp(-2N|U_{\mathbf{h}}||U_{\mathbf{k}}||U_{\mathbf{l}}||U_{\mathbf{m}}| \cos \Phi)I_0(2N|U_{\mathbf{h}+\mathbf{k}}||U_{\mathbf{h}+\mathbf{k}}^{\mathrm{ME}}|)$$

Similar terms arise from the knowledge of $|U_{\mathbf{h}+\mathbf{l}}|$ and $|U_{\mathbf{h}+\mathbf{m}}|$. If the initial distribution $p_0(\Phi)$ is now multiplied by these three likelihood factors, in accordance with Bayes's theorem, then the alert reader will immediately recognize the result as being Hauptman's quartet formula (Hauptman, 1975). The content of this complex formula is immediately obvious in the present derivation:

— if $|U_{\mathbf{h+k}}|$ is large, it is likely that the two contributions to the offset $U_{\mathbf{h+k}}^{\mathrm{ME}}$ interfere constructively, i.e. $\Phi \approx 0$;
— if $|U_{\mathbf{h+k}}|$ is small, it is likely that the two contributions interfere destructively, i.e. $\Phi \approx \pi$.

Of course, in the maximum-entropy setting, we need not analyze the situation in terms of individual phase invariants: different phase choices $\mathbf{U}^*$ may be ranked by using as a figure of merit the global log-likelihood criterion $L(\mathbf{U}^*)$ given by (6), and they may be refined by numerically optimising this criterion with respect to these phases. This is equivalent to simultaneously using a very large set of phase relations (call it an 'orchestra', by extrapolation of the musical progression of quartets, quintets, etc...), without having to explicitly enumerate them or obtain an analytical formula for them: all this is done implicitly, directly in numerical form, through the process of maximum-entropy extrapolation (which leads from $\mathbf{U}^*$ to $\mathbf{U}_\perp^{\mathrm{ME}}$) followed by the use of criterion (6).

For phase refinement by optimisation of $L(\mathbf{U}^*)$ with respect to the phases in $\mathbf{U}^*$, the Jacobian matrix $\partial(\mathbf{U}_\perp^{\mathrm{ME}})/\partial(\mathbf{U}^*)$ must be obtained; this is a simple calculation, whose result will be reported elsewhere. It is also possible to modify the likelihood criterion so as to take into account measurement errors in the observed values of the moduli.

5 Multisolution strategy: tree-directed phase permutation followed by likelihood optimisation

When tackling a structure determination *ab initio* by the method proposed here, the familiar strategy of direct methods has to be modified. Indeed, we no longer want to list phase relations leading to 'estimates' of individual phase invariants: instead, we want to handle at all times the saddlepoint approximation P^{SP} (perhaps $P^{\mathrm{SP}}_{\mathrm{diag}}$) to the joint distribution of *all* structure factors, corresponding to initial phase choices contained in a vector $\mathbf{U}^*$; use likelihood criteria such as (6) to refine these phases; look at the conditional distribution $P^{\mathrm{SP}}(\mathbf{U}_\perp|\mathbf{U}^*)$ as a function of the phases in $\mathbf{U}_\perp$; and enlarge the basis set $\mathbf{U}^*$ by those reflexions in $\mathbf{U}_\perp$ for which strong indications are given.

The hall-mark of this approach is the constant updating of the prior distribution of atoms $q(\mathbf{x})$ to the maximum-entropy distribution compatible with the phase choices made at each stage. To implement these ideas computationally, it is convenient to use as a book-keeping device *a multisolution tree* representing the various phase choices made and the parentage relations between them. Serendipitously, this extra complication allows one to also bring under control a troublesome phenomenon, the 'branching problem' (MEFDM, ∮ 4.3), which arises from the fact that the conditional distributions $P^{\mathrm{SP}}(\mathbf{U}_\perp|\mathbf{U}^*)$, viewed as functions of the phases in $\mathbf{U}_\perp$, are strongly multimodal so that the process of propagating phase information from $\mathbf{U}^*$ to $\mathbf{U}_\perp$ is inherently ambiguous. This phenomenon is not restricted to the use of entropy: it is present in standard direct methods as well, and is handled through the use of magic integers and other tricks. In the present setting, the branching problem can be solved locally by casting it in the form of a generalised eigenvalue problem (MEFDM, ∮ 7.2); the multisolution tree can then be used to record the branching explicitly and to organise the global search for an optimal set of phases so as to avoid becoming uncontrollably trapped around false solutions.

The overall strategy now goes as follows (MEFDM, ∮ 8.1). The root node of the tree consists of the origin-fixing phases, and its first ramification occurs with the choice of enantiomorph. The subsequent growth of the tree is governed by a combination of the following computations at each node:

(a) update the prior distribution of atoms $q(\mathbf{x})$ to the maximum-entropy distribution $q^{\mathrm{ME}}(\mathbf{x})$ compatible with the phase choices (hence the structure factor values $\mathbf{U}^*$ attached to that node;

(b) construct the conditional distribution $P^{\mathrm{SP}}(\mathbf{U}_\perp|\mathbf{U}^*)$ of the yet un-phased structure factors on the basis of $q^{\mathrm{ME}}(\mathbf{x})$;

(c) construct the likelihood function $\wedge(\mathbf{U}^*)$ by integrating the conditional distribution over the unknown phases;

(d) refine the basis phases in $\mathbf{U}^*$ by maximizing $\wedge(\mathbf{U}^*)$ with respect to these basis phases;

(e) solve the local branching problem for (i.e. identify the local maxima of) the conditional distribution with respect to a subset of unknown phases when the values of the corresponding moduli are introduced;

(f) expand the current node by creating a branch leading to a new tip node for each of these maxima; if the conditional distribution is too flat for (e) to be meaningful, just do some phase permutation; each such node has an enlarged set of constraints $\mathbf{U}^*$, and process (a) may be reiterated for each of them.

The multisolution tree would quickly grow to an unmanageable size, and its growth must be supervised so as to maximize the chance of finding the correct set of phases without having to develop too many of its spurious branches. For this purpose, we may use two criteria:

– the loss of entropy in going from a uniform prior $q(\mathbf{x}) = 1/V$ to the current non uniform prior $q^{\mathrm{ME}}(\mathbf{x})$; according to Shannon, this measures the shrinkage of the population of reasonably probable structures when the constraints $\mathbf{U}^*$ are activated;

– the likelihood $\wedge(\mathbf{U}^*)$ calculated from the observed values of the yet unphased moduli; it measures the degree to which the phase choices made in $\mathbf{U}^*$ are able to anticipate correctly, through their maximum-entropy extrapolation, some of the information present in the yet inactive constraints, and hence measures the 'chances of survival' of the current node in the face of these forthcoming constraints.

Intuitively, the entropy of a node measures the 'girth' of the branch directly leading to it from the root node, while its likelihood is an estimate of the sum of the girths of the branches bifurcating away from that node once the values of the unphased moduli are taken into account. One would thus expect likelihood (which looks 'ahead' at the degree of pre-established harmony between its predictions and the actual observations for the yet unphased data) to be a much more versatile criterion than entropy (which merely looks 'back' at the cost of accommodating the current phase assumptions but is blind to the rest of the data). Of course this is not an indictment of entropy, since log-likelihood is in fact another kind of relative entropy.

A suitable combination of these two criteria can then be used as a score which determines the priority with which each node will be allowed to 'ramify' in step (f) to achieve an optimal scheduling of the growth of the multisolution tree.

The approach just described has been actually programmed in collaboration with Chris Gilmore. Applications to some of the most difficult structures in Sheldrick's data bank (Winter2 and sucrose octa-acetate) are most encouraging and show that, even with a diagonal approximation to P^{SP}, the likelihood criterion (6) is extremely powerful. Full details of this work will be reported elsewhere.

6   Generalisation and perspectives: an integrated phase determination method based on likelihood functions

A choice procedure for pruning the multisolution tree would obviously be to bring to bear on this operation any independently available source of phase information, such as that provided by data (however poor) on an isomorphous derivative or by anomalous scattering measurements: even if this information is in no way sufficient on its own to lead to an interpretable electron density map, it will have considerable 'pruning power'. Similarly, if the structure is known to obey non-crystallographic symmetry or molecular envelope constraints, it is possible to select only those nodes of the tree which best obey these constraints. In other cases, one may be able to use the presence, in recognizable form, of a known molecular fragment as a criterion for ranking the competing solutions.

One would feel, however, some degree of disappointment at having to abandon so highly structured an approach as the tree-directed heuristic search described above, and having to resort, through a variety of *ad hoc* hybrids, to the usual disparate collection of gimmicks used so far to circumvent the phase problem.

Fortunately, this capitulation can be avoided. The formalism presented in MEFDM in the case of one unknown structure built (for simplicity) from point atoms of unit weight has now been extended (Bricogne, to appear) to the case of *families of related structures* made from several types of atoms, each atom type having an arbitrary (complex) scattering factor which can be different in each structure of the family, and being distributed with its own non-uniform prior *simultaneously* in all these structures. This setup is general enough to include heterogeneous structures, isomorphous replacement, anomalous scattering at one or several wavelengths, and contrast variation. The joint probability distribution of any set of structure factors taken from the different structures (for instance, 'cylindrical' sets comprising a given collection of reflexions considered simultaneously across all members of the family of structures) can then be obtained, extending in particular the recent results of Hauptman and of Karle on the incorporation of isomorphous replacement and anomalous scattering into direct methods. The equivalence between the maximum-entropy method and the saddlepoint approximation continues to hold in this generalized setting. This derivation of statistical phase relations for arbitrary complex-valued scattering factors shows clearly that the source of such relations is the positivity of the prior probability distributions of the various type of atoms, *not* the positivity of the electron density as has been claimed since 1950.

In the same work, the maximum-entropy formalism has also been extended into a statistical formulation of the molecular replacement and molecular averaging methods, by deriving joint distributions of structure factors in the presence of known structural fragments, of solvent regions, of non-crystallographic symmetries, and even of multiple crystal forms (the latter allowing one to deal with non-isomorphous heavy-atom derivatives). These extensions are readily merged with those concerning the treatment of families of related structures.

Furthermore, the optimal Gaussian approximation of the conditional distributions given by the saddlepoint method have been used systematically to construct likelihood functions from all the observed data available (including their error estimates) which measure the adequacy of all the ingredients of the statistical model used to derive the conditional distribution in the first place. The numerical optimisation of these likelihood functions can be used, as before, to refine initial phase assignments; but it can also be used to detect heavy atoms or known fragments, or to refine a description of molecular boundaries or of local symmetry elements. In this way, all the distinct sources of phase information which had hitherto been treated via separate — or at best loosely coupled — procedures may now be first identified, then optimally combined, through a single basic computational device.

This scheme does more than simply add the strengths of the

individual sources of phase information, since it introduces new and powerful interactions among them, namely the statistical relationship between the phases of different reflexions from a family of related structures, as they are specified by the joint distribution of all structure factors for which data are available.

Finally, the likelihood functions have been obtained in a sufficiently general form to allow one to consult not only single crystal data, but also partially corrupted data such as those originating from fibre and powder diffraction experiments: they can thus serve to extend the use of direct methods to these types of data.

Conclusion

I hope to have given not too inadequate an outline of an attempt at formulating a comprehensive statistical treatment of the phase problem, incorporating all known types of phase information and yet treating them all through a single computational process: the maximisation of likelihood functions calculated from joint probability distributions of structure factors in which all contributions from randomly placed atoms are handled through the saddlepoint approximation, i.e. by always using maximum-entropy prior distributions of atoms. It is the latter method which allows one to deal optimally with strongly non-uniform distributions of atoms, such as those encountered in macromolecular crystal structures, which had so far been thought to be one obstacle (among many) to the application of direct methods to macromolecular crystallography: indeed, the saddlepoint method turns non-uniformity from a handicap into a strength.

REFERENCES

Bertaut, E. F. (1955a). *Acta Cryst.*, **8**, 537–543.
Bertaut, E. F. (1955b). *Acta Cryst.*, **8**, 544–548.
Bertaut, E. F. (1955c). *Acta Cryst.*, **8**, 823–832.

Bricogne, G. (1984). *Acta Cryst.*, **A40**, 410–445.

Cochran, W. (1955). *Acta Cryst.*, **8**, 473–478.

Cochran, W. and Woolfson, M. M. (1955). *Acta Cryst.*, **8**, 1–12.

Hauptman, H. and Karle, J. (1953). *The Solution of the Phase Problem: I. The Centrosymmetric Crystal.* A.C.A. Monograph No. 3. Pittsburg: Polycrystal Book Service.

Jaynes, E. T. (1957). *Phys. Rev.*, **106**, 620–630.

Klug, A. (1958). *Acta Cryst.*, **11**, 515–543.

Peschar, R. and Schenk, H. (1987). *Acta Cryst.*, **A43**, 513–522.

Shannon, C. E. and Weaver, W. (1949). *The Mathematical Theory of Communication.* Urbana: University of Illinois Press.

6

MOLECULAR REPLACEMENT: THE METHOD AND ITS PROBLEMS

E. J. Dodson

1   Introduction

Many protein crystals grow with more than one molecule in each assymetric unit; these molecules are related by short-range, noncrystallographic symmetry, and indeed we can often see the pseudo symmetry by inspecting the diffraction patterns. It was an obvious path of study to try and characterise and use this information.

The early work by David Blow and Michael Rossmann was concerned with exploiting this, and they established methods for describing these relationships, although they are unable to generate phases from the relationships alone as they had hoped. So far all the primary achievements in determining the structure of proteins still rest on defining phases by the isomorphous substitution of heavy atoms into the protein crystal, but when there are several molecules in each assymetric unit the maps calculated from these isomorphous phases contain duplicate information, and Gerard Bricogne has shown how to exploit this information to reduce the isomorphous phase error. He developed methods for efficiently averaging the density of the molecular repeats, and transforming this density back into phase information. The recent determination of virus structures are a triumph for this technique. (Rossmann (1985)) (I shall give a brief summary of his procedure at the end of the paper.)

It was also found that many proteins form different crystal forms and crystallise under different conditions; these are often in different space groups, and it was realised that the structural information from one crystal form could be used to help solve the other.

The techniques developed to do these studies in both the *ab initio* case and when there is a known structure are known collectively as the Molecular Replacement Method. It depends on some of the most pleasing aspects of crystallography; symmetry, molecular transforms, and the convolution theorem.

2 The Method

The idea is simple but the practice can prove extremely frustrating.

Questions that must be answered are listed below.

1. How can one detect whether there is structure conservation within the one crystal, or between related crystal forms?

2. How can we describe the relationship if it exists?

3. How can we use the relationship ... if it exists?

First let us recap what we know;

A) We have as data the Structure amplitudes and the information encapsulated in the Patterson function they generate. This function is the sum of the pattern of vectors between atoms within each subunit (the self vectors) and of the vectors between atoms belonging to different subunits (the cross vectors) — the cross vectors can result from either crystallographic or from noncrystallographic symmetry.

B) We also know that a rigid body transformation can be described by six parameters; three angles needed to define the rotation, and three translational components.

It is simpler to solve the problem in two stages, the rotation angles can be obtained from inspection of the self vectors in the Patterson which are centred at the origin, and then the search for the translation components can use the correctly orientated parent molecule to predict patterns.

First let us consider methods for finding the three angles which define the rotation matrix [Q].

3   The Rotation Overlap Function

Rossmann and Blow (1962) defined a Patterson overlap function

$$R[Q] = \int_V P_1(u')P_2(u)dV \tag{1}$$

where $P_1(u')$ and $P_2(u)$ are the values of the Patterson functions at the points u' and u where $u' = [Q]u$ (see below for detailed definition of this equation) and V is the volume over which matching is expected. When [Q] is correct the self vectors of the Patterson functions will overlap and $R[Q]$ will be large. This is true when $P_1(u')$ and $P_2(u)$ are taken from the same Patterson (self search) or are derived from different structures.

Rossmann and Blow showed that if $P(u)$ is expanded in the usual way as a Fourier series

$$p(u) = \sum_{hkl} F_{hkl}^2 exp(-2\pi i \mathbf{h}.\mathbf{u}) \tag{2}$$

$R[Q]$ simplifies to

$$R[Q] = u/V^3 \sum_{\mathbf{h}} \sum_{\mathbf{h'}} F_{\mathbf{h}}^2 F_{\mathbf{h'}}^2 G_{\mathbf{h}.\mathbf{h'}} \tag{3}$$

where $\mathbf{h}$ and $\mathbf{h'}$ range over reciprocal space
and

$$[h'] = [h][Q].$$

$G_{\mathbf{h}.\mathbf{h'}}$ is a shape function which can be defined for simple volumes of integration. In fact $G_{\mathbf{h}.\mathbf{h'}}$ is virtually zero unless $\mathbf{h}$ is close to $\mathbf{h'}$.

Tony Crowther showed that by expanding the Patterson density within a spherical volume in terms of spherical harmonics it was possible to use fast Fourier transforms to calculate values of the overlap function; this meant that the calculations ran up to 100 times faster. This allows the user to run the program with different combinations of the control parameters, using different shells of data.

3.1 Expansion of $[Q]\mathbf{x}$ and $[Q]\mathbf{u}$

Molecular replacement assumes that relative to some orthonormal axial system $[I\ J\ K]$ we can generate the new coordinate set $\mathbf{x}_1$ from the coordinate set $\mathbf{x}_0$ and similarly, the new vector set $\mathbf{u}_1$ from $\mathbf{u}_0$.

To write this in full, the vector

$$[I\ J\ K]\begin{bmatrix} x \\ y \\ z \end{bmatrix} = [I\ J\ K][Q]\begin{bmatrix} x \\ y \\ z \end{bmatrix} + \begin{bmatrix} tx \\ ty \\ tz \end{bmatrix} \tag{4}$$

where $[I\ J\ K]$ are the orthonormal axial system

$$\begin{bmatrix} x \\ y \\ z \end{bmatrix}$$ the coordinates of individual atoms in this axial system

$$\begin{bmatrix} tx \\ ty \\ tz \end{bmatrix}$$ a translational vector in this axial system

In practice we assume the axial system for both sides of equation 4 are the same, and simply write

$$\mathbf{x}_1 = [Q]\mathbf{x}_0 + \mathbf{t} \tag{5}$$

3.2 Expressions for $[Q]$

Any matrix with real elements and determinant equal to unity represents a rotation in real space, and any rotation can be expressed in terms of three independent angles. There are amny ways of doing this, but as long as the user knows how his/her matrix has been defined, the matrix ELEMENTS should be the same for each and all methods. It is absolutely necessary to make sure all the programs used have the same conventions built into them! However different authors of papers about molecular replacement seem to feel no compulsion to standardise their definitions with those of other authors – the authors in the book edited by Michael Rossmann illustrate this! (Rossmann (1972)).

One useful form is given in International Tables Volume 2 p. 63 (1952).

1. Spherical Polar Angles

If $[Q]$ describes a rotation of ω about a vector with direction cosines $l_1\, l_2\, l_3$, then

$$[Q] = \begin{bmatrix} cos\omega + l_1(1 - cos\omega) & l_1l_2(1 - cos\omega) & l_3l_1(1 - cos\omega) \\ & -l_3sin\omega & +l_2sin\omega \\ l_1l_2(1 - cos\omega) & cos\omega + l_2(1 - cos\omega) & l_2l_3(1 - cos\omega) \\ +l_3sin\omega & & -l_1sin\omega \\ l_3l_1(1 - cos\omega) & l_3l_2(1 - cos\omega) & cos\omega + l_3(1 - cos\omega) \\ -l_2sin\omega & +l_1sin\omega & \end{bmatrix} \quad (6)$$

If the rotation axis makes an angle of ω with the K_0 axis, and its projection in the I_0J_0 plane makes an angle of θ to the I_0 axis, (measured as positive towards the J_0 axis), the direction cosines are

$$[l_1\, l_2\, l_3] = [sin\omega cos\theta \quad sin\omega sin\theta \quad cos\omega] \quad (7)$$

and the substitution of these values gives $[Q]$ in its spherical polar form. It is often useful to use this form when searching for non-crystallographic symmetry within the assymmetric unit, where we might expect to find a two–fold or a three–fold or even a seventeen–fold axis, and therefore know the expected value of ω.

There is a version of the rotation program which searches in spherical polar angles. It is called POLARRFN, written by W. KABSCH and has been installed into the UK protein crystallographic program suite by Phil Evans of the MRC Cambridge. (CCP4)

2. Eulerian Angles.

The program originating from Tony Crowther works in Eulerian angles which he defines with respect to rotation of AXES. (I find this difficult to visualise, but it is possible to show that rotation of axes by α , then β then γ is equivalent to rotating coordinates relative to fixed axes by γ then β then α .) He defines [Q] using the system given below.

If the two sets of coordinates are defined relative to an orthonormal axial system I, J, K, the matrix [Q] can be written as a function of three Eulerian angles, α, β, and γ.

If γ is a rotation about the initial direction of K_0, β is a rotation about the subsequent direction of J_1, and γ is a rotation about the final direction of K_2, then

$$[Q] = \begin{bmatrix} \cos\alpha\cos\beta\cos\gamma & -\cos\alpha\cos\beta\sin\gamma & \cos\alpha\sin\beta \\ -\sin\alpha\sin\gamma & -\sin\alpha\cos\gamma & \\ \sin\alpha\cos\beta\cos\gamma & -\sin\alpha\cos\beta\sin\gamma & \sin\alpha\sin\beta \\ +\cos\alpha\sin\gamma & +\cos\alpha\cos\gamma & \\ -\sin\beta\cos\gamma & \sin\beta\sin\gamma & \cos\beta \end{bmatrix} \tag{8}$$

Other programs use Eulerian angles defined relative to different axial systems.

(Note that the actual multiplication by [Q] are sometimes done in real space, sometimes in reciprocal space, and it is important to be clear whether [Q] is being used to pre-multiply the column vector of coordinates, or to post-multiply a row vector of indices.)

3.3   Expressions for Orthogonalising Matrix [RO]

So far all these definitions have referred to ORTHONORMAL axes, and it is obviously essential that we are clear how we have defined the orthonormal axes relative to the crystal axes. The program ALMNFR,(CCP4) based on Tony Crowther's fast Fourier summation for calculating the Rossmann/Blow rotation overlap allows three ways for defining these.

1. I along a, K along c^*.

2. I along b, K along a^*.

3. I along c, K along b^*.

The atomic coordinates x_0 relative to these axes are derived from the fractional coordinates relative to the crystal axes by $x_0 = [RO]x_f$ and the elements of [RO] for each system are given in below.

It reduces the amount of calculation to have the highest symmetry along K, so option 3 is used for monoclinic spacegroups, option 1 for all high symmetry spacegroups with 3,4 or 6 fold axes along c. For orthorhombic spacegroups the choice is arbitrary.

Option 1. (Remember this α, β, and γ are the cell angles).

$$[RO] = \begin{bmatrix} a & b\cos\gamma & c\cos\beta \\ 0 & b\sin\gamma & -c\sin\beta\cos\alpha^* \\ 0 & 0 & c\sin\beta\sin\alpha^* \end{bmatrix} \tag{9}$$

Option 2.

$$[RO] = \begin{bmatrix} a\cos\gamma & b & c\cos\alpha \\ -a\sin\gamma\cos\beta^* & 0 & c\sin\alpha \\ a\sin\gamma\sin\beta^* & 0 & 0 \end{bmatrix} \tag{10}$$

Option 3.

$$[RO] = \begin{bmatrix} a\cos\beta & b\cos\gamma & c \\ a\sin\beta & -b\sin\alpha\cos\gamma^* & 0 \\ 0 & b\sin\alpha\sin\gamma^* & 0 \end{bmatrix} \tag{11}$$

4   Practical Difficulties

I shall refer for examples to two papers in *Acta Cryst*; one on Haemoglobin, (Derewenda *et al.*, (1981)) and the other on des-pentapeptide insulin (DPI) (Bi Ru-Chang *et al.*, (1983)). the haemoglobin study was very straightforward, but the DPI one was often misleading.

The two principal pitfalls for the user are obvious when you consider the nature of the overlap function.

1. The overlap search is done by matching SPHERICAL SHELLS of Patterson space. This shell should ideally contain most of the intra-molecular vectors, but few of the inter-molecular ones resulting from crystal packing. This is possible if the molecular is roughly globular, and in a large volume of solvent, like haemoglobin. The choice of radius for the sphere of integration is determined by the dimensions of the molecule, its shape, and its packing in the crystal. If the molecule is markedly ellipsoidal vectors will be omitted if the radius is limited to the molecule's smallest dimension; on the other hand, if the radius is set to the largest dimension, the spherical volume will be grotesquely larger than the molecular volume.

When there is close crystal packing with small solvent volume and an irregularly shaped molecule as with des-pentapeptide insulin, it is impossible to separate the inter- and intra-molecular vectors within a spherical volume, and any change of radius can alter the appearance of the overlap map considerably. Changing the radius from 11Åto 13Åincreased the height of the true peak relative to the spurious ones. Increasing the radius to 15Åincreased the noise level.

Obviously if the sphere radius is greater than half a cell edge the same Patterson density will be included twice, and if the radius were set greater than a cell edge the origin peak would be included again. The program now grinds to a halt if you attempt to do this.

The inner radius is used in a function to modify the F^2 terms to remove the origin peak. It must be at least equal to the resolution limit.

This problem of non-spherical molecules is an insoluble one when using the Crowther-Blow method and I feel it is the main reason why some overlap maps look like the haemoglobin one and some like the DPI ones.

2. The overlap function is a Patterson function using squared and sharpened values of F, and is therefore likely to be distorted by a few large terms. I find it is always necessary to sharpen the input intensities to give an even distribution of large F^2s against resolution; the program lists the number of terms in each $sin\theta/\lambda$ shell after sharpening thus allowing you to check whether the chosen sharpening factor is sensible. Ian Tickle suggests using E values.

3. Another problem arises when comparing an observed Patterson with one generated from calculated F^2 derived from a model structure; obviously differences between the two structures will affect the degree of overlap. Some differences are inevitable; usually the known structure will have a different sequence or crystal form, so it may seem safer to omit many sidechains and external regions. However the DPI example shows that caution can be overdone – a better result was obtained when more atoms were included even though many were not correctly placed in the new form. Presumably their presence gave a more realistic distribution of the calculated amplitudes. The best way round this problem is to repeat the overlap search several times using different shells of intensities. Any true solution should be at least positive in all such searches, even if it is not the local maxima. There seems no way of predicting which shell will be the best for your problem; if there has been a good deal of movement between different protein domains the best result may come from low resolution matching; but if there are packing difficulties then the solution may only show up when high resolution data is used allowing matching of more precise features. It is not possible to generalise about which parameters will work for any given problem.

4. Crystal symmetry can make interpretation of results very complex. There are formulae for predicting symmetry equivalents of Eulerian angles, and the equivalent rotation matrices can be generated.

If $\quad \mathbf{x}_1 = [Q_{11}]\mathbf{x}_0$, then there is a rotation matrix $[Q_{ij}]$ such that

$$[S_i]\mathbf{x}_1 = [Q_{ij}][S_j\mathbf{x}_0 \tag{12}$$

so

$$[S_i][Q_{11}] = [Q_{ij}][S_j] \tag{13}$$

and

$$[Q_{ij}] = [S_i][Q_{11}][S_j]^{-1} \tag{14}$$

(Remember that the symmetry matrices here apply to ORTHOGO-NAL coordinates and are the product of the crystal symmetry AND the orthogonalising matrix.)

The ALMNFR version of the fast rotation program now tabulates all equivalent sets of Eulerian angles for any maxima, with the spherical polar equivalents and the direction cosines of the axis of rotation.

In the DPI study the difficulty was not that the complexity of the solution was disguised by the crystal symmetry, but that there were high noise levels and ambiguous peaks. We were however able to crosscheck all the results from the four crystal forms by using the suitable symmetry equivalents of the maxima to find equivalent matrices matching form A to for B, form B to form C and Form A to form C, etc.

5. There is another insoluble problem which can arise from crystal symmetry which cannot be avoided. If the non-crystallographic symmetry axis is approximately aligned with a crystal symmetry axis, any maxima of the search function will tend to smear into its symmetry equivalent, and it is often not possible to be sure where the true peak is. This hazard is obviously more likely to arise with higher symmetry space groups.

5 Positioning the Molecule in its Cell

5.1 Translation Functions

There is further information in the Patterson which can give the translation components to position a parent molecule in the new cell. (I shall not consider the case where there is no set of known coordinates — the translation searches within the same Patterson are not accurate, and most workers who quote the translation components have obtained them from their heavy atom coordinates.) The pattern of self vectors will overlap the observed Patterson when oriented correctly and centred at the Patterson origin. The cross vectors from one molecule to its symmetry equivalents will produce a similar pattern centred about a point $[S_i]\mathbf{x}_c$ - $\mathbf{x}_c$ where $\mathbf{x}_c$ is the molecule's centroid. (In fact this is the Harker vector of the centroid.) The translation functions look for this correlation.

Ian Tickle gives a very clear summary of the mathematics of the Crowther and Blow T2 functions. (Tickle 1985)

$$T2(\mathbf{t}) = \int_V (P_o(\mathbf{u}) - P_m(\mathbf{u})) \cdot P_c(\mathbf{u},\mathbf{t}) du \qquad (15)$$

where $P_0(\mathbf{u})$ is the observed Patterson

$P_m(\mathbf{u})$ is the 'self Patterson' for the given

coordinates and symmetry

and $P_c(\mathbf{u},\mathbf{t})$ is the calculated Patterson for the

correctly oriented molecules with

an unknown translation.

We use a program developed by Ian Tickle. the translation terms are calculated then passed to the Fast Fourier program to calculate a map which should give maxima at the Harker positions for the centoid.

5.2   R Factor Search Function

An alternative method for finding the translation parameters is by searching for a minimum R value between the observed intensities and the different sets of calculated amplitudes generated as the molecule was moved through the unit cell. This method was used by Cutfield *et al.* (1974) and described fully by Nixon and North (1976). The program used (called SEARCH (CCP4)) uses the fact that if the partial Fc's, Fc_1, Fc_2,...are calculated for all the symmetrically equivalent model molecules, orientated according to the appropriate rotation matrix, then the value of Fc for any set of translation, t_1, t_2, ..., where the t_i are symmetrically equivalent, will be given by

$$\mathbf{F}c(hkl) = \mathbf{F}c_1(hkl)exp(-2i\pi\mathbf{h}\cdot\mathbf{t}_1) + \mathbf{F}c_2(hkl)exp(-2i\pi\mathbf{h}\cdot\mathbf{t}_2) + \dots \quad (16)$$

So, once the $\mathbf{F}c_i$ are calculated it is only necessary to sum them together with appropriate phase modifications to generate the different sets of $\mathbf{F}c$'s.

It is possible to use the inverse Fast Fourier transform for spacegroup P1, using the cell dimensions of the observed intensity set to calculate all the required $\mathbf{F}c_i$ relatively inexpensively, although this method is much slower than the translation function.

The appearance of any translation parameter search gives some verification to the quality of the rotational parameters. If these are inaccurate, or if there is a poor match between the model and the new structure, it is impossible to find consistent clear solutions for the translation parameters. The two functions are different; one works with F^2 terms and the other with F.

6   Real Space Molecular Replacement

I have not mentioned molecular replacement in real space; programs are available which work directly with the Patterson maps. The great advantage of these programs is that a non-spherical masking map can be defined which screens for points to be included in the overlap. The disadvantage is that the procedure is orders of magnitude slower.

Manfred Buehner has a useful comparison of his experiences with reciprocal and real space molecular replacement. (Buehner (1985)).

7   Verifying the Results

We use two procedures to check the quality of our fitting. The first necessary but not sufficient check is to run a distance/angles calculation to see that the suggested fit is not causing different symmetry models to collide. The program, PRJANG, (a descendent of DJANGO), does this very quickly. (It can also be used to step through the unit cell with a reduced set of atoms, usually the $C'_\alpha s$, to chart possible and impossible locations within the cell.)

The best check is to generate phases from the model in the new cell and see if these phases can be used in any way. If there is heavy atom derivative data, then these phases should show up the heavy atom sites.

In DPI they were sufficiently accurate to show sulphur atoms which had been excluded from the phasing. But a word of warning; if the orientation angles are correct the phases will have some truth in them, and if the translation is out by some fraction of the symmetry operations a great many of the phases are 'right' and may produce ghosting.

The ultimate test of the model is: is it good enough to allow us to refine it to an accurate structure? There are bound to be problems, there will be some error in the rotation and translation parameters which will become more serious at greater distance from the molecule's centre. And there will be some gross errors where the model structure and the new form are different. Nevertheless, in our experience if the refinement of some parts of the new form can proceed smoothly and generate improved phases sufficiently accurate to allow rebuilding of the different fragments, then the search for isomorphous derivatives is bypassed, and a useful accurate crystal structure can be quickly obtained. This has been done for both Haemoglobin and DPI, as well as many other structures.

We have found initial R factors of about 60% for 2Å data fall to 30%–35% automatically when the initial rotation and translation parameters have been substantially correct. At that point rebuilding and extension of the model has been fairly straightforward. A work of warning; incorrect structures have 'refined' too. The refinement procedures will ALWAYS lower the R factor whether the model is nonsense or not, and other biochemical criteria are necessary in deciding on the correctness of any solution.

A sensible plan of approach is to use a rigid body refinement to correct some of the larger errors (Leslie 1984). The program CORELS (Sussmann 1977) refines the parameters of restrained groups of atoms. It has a larger radius of convergence than any atom-by-atom refinement and can be used to both improve the transformation parameters, and to shift parts of the structure relative to other bits. Its disadvantages are: it is rather difficult to set up — 'rigid' groups have to be defined before the procedure starts, and it is very slow.

Zygmunt Derewenda refined the haemoglobin tetramer in $P2_12_12$ beginning from coordinates of the dimer in $P2_1$ (Shaanon 1983). He found that after the initial cycles of automatic atomic refinement he was able to improve the rotation and translation parameters by recalculating the fit of the inner core of the parent molecule and the roughly 'refined' model. This gave corrections to the Eulerian angles of the order of 2 degrees, and to the translation parameters of the order of 0.2Å. By repositioning the parent molecule with these new parameters and beginning the refinement cycles again he was able to go farther with the automatic procedure, and to reduce the work of rebuilding and correcting the model, especially in the outer shell.

8   Use of non-Crystallographic Symmetry for Phase Improvement

The phase constraints implied by duplicated structural features were first derived by Rossmann and Blow (1963) and used in reciprocal space by Jack (1973) in work on the TMV coat protein disk. However, the computing requirements were proportional to N^2 where N was the number of independent reflections.

Bricogne (1974) reformulated the theory in real space, and showed that this procedure was equivalent to the general reciprocal space technique, and was a great deal more economical. He set up programs to use this on the IBM and now there are versions of these programs available for VAX machines.

To use this method the rotation and translation parameters relating the different molecules have to be known. This requires a good solution to the self rotation, and translational parameters. In some cases these have been obtained from the matching of heavy atom sites (Jack 1973) or from the known symmetry of the molecule, for example, the viruses. The programs can make some correction to orientation parameters, but they must be substantially correct, and the noncrystallographic symmetry must be fairly exact.

The necessary steps are:

1. trace an envelope for one independent subunit,

2. rotate the other subunits to this orientation by the previously acquired matrices and translation vectors,

3. average the density of all the subunits,

4. set the density outside the molecular boundary to an average value,

5. reconstruct an improved electron density map by reversing the rotations and translations,

6. transform this map to generate phases and amplitides (which could extend past the initial resolution limit),

7. recombine these modified phases with the initial M.I.R set, weighting them appropriately,

8. recalculate an improved map.

Steps 2 and 8 are recycled and the phase shifts converge quite rapidly after 3 or 4 cycles.

NOTE: This technique has something in common with the Wang procedure for improving phases. That uses steps 1, 4, 6, 7 and 8 and it is clear why the Wang procedure has to be much less successful than the molecular averaging.

REFERENCES

Bi Ru-Chang, Cutfield, S. M., Dodson, E. J., Dodson, G. G., Giordano, F., Reynolds, C. D., Tolley, S. P., (1983). *Acta Cryst.*, **B39**, 90–98

Blow, D. M., (1962). *Acta Cryst.*, **15**, 24–31

Bricogne, G., (1973). *Acta Cryst.*, **A30**, 395–405

Buehner, M., Hecht, H. J., (1985). In *Molecular Replacement. Proceedings of the Daresbury Study weekend*, (compiled by P.A. Machin), pp 62–69. SERC Darebury Laboratory, Warrington, U.K.

Crowther, R. A., Blow, D. M., (1967). *Acta Cryst.*, **23**, 544–548

Crowther, R. A., (1972). In *The Molecular Replacement Method. A Collection of Papers on the Use of Noncrystallographic Symmetry*, (ed. M. G. Rossmann), pp 173–178, New York: Gordon and Breach

Derewenda, Z. S., Dodson, E. J., Dodson, G. G., Brzozowski, A. M., (1981). *Acta Cryst.*, **A37**, 407–413

Jack, A., (1973). *Acta Cryst.*, **A29**, 545–554

Leslie, A. G. W., (1984). *Acta Cryst.*, **A40**, 451–

Nixon, P. E., North, A. C. T., (1976). *Acta Cryst.*, **A32**, 320–325

Program Suite for Protein Crystallography (CCP4) SERC Daresbury Laboratory, Warrington, U.K.

Rossmann, M. G., Arnold, E., Erickson, J. W., Frankenberger, E. A., Griffith, J. P., Hecht, H. J., Johnson, J. E., Karmer, G., Luo, M., Mosser, A. G., Rueckert, R. R., Sherry, B., and Vriend, G., (1985). *Nature*, **317**, 145

Rossmann, M. G., Blow, D. M., (1962). *Acta Cryst.*, **15**, 24–31

Rossmann, M. G., Blow, D. M., (1963). *Acta Cryst.*, **16**, 39–45

Rossmann, M. G., (1972). In *The Molecular Replacement Method. A Collection of Papers on the Use of Noncrystallographic Symmetry*, (ed. M. G. Rossmann), pp 173–178, New York: Gordon and Breach

Shaanan, B. J., (1983). *J. Mol. Biol.*, **171**, 31–51

Sussman, J. L., Holbrook, S. R., Church, G. M., Kim, S. H., (1977). *Acta Cryst*, **A33**, 800

Tickle, I.J., (1985). In *Molecular Replacement. Proceedings of the Daresbury Study Weekend*, (compiled by P. A. Machin), pp 22–26, SERC Daresbury Laboratory, Warrington, U.K.

Tollin, P., (1966). *Acta Cryst.* **21**, 613–614

Wang, B.C., (1985). In *Methods in Enzymology, Vol. 115, Diffraction Methods for Biological Macromolecules*, (ed. H. W. Wyckoff, C. H. W. Hirs and S. N. Timasheff), pp 90–111, Academic Press, Inc.

7

STRUCTURAL ANALYSIS BY THE METHOD OF MULTIWAVELENGTH ANOMALOUS DIFFRACTION

Wayne A. Hendrickson

1. INTRODUCTION

Multiwavelength anomalous diffraction (MAD) analysis has emerged
in recent years as an advantageous alternative to the multiple
isomorphous replacement (MIR) method for ab initio determination
of macromolecular structure. This advance has been spurred by
the availability of readily tunable x-rays from synchrotron
sources and by developments in methodology for analyzing the
diffraction measurements. The data are measured at a few
wavelengths, typically three or four, in the vicinity of an
absorption edge of an appropriate anomalous scatterer.
Absorption edges for elements of atomic numbers in the range
from 20(Ca) to 47(Ag) and from 50(Sn) to 92(U) are accessible to
x-ray wavelengths of from 0.5 to 3.0 A. Intense beams in this
range are available at most hard x-ray synchrotron sources. It
has long been recognized that multiple wavelength measurements
from crystals containing anomalous scatterers should suffice for
definitive phase evaluation. Indeed, the multiwavelength
experiment can be viewed as analogous to multiple replacements
that are strictly isomorphous. While data analysis by MIR
procedures is possible, greater power is found in a fundamental
reformulation of the problem.

The MAD procedure is beginning to have appreciable success. We have described a test with lamprey hemoglobin (Hendrickson, 1984) and the structure determination of selenolanthionine (Hendrickson, Troup, Swepston and Zdansky, 1986). Kahn et al. (1985) have reported the analysis of a parvalbumin and Harada et al. (1986) have described an application to a cytochrome c'. Korzun (1987) has carried out the analysis of an azurin structure based on two wavelengths. Recently, we have determined the structures of a bacterial ferredoxin at 5A resolution and of core streptavidin at 3.1A resolution. Applications to other problems are in progress by us and others.

The mathematical basis and experimental procedures for MAD phasing analysis have been reported elsewhere (Hendrickson, 1985a; Hendrickson, Smith and Sheriff, 1985; Hendrickson, 1987) and detailed reports of specific applications that we have made are in preparation. This article focuses on the system of computer programs, MADSYS, that we have developed to implement phase determination by the MAD method. A brief recounting of the theoretical foundation of the method is also given as a framework for this discussion of computational procedures.

2.   THEORETICAL FOUNDATION

The basis of phase determination from anomalous diffraction measurements lies in the distinctiveness of scattering from different kinds of atoms that can arise from resonance effects. Classical Thomson, or "normal" scattering relates to the interaction of x-rays with free electrons. Thus, induced vibrations respond directly to the incident frequency and the scattering strength is independent of wavelength. In reality, the electrons in atoms are in bound orbitals and vibrations induced by the incident x-ray wave resonate with natural frequencies of bound electrons. The resulting scattering is perturbed from the "normal" value, and this "anomalous" correction obviously varies with wavelength and the atom type. Moreover, whereas the normal scattering f^o from an atom, relative to that from a free electron, is purely real; the anomalous scattering includes a phase shift and thus it is complex with real and imaginary components denoted by f' and f''. The total atomic scattering factor is then

$$f = f^o + f'(\lambda) + i\, f''(\lambda). \tag{1}$$

The separate terms of the atomic scattering factor lead to corresponding terms in the structure factor equation. This then also separates wavelength-dependent and wavelength-invariant factors. Karle (1980) has shown how these factors can also be separated in analyzing the diffracted intensities. We use a somewhat different division of components (Hendrickson, 1985a) that lends itself to direct analysis of the crystal structure at hand. The relationships increase in complexity with the number of kinds of anomalous scattering centers, but in the commonly occurring case of a single kind of anomalous center they take on a particularly simple form. If we use $|^{\lambda}F(h)|$ to denote the complete structure factors for reflection h at a particular wavelength λ, then the diffracted intensity, $I = K|F|^2$, is directly proporation to

$$|^{\lambda}F(\pm h)|^2 = |^oF_T|^2 + a(\lambda)\ |^oF_A|^2 \tag{2}$$
$$+ b(\lambda)\ |^oF_T||^oF_A|\cos(^o\varphi_T - {}^o\varphi_A)$$
$$\pm c(\lambda)\ |^oF_T||^oF_A|\sin(^o\varphi_T - {}^o\varphi_A).$$

Here, $^oF_T = |^oF_T|\exp(i\,{}^o\varphi_T)$ is the normal scattering (f^o) contribution from all atoms in the structure and $|^oF_A|$ is the corresponding contribution just from the anomalous centers. These factors are, of course, wavelength invariant. All

wavelength dependence is encoded in a, b, and c -- factors that
are ratios of anomalous to normal scattering components:

$$a(\lambda) = (f'^2 + f''^2)/\ f^{o2}, \tag{3a}$$

$$b(\lambda) = 2\ (f'/f^o), \tag{3b}$$

$$c(\lambda) = 2\ (f''/f^o). \tag{3c}$$

The formulation given by Eq. (2) is the basis for
structural determination based on anomalous diffraction
measurements. From the data measured at a few appropriate
wavelengths and at both Bijvoet mates it is possible to evaluate
$|^oF_T|$, $|^oF_A|$ and $\Delta\phi = {}^o\phi_T - {}^o\phi_A$ for each reflection. Then,
using the magnitudes $|^oF_A|$, the structure of anomalous
scatterers can be determined and used to calculate ${}^o\phi_A$. Given
$\Delta\phi$, this yields the desired phase, ${}^o\phi_T$, needed to produce a
Fourier synthesis for the total structure.

3.   COMPUTATIONAL PROCEDURES

Analysis of diffraction measurements at multiple wavelengths
from crystals that contain anomalous scatterers involves a
variety of calculations. The program, MADLSQ, that solves the
system of equations (2) lies at the heart of the procedure, but
programs used to evaluate scattering factors, to scale the data,
to locate anomalous scatterer positions, to evaluate
macromolecular phases, and to compute Fourier syntheses are also
involved. We have collected the necessary programs together in
a system named MADSYS. Some of the programs in this system
derive from other problems and some were acquired from other
investigators, but many have been developed expressly for this
problem.

3.1 Determination of Scattering Factors

As is obvious from Eqs. (2) and (3), knowledge of the
atomic scattering factors is an essential ingredient for the
multiwavelength data analysis. The normal scattering factors
are simply derived from quantum calculations of atomic electron
density distributions and these have been tabulated. Cromer and
Liberman (1970) have also shown how the anomalous scattering
factors can be computed from cross sections derived from wave
functions for the free electrons. These have been tabulated for
certain characteristic X-ray lines, but they can also be
calculated at arbitrary wavelengths with a program by Cromer
(1983). We have incorporated this into a program, SPECTRUM,
used to produce a complete wavelength profile for any atomic
species.

While SPECTRUM is very useful for points somewhat remote
from the absorption edges, the theoretical values from free
atoms do not apply very near the edge. (Templeton, Templeton,
Phizackerley, and Hodgson, 1982). Both the edge position and
the magnitude of scattering are affected changes in wave
function and atomic energy levels due to the molecular bonding
environment. While these effects could in principle be taken
into account in molecular orbital calculations, this would
require prior knowledge of the atomic structure. Fortunately,
sufficient experimental information is available. The imaginary
part of the anomalous scattering factor, f", is directly related
to the atomic absorption coefficient, and the real part, f', can
be obtained from the f" spectrum by Kramers-Kronig
transformation (James, 1948). The atomic absorption data can be
measured by fluorescence detection from the very crystal used
for diffraction measurements. We have developed a program
XASFIT to reduce such x-ray absorption spectra to f" values and
another program KRAMIG performs the Kramers-Kronig
transformation to produce f' values. Both of these calculations

are typically made subject to constraints to values from
SPECTRUM at points relatively remote from the edge.

    A further complication arises due to anisotropy in
anomalous scattering factors. The pleiochroism in diffraction
described by Templeton and Templeton (1985) extends to the local
environment of individual atoms. Thus, ultimately the
scattering factors must be treated as refinement variable as in
the program MADDST that we have planned (Hendrickson, 1985a).
Values from XASFIT and KRAMIG would be used as the starting
point for this refinement. Dr. Yoshinori Satow (personal
communication) has already written a program that can refine
isotropic scattering factors.

3.2 Scaling

    As usual, diffraction intensities must be processed to
account for a number of factors including the Loventz factor,
polarization effects, absorption, radiation damage and source
intensity variation. We assume that these have been made with
whatever data acquisition software has been used to process the
measurements. There generally remain certain systematic errors
that can be reduced by local scaling procedures, and it is
important to make these corrections for problems with relatively
weak signals as can happen for protein structure determination
based on K-edge scatterers.

    The special scaling procedures that we use are based on
those developed in the course of our analysis of crambin
(Hendrickson and Teeter, 1981). These have been described at
earlier computing schools (Smith and Hendrickson, 1982;
Hendrickson, 1985b). ANOSCL and SCALE2 are used to determine
and apply parameterized local scaling factors that bring sets of
data related by Friedel symmetry into correspondence. Various
auxiliary routines, such as ANOCHK, RMSANO and ANORES, are used
to support these calculations. The data from separate

wavelengths must also be brought to a common scale. This is
first done by bringing the various sets into least-square
agreement by the parameterized local scaling algorithm used in
ANOSCL. The program used to accomplish this, WVLSCL, was
written by Janet Smith (personal communication). Additional
scale factors that account for variation in total scattering
strength with wavelength must also be applied. These factors do
not differ appreciably for dilute, weak scatters (as for iron in
lamprey hemoglobin) but they can become very important if
anomalous scattering is strong (as in selenolanthionine).
Finally, although it may not be essential to have data placed on
an absolute scale this can be very useful. We accomplish this
by a program, KCURVE, that is based on Wilson's statistics as
employed by Karle and Hamptman.

It is worth mentioning that data sorting and merging
routines are also needed, e.g. LEXICO and OMERGE (written by
Janet Smith) and that in some instances it is advantageous to
defer merging of equivalent data until after phase determination
for separate sets.

3.3 Structure of Anomalous Scatterers

Once the diffraction data have been reduced to a common
scale and appropriate scattering factor values have been
determined, the system of MAD equations can be solved for each
reflection. MADLSQ accomplishes this in two steps described
previously (Hendrickson, 1985a). First, given a, b and c, the
cofactors of the four terms in Eq. (2) are treated as variables
in a linear least-squares fitting. Then the trigonometric
constraint that $\cos^2\theta + \sin^2\theta = 1$ is introduced as a LaGrange
constraint in a non-linear refinement. A similar procedure has
been encountered by Chapuis et al. (1985), but the slow
convergence described by those investigators is contrary to our
experience. Thus MADLSQ procedure produces, among other

results, a set of $|^{O}F_{A}|$ values for use in determining the
structure of anomalous scattering centers.

The anomalous scatterer structure amplitudes from MADLSQ
have the advantage of being limited in accuracy by measurement
errors rather than by factors such as approximations in theory,
intercrystal scaling or lack of isomorphism. Hence, these
values can be used directly in Patterson syntheses (PATTER) or
direct methods (MULTAN) for interpretation of the structure.
Both techniques have proved effective in our applications.
However, in actual practice to date, especially with
area-detector data, we find that the least squares fittings for
some reflections give unrealistic values for $|^{O}F_{A}|$. Poor
fittings can often be detected from the statistical output from
MADLSQ, but it remains important to eliminate outlier values in
computing Pattersons and in producing and selecting normalized
structure amplitudes for direct methods.

Once a trial structure has been deduced for the anomalous
centers, this must be refined. The program for doing this,
ANOLSQ, also derives from the crambin study. Even though the
"observed" data for this refinement as produced from MADLSQ may
still contain substantial error, it is noteworthy that the
positional parameters can be determined rather accurately since
there is a great degree of over determination. When a
satisfactorily refined model has been obtained, FACALC is used
to compute phases and structure amplitudes corresponding to the
normal scattering from the anomalous centers.

3.4 Phase Determination

The actual phase calculation is in principle trivial.
Given the $\Delta\phi$ from MADLSQ and $\phi_{A}(calc)$ from FACALC, the desired
phase is simply $^{O}\phi_{T} = \Delta\phi + \phi_{A}(calc)$. All space group
symmetry restrictions are inherent in $\phi_{A}(calc)$. One

complication that arises concerns the estimation of errors.
Since measurements are usually made at only a few wavelengths,
the variance estimates from individual MADLSQ refinements are
not very reliable. An alternate phase probability procedure has
been devised by Arno Pähler and it has been implemented in a
program MADABCD by Pähler and Smith (personal communication).
This program uses $|F_A(calc)|$ and $\phi_A(calc)$ values in a
refinement to produce $|{}^oF_T|$ magnitudes and new ${}^o\phi_T$ estimates.
Then the A, B, C and D coefficients from a lack-of-closure
analysis (Hendrickson and Lattman, 1970) are generated and used
to calculate centroid phases and figure-of-merit weights. The
probability distribution are particularly useful for combination
with other phase information as from molecular averaging or
solvent flattening.

One essential ambiguity generally remains at this stage of
the determination; namely, the choice of hand. The structure of
anomalous centers, if handed, would equally well explain the
observed magnitudes either as deduced from Patterson or direct
methods calculations or as the enantiomorphous structure. (If
the anomalous structure is centric, there is no ambiguity and no
loss of generality in the solution.) The phase difference from
MADLSQ will, however, only be compatible with the absolute
configuration. Usually, Fourier syntheses (FORIER) computed
with phases deduced from the alternative enantiomorphs will
distinguish the correct hand by chemical and physical
reasonableness. If non-crystallographic symmetry is present,
the correlation coefficients relating the equivalent subunits
provide a useful monitor. Chemically interpretible features are
the ultimate test.

4.   PROSPECTS

The applications that we have made to date convince us that the
analytical procedures described here are effective. Moreover,
the prospective use of selenomethionyl proteins as a general
phasing vehicle (Hendrickson, 1985a; Hendrickson, 1987) and the
construction of synchrotron beamlines dedicated to these
experiments make MAD phasing a promising future direction for
macromolecular crystallography. Improved experimental facilites
should yield more accurate data, but many future problems are
likely to be inherently more challenging than those attacked to
date. Thus, continued development in methods is essential.

     Several prospective development come to mind. One is to
realize the MADDER and MADDST programs envisioned earlier
(Hendrickson, 1985a) whereby atomic parameters and scattering
factors for the anomalous centers will be incorporated directly
as global variables in the phasing calculations. Another is to
further develop the error analysis. A third prospect is to
treat properly the anisotropy of anomalous scattering for some
species. The Templetons have observed substantial pleiochroism
from our selenolanthionine crystals and we see similar behavior
with selenobiotinyl streptavidin. Although the streptavidin
analysis worked well without treating this effect, improvements
can be expected when anisotropic scattering factors are used.
Yet another important area concerns methods for determining
complicated anomalous scatterer structures from MADLSQ results.
While our simulations of data for MULTAN are encouraging, real
problems will not produce strictly random errors. Nevertheless,
since the full set of macromolecular data gives a vast
overdetermination of anomalous center parameters, and these
centers are well resolved, appropriately designed direct methods
should succeed in spite of the limited accuracy of the data.

ACKNOWLEDGEMENTS

I thank Janet Smith, Arno Pähler, Krishna Murthy and John Horton for their participation in the development and application of MAD phasing methods, and I am very grateful to Ethan Merritt and Paul Phizackerley of the Stanford Synchrotron Radiation Laboratory and to Yoshimori Satow of the Photon Factory for help with the synchrotron experiments that sustain the effort. This work was supported in part by a grant, GM-34102, from the U. S. National Institutes of Health.

REFERENCES

Chapuis, G., Templeton, D. H., and Templeton, L.K. (1985). Acta Cryst. A41, 274-278.

Cromer, D. T. (1983) J. Appl. Cryst. 16, 437

Cromer, D. T., and Liberman, D. (1970). J. Chem Phys. 53, 1891-1898.

Harada, S., Yasui M., Murakawa, K., Kasai, N., and Satow, Y., (1986) J. Appl. Cryst. 19, 448-452.

Hendrickson, W. A. (1984). Acta Cryst. A40, C-3.

Hendrickson, W. A. (1985a). Trans. Amer. Cryst. Assn. 21, 11-21.

Hendrickson, W. A. (1985b). In Computational Crystallography, G. M. Sheldrick, C. Krüger and R. Goddard) Oxford Univ. Press,
Oxford, pp. 277-285.

Hendrickson, W. A. (1987). In Crystallography in Molecular Biology, (D. Moras et al. Eds.) Plenum Press, New York, pp 81-87.

Hendrickson, W. A. and Lattman, E. E. (1970) Acta Cryst. B 26, 136-143.

Hendrickson, W. A., Smith, J. L., and Sheriff, S. (1985). Methods in Enzymology 115, 41-55.

Hendrickson, W. A., and Teeter, M. M. (1981) Nature 290, 107-113.

Hendrickson, W. A., Troup, J. M., Swepston, P. N., and Zdansky, G. (1986). Abstracts of the American Crystallographic Association Abstracts, Series 2, Vol. 14, p. 48.

James, R. W. (1984). The Optical Principles of the Diffraction of X-rays, Bell, London

Kahn, R., Fourme, R., Bosshard, R., Chaimdi, M., Risler, J. L., Dideberg, O., and Wery, J. P. (1985) FEBS Lett. 179, 133-137.

Karle, J. (1980). Int. J. Quant. Chem. 7, 356-367.

Korzun, Z. R. (1987). J. Mol. Biol. 196, 413-419.

Smith, J. L., and Hendrickson, W. A. (1982) In Computational Crystallography, (D. Sayre, Ed.) Oxford Univ. Press, Oxford, pp. 209-222.

Templeton, D. H., Templeton, L. K. , Phizackerley, R. P.and Hodgson, K. O. (1982). Acta Cryst. A38, 74-78.

Templeton, D. H. and Templeton, L. K. (1985). Acta Cryst. A41, 365-371.

Structure refinement

8

RESTRAINTS AND CONSTRAINTS IN LEAST SQUARES REFINEMENT OF SMALL MOLECULES

David Watkin

Very few crystallographers are now engaged in writing refinement programs for small molecules, but most will be forced at some stage to use them. In these programs, as in others,

$$\text{Garbage in} = \text{Garbage out}$$

Increasing sophistication in data collection procedures has reduced the Garbage coefficient of the experimental data, but there remains another fruitful source of garbage — the Users of the program. Mis-use of a program, perhaps through failure to understand the principles of the technique, can have a very high Garbage coefficient. To use a program well, you don't necessarily need a deep understanding of the maths, but you must at least have a feeling for the background to the techniques involved.

1 Background

An aim of X-ray structural analysis is to provide some model for the periodic pattern of electron density which is a currently acceptable way of describing the internal structure of a crystal. There are well established hypotheses about the nature of the interaction between X-rays and this periodic density which enable us, given the amplitudes, to build models of the electron density.

If we have reliable observations of amplitudes and estimates of the phases (which might well be the case in a centro-symmetric structure) then, neglecting such effects as diffuse scattering and so on, a three dimensional Fourier summation enables us to compute the electron density at points in the crystal. Such a representation is perhaps the best we can achieve, and by plotting out the density at many points as figure fields or contour maps, we produce diagrams providing valuable insight into the 'real' electron density.

Such diagrams are particularly valuable when there are effects like disorder, but are rather impractical for archiving, for representing concisely, and for comparing one structure with another. For these purposes we need to condense the available information, and this can be done by postulating a mathematical model which can be used to represent the density. It is in the definition and refinement of this model that we need to keep our wits about us, and be aware of the effect of constraints and restraints.

References for full descriptions of least-squares (LS) techniques are given later (Press, 1986), and so I will give here just a brief reminder. We have an expression for Y_c as a function of some variables x whose values we seek. We also have some observed (experimental) values for Y_o. We can expand Y_c as a Taylor series in x (even if the expression is linear) to find shifts which must be applied to initial values of x to improve Y_c.

$$Y_{old} + (\partial Y/\partial x_1) \cdot \delta x_1 + (\partial Y/\partial x_2) \cdot \delta x_2 \ldots = Y_{new}$$

The Y_{new} we want is the same as Y_o, giving the observational equations

$$(\partial Y/\partial x_i) \cdot \delta x_i \ldots = Y_o - Y_c$$

in which we compute Y_c and the derivatives from the current model. We can write the whole set of equations as

$$A \cdot \delta x = y$$

Note that if the observations are of different precision or their errors are correlated, we need to weight these equations in a more or less complex way. Pre-multiplying both sides by A transpose (A′) gives the Normal equations

$$H \cdot \delta x = A' \cdot y = r$$

These we solve for the corrections (δx) to be applied to the current model. The normal matrix H is square and composed of terms of the form

$$h_{ij} = \sum (\partial Y_m / \partial x_i) \cdot (\partial Y_m / \partial x_j)$$

The right hand side is a vector of terms of the form

$$r_i = \sum (\partial Y_m / \partial x_i) \cdot (Y_{obs} - Y_{calc})_m$$

and the solution is $\partial x = H^{-1} r$. It should be noted that not every possible parameter need contribute to every Y, i.e. the derivative can be zero.

In crystal structure analysis, the bulk of the observations Y_{obs} are structure amplitudes, and the Y_{calc} are computed by the structure factor expression from the atomic and over-all parameters. In general, there is a one-to-one correspondance between the physical parameters and those being refined in the LS process. However, it is not always possible or appropriate to refine all possible physical parameters, and in some cases physical parameters may be combined to form more complex LS parameters. This information must be coded in some way, and can be represented in a table.

The table can be used not only to relate physical parameters to simple LS parameters (as in x,y,z for atom 1), but also to control the contributions to complex LS parameters, sometimes called 'free variables', if the entries are the derivatives of the physical parameters *wrto* the LS parameters (Larson, 1980). Thus the entries for parameter 5 indicate that the site occupancies for atoms 1 and 2 are strictly inversely related (anti-riding), and the entries for 6 and 7 define two new LS parameters

$$ls6 = x2 + x3, \quad \text{and} \quad ls7 = x2 - x3$$

Entries represented by '.' mean that there is no contribution from the physical parameter to the corresponding LS parameter. Physical parameters with only '.' below them are thus not being refined, though

Restraints and Constraints

TABLE 1

LS prm	atom 1							atom 2				atom 3			atom 4			atom 5		
	x	y	z	Occ	Uis	U11	U22..	x	y	z	Occ..	x	y	z...	x	y	z.	x	y	z
1	1	.	.	.	.	.	.	.	.	.	.	.	.	.	.	.	.	.	.	.
2	.	1	.	.	.	.	.	.	.	.	.	.	.	.	.	.	.	.	.	.
3	.	.	1	.	.	.	.	.	.	.	.	.	.	.	.	.	.	.	.	.
4	.	.	.	.	1	.	.	.	.	.	.	.	.	.	.	.	.	.	.	.
5	.	.	.	1	.	.	.	.	.	.	-1	.	.	.	.	.	.	.	.	.
6	.	.	.	.	.	.	.	1	.	.	.	1	.	.	.	.	.	.	.	.
7	.	.	.	.	.	.	.	1	.	.	.	-1	.	.	.	.	.	.	.	.
8	.	.	.	.	.	.	.	.	.	.	.	.	.	.	1	.	.	1	.	.
9	.	.	.	.	.	.	.	.	.	.	.	.	.	.	.	1	.	.	1	.
10	.	.	.	.	.	.	.	.	.	.	.	.	.	.	.	.	1	.	.	1

they CAN contribute to Y_{calc}, and hence influence the refinement if their current values are used in the structure factor expression.

2   Constraints and Restraints

Refinements with Constraints and Restraints are closely related techniques, early literature often calling restraints 'slack constraints', though their mathematical implementations are rather different. A constraint or restraint is a hypothesis which is made about the solution to the problem. The solution is restrained to satisfy the hypothesis, the restraint being enforced more vigorously as our confidence in the hypothesis increases, and is enforced absolutely as a constraint when we admit no possibility of doubt. Table 2 lists three classes of hypothesis with examples.

3   Implied constraints

These are those things which are generally regarded as being computable without appreciable error with respect to their 'true' values. This implies that we have a good understanding of the effect, and a good mathematical

TABLE 2

IMPLIED CONSTRAINTS	EXPLICIT CONSTRAINTS	RESTRAINTS
Lp correction.	Space group.	Geometry
Analytic absorption correction.	Temperature factor.	Temperature factor
Ideal imperfection.	Extinction.	Shift limitation
Atomic form factors.	Atomic groups.	parameter sums
	Occupancy.	parameter averages
	Special positions.	Symmetry
	Floating origins.	Special positions
		Floating origins

representation of it. Our structural model is going to be constrained in some way by the equations of these corrections or tabulated data, and we normally dismiss the thought that these corrections are part of our model. Occasionally, in less than routine analyses, workers are forced to look futher than Vol 4 for form factors, and need to devise directed or molecular form factors. All of these items are described by parameters which should appear in Table 1, and the equations using them be involved in the formation of Y_{calc}.

4   Explicit constraints

These are some items which the analyst may care to think about, and about which decisions have to be made. Sometimes several different hypotheses seem to explain the observations equally well, in which case it is customary to differentiate between them by applying Occams Razor, though it is not always easy to decide which of two very dissimilar assumptions IS minimal. An additional or more careful experiment may resolve the problem, but this is not always possible. Under this latter condition, we have to postulate characteristics for the solution, and then keep alert for possible consequences. These postulates are constraints.

For example, we may postulate that a heavy atom has an isotropic atomic displacement parameter (temperature factor). If this is valid, then we do no harm and it should have no ill effects on the rest of the analysis. However, if it is wrong then we are not adequately modelling the electron density distribution, and something else in the analysis may become distorted as a consequence — in particular, light atoms close to the heavy atom might take on unreasonable positions. In this case a solution may be to use a more complex model, with anisotropic temperature factors (Uaniso). More complex models are not a universal cure. It used to be fashionable to see if a group of atoms in a structure could be described as a 'rigid group' by fitting a TLS model to the individual Uaniso. At the end of the computation you get translation, libration and a screw coupling tensor for the group, and an R factor based on the discrepancy between the atomic Uaniso, and those computed from TLS. Every generation of students would find an example where R(U) was, say, 4%, and a diagonal element of L was -200. Since these terms represent libration squared, the value is worthless, and throws doubt onto the value of any element of TLS. The problem is over-parameterised for the data content of the Uaniso. Either a simpler model must be used, or extra information be introduced, by saying for example that the components of T for the fragment are the same as those for a larger fragment (whole structure?), and then finding L and S for the sub-fragment. The constraint here is that we have T, from the larger fragment.

A similar situation can occur in structure analysis when diffraction data are poor or scarce, in which case we might find C-C distances in a phenyl group in the range 1.2 – 1.6Å. Such values are improbable, and if we observe them we must wonder what other distortions are present. Because the refinement is non-linear, we are doing refinement against $(F_o\text{-}F_c)$, which must be incorrect if there is a material error in the model. One solution is to replace the atomic parameters by parameters describing rotation and translation of the group as a whole. The atoms concerned are initially 'regularised' to some ideal geometry (not necessarily a regular hexagon —

the user is free to choose any geometry his experience indicates should be suitable). This geometry is preserved during the refinement. The individual atomic parameters are replaced by 6 LS parameters, 3 defining a rotation, and 3 a translation. Some implementations put the ideal model at the origin of a local coordinate system and compute shifts needed to position this in the cell. I prefer to place the ideal model at approximately the crystallographic position, and compute corrections. This leads to a well conditioned matrix under most conditions, including those where atoms in the group are also the subject of restraints. Table 1 would now contain items which are the derivatives of physical parameters *wrto* the group parameters, and so must be re-computed between each round of refinement. The reduction in the number of variables (from 3n to 6 for n atoms) reduces the work done in forming the normal matrix, but this is to some extent offset by having to use the chain rule for partial derivatives, and in any case the really important thing is that the solution is kept within reasonable bounds. A related technique is that of putting hydrogen (or other) atoms into structures either at positions geometrically computed at the end of each round of LS, or by making the parameter shifts 'ride' upon those of an adjoining atom. In this approach, a single LS parameter is made to represent 2 (or more) physical parameters (LS parameters 8,9,10). The derivatives for all the physical parameters should be added into the normal matrix, though this is sometimes omitted for hydrogen atoms to save time. In a general implementation it should be possible to make any parameter ride on any other, and thus provide a mechanism for relating parameters like temperature factors or site occupancy. Equating all the x, the y and the z parameters to 3 LS variables is a mechanism for group translational refinement.

The analyst needs to be clear about the difference between rigid body group and riding refinements, and rigid body and riding models for thermal motion. In refinement, the analyst is making postulates about the relationships between parameter shifts. In thermal analysis he is making postulates about the correlation of the thermal motion of groups of atoms.

These things are not necessarily related at a mathematical level, though may be, by being applied to the same groups of atoms.

Care must be taken in computing the esds of molecular parameters when there is not a simple 1–to–1 relationship between the LS and physical parameters. For example, a hydrogen atom riding on a carbon will have the same positional esds as the carbon, but a bond length esd computed from the atomic esds alone is nonsense, since in a riding model the bond length is a constant. A proper esd computation using the full variance-covariance (VCv) matrix will produce a zero esd, since the covariance matrix carries the information that the parameters are not independant. Note also that in a group refinement, atoms near the periphery of the group will have larger esds than those near the centre, but properly computed bond lengths for atoms in the group have zero esd. The problem is most subtle when some atoms are in groups, and the others are not. BEWARE of molecular geometry programs that only use atom parameter esds.

A different sort of constraint is necessary in those space groups where the origin is not fixed by symmetry in one or more direction, e.g. P1, Pc, P2. Here, the normal matrix will be singular since there is no crystallographic data to fix the origin. A traditional solution is not to refine the corresponding coordinates of one atom (e.g. y in P121). However, after refinement the esd of the unrefined parameter must be zero, and in order to preserve the relative uncertainties between (in this case) the y coordinate of the fixed atom and those of the other atoms, the y esd of those atoms must increase. If molecular parameters are computed, their esds will be too high unless the VCv matrix is used. The problem is less severe if a heavy atom is used to fix the origin, but the situation is most conveniently handled using restraints.

It is natural to hope that crystal structures fall into one of the 230 regular spacegroups, but in practice structures are found which cannot be completely described in a high symmetry space group, but for which the use of a lower symmetry one introduces so many extra variables (which are probably highly correlated anyway) that the refinement is unsatisfactory.

For example, all but a few atoms may be related by a good pseudo mirror plane. The low symmetry refinement is unstable, and the high symmetry refinement means accepting disordered atoms. One solution is to perform the refinement in the low symmetry space group, but apply symmetry constraints to parameters related by the pseudo mirror. The shifts for corresponding pairs of parameters are equivalenced to single LS parameters, though with opposite signs to preserve the mirror. The structure is thus partly in one space group, and partly in another. Similar strategies can be used to preserve the total occupancy of two partially occupied sites, or one site occupied by two different species of atom. If three sites or species are involved, linear constraints of this form become difficult to implement, and it is more convenient to use restraints.

When there is no serious ambiguity about the space group, there is little difficulty in recognising the constraints to be imposed on the parameters of atoms in special positions. These are handled by not refining a parameter, or by equivalencing several atomic parameter shifts (perhaps with multiplicities) to a single LS parameter (e.g. the x and y parameters of an atom at (x, x, z) on a diagonal mirror).

5 Restraints

So far, the Y_{obs} have been regarded as being structure factors or amplitudes, and constraints have been imposed on the model by manipulating the parameters to be refined. These constraints MUST be obeyed even if they are nonsensical, because there is no mechanism in the mathematics (except failure of the program) for them not to be.

A more democratic approach to influencing the refinement is to admit other types of Y_{obs}. All that are needed are expressions relating Y_{obs} to the structure parameters, and values for the Y_{obs}. The derivatives then can be found, as with the structure factor expression, and the terms added into the normal matrix and right hand side vector. Sometimes additional variables need to be introduced, e.g. a dummy atom for some implementations of the planarity restraint, or perhaps even real parameters as when one co-refines X-ray and neutron data.

Thus, in restrained refinement, a value is postulated for some function of the parameters, and this is used in exactly the same way as the X-ray data. These supplemental observations may fit in nicely with the X-ray data, or resolve an ambiguity left by the X-rays, or may be in conflict with them — in which case something needs to be reconsidered. In an earlier section, we noted the need to weight the X-ray observations. In general, these are measured on a common scale and to a similar precision, so that we have means for estimating the relative weights to be applied. Supplemental observations are of an entirely different kind, and need to be put on the same common scale. If there are only relatively few supplemental observations, then we can use some convenient mechanism to indicate their relative precision between them selves (e.g. an expected e.s.d.), and then scale this to the weighted residual for the X-ray observations (Rollett, 1965). When the bulk of the data is supplemental, for example in the refinement of powder diffraction data of complex materials, there seems to be no well established practice.

We noted above that for the X-ray data we have a function of the form

$$F_{cal} = \sum (f \cdot \exp(t) \cdot \exp(2\pi \cdot i \cdot hx))$$

where t is a temperature factor and x the atomic positions, so that every atom coordinate has a contribution to F_{calc}. For restraints, the Y_{calc} are usually a function of only a subset of the parameters, so that an observational equation consists largely of zeros. Algorithms must be used for efficiently finding those matrix elements which are a function of the parameters involved.

A commonly used restraint is to set an interatomic distance (bonded or non-bonded) to a given value, so the function for Y_{calc} is

$$(X_1 - X_2)^2 + (Y_1 - Y_2)^2 + (Z_1 - Z_2)^2 = D^2$$

where X, Y, and Z are orthogonal coordinates. Six parameters are linked by this restraint, the fractional positional coordinates of the two atoms.

If we recall that $D^2 = \Delta x' \cdot G \cdot \Delta x$, where G is the metric tensor and Δx is the vector of fractional coordinate differences, this expression can be differentiated directly to give terms like

$$\partial D / \partial x_1 = (g_{11}\Delta x + g_{12}\Delta y + g_{13}\Delta z) \ / \ D$$

and

$$\partial D / \partial x_2 = -(g_{11}\Delta x + g_{12}\Delta y + g_{13}\Delta z) \ / \ D$$

Note that these expressions make no use of the absolute position in the cell (unless x_2 is x_1 operated on by a suitable symmetry operator), so that the use of molecular geometry restraints alone as a means for regularising a structure usually leads to a singular matrix - the position in the cell is not fixed.

The value for D could be one taken from the literature, if that is believed to be appropriate, or, in a case where it is expected that several bonds should have the same length (by molecular symmetry), D could be set to the current average value. This must be computed between each round of refinement. The effect of restraints is to add contributions to terms dotted about the matrix. The matrix inverter may warn that parameters are now highly correlated, but this is not unexpected since we are deliberately correlating positions through the restraints. The form of the distance restraint has a symmetrical effect whether the actual separation is larger or shorter than the target value, and so is suitable for distances defined by fairly rigid molecular geometry. For softer (Van der Waals) contacts, we expect distances shorter than the target to be unlikely, but know less about longer ones. These contacts have usually been handled with a restraint in the form of an energy penalty function, e.g. part of a Buckingham potential. I prefer to use an ordinary distance restraint, but with a weight which is itself a function of the observed interatomic distance. When this is shorter than the target VdW distance, the restraint has a large weight, and tries hard to force the atoms apart; when the distance is greater, the weight is reduced and so tries only gently to pull the atoms together, with an effect that decreases as the separation increases. The fact that D_o-D_c now appears in both the y vector and the weight means that the process is no longer strictly LS. The convergence properties of these techniques are mentioned in articles on robust-resistant refinements (Prince, 1982).

The restraints described above are of an essentially chemical nature, but other types of observational equation are possible, e.g.

$$x_1 + x_2 + \ldots = constant$$

If x_1 and x_2 are site occupancies of two atoms, and the constant is 1, the restraint is equivalent to the constraint applied to atoms 1 and 2 via ls5 above. However, the restraint has advantages. The target value (1) may not be achieved, indicating partial occupancy or problems with the temperature factors. In addition, the sum is not limited to two atoms, so that quite complex models for site or compositional disorder can be devised.

If the x_i are atomic coordinates along a polar direction, and the constant is the sum of their current values, then the restraint (for b polar)

$$y_1 + y_2 + \ldots y_n = \sum(y_i)$$

will keep the centre of gravity along b fixed, and thus fix the origin. This is a better technique than fixing one atom because of the way the parameters errors become correlated.

If we use only one x_i, and use the current value as the target, we have a shift limiting restraint, which tries to keep the coordinate at its current value. Note that this is only one observation added in with all the others, and if they strongly influence x_i, it will shift. However, if they contain little information about x_i, it should not be influenced by accumulated errors in the matrix work.

We can also have observations of the form

$$x_i = constant, \text{ or } x_i - x_j = constant$$

An application might be U[iso] = 0.10, for a water of solvation. Space group symmetry requirements of atoms on special positions can also be introduced as restraints, the constraint for the atom on (x,x,z) given above being replaced by the restraint

$$x_1 - y_1 = 0.0,$$

or for an atom on $(x,\ 1/2,\ z)$

$$y_1 = 0.5$$

For an ordinary free refinement of atomic parameters, there is not much advantage in applying symmetry conditions as restraints, though it does make programming simpler at the cost of a slightly larger matrix — an advantage which may become important. However, when atomic parameters are replaced by group parameters, the direct organisation of the matrix for groups containing atoms on special positions becomes very difficult to implement in a problem independent way, because special relationships must be imposed on the group parameters. If the group refinement is set up in the normal (unconstrained) way and special conditions imposed on atoms as restraints, the correct solution emerges without additional problems. We have not yet programmed group thermal parameters (TLS), but for all the constraints and restraints outlined above it seems that the user can mix them in any sensible fashion without risk of numerical instability.

6 Singular Value Decomposition

This is the recommended method for solving LS problems because of the information it gives about ill- or un-defined parameters. However, it requires storage of the full observational matrix together with another the size of the normal equations, and so is currently impractical. None the less, the technique can be applied to the solution of the normal equations (in the forming of which some information may have been lost) and still give useful diagnostics. In this context it is sometimes called eigenvalue filtering. The normal matrix is decomposed into a rotation matrix V (square but not symmetric, so we must store it all) and a diagonal matrix D, whose elements are the squares of the singular values.

$$A = V \cdot D \cdot V' \text{ and } A^{-1} = V' \cdot D^{-1} \cdot V$$

In effect, the rotation matrix maps the actual parameters into a space in which the new parameters are uncorrelated. Any zero diagonal elements of D mean that A contains no information about the model in that particular direction in this new space, — we have less real degrees of freedom than we thought, or we have over parameterised the real model. In fact, any small value of d_{ii} indicates incipient singularity in A. D is inverted by taking the reciprocal of the diagonal elements. Setting the reciprocal of any small or zero element to zero rather than a large value removes the parameter from the intermediate space, and hence its influence on the real parameters. This technique was coded up in Oxford many years ago, but fell into dis-use because of the computational cost. It seems it provided an effective way for dealing with atoms on special positions or floating origins, and I feel it should be re-introduced as a way of dealing with unexpected correlations.

7   Conclusions

Constraints and restraints are an inevitable feature of structure refinement. Their proper use makes the difference between getting a significant physical solution, and just getting numbers out of the program. The major advantage of constraints is that they reduce the size of the computation. Restraints offer the advantage of great flexibility, and an indication of when they are inappropriate. A 'reasonable' restraint which is strongly disobeyed (Y_{calc} very different from Y_{obs}) after refinement is a good indication of either some serious problem with the data, or that the restraint is inappropriate, in which case you have mis-understood the model.

The material covered in this brief note has been strongly influenced by discussions over many years with John Rollet and Bob Carruthers, to both of whom I owe a great debt; and to experience with CRYSTALS, which implements most of the techniques described.

REFERENCES

Larson, A. C. (1980). In *Computing in Crystallography* (ed. R. Diamond, S. Ramaseshan, and K. Venkatesan) Indian Academy of Sciences, Bangalore.

Press, W. H. (1986). Numerical Recipies - The Art of Scientific Computing. Cambridge University Press.

Prince, E. (1982). Mathematical Techniques in Crystallography and Materials Science, Springer-Verlag.

Rollett, J. S. (1965). In *Computing Methods in Crystallography* (ed. J. S. Rollett) Pergamon Press, Oxford.

9

CRYSTALLOGRAPHIC REFINEMENT BY SIMULATED ANNEALING

Axel T. Brünger

1 Introduction

Conventional refinement of biological macromolecules involves a series of steps, each of which consists of a few cycles of least-squares refinement with stereochemical and internal packing constraints or restraints (Sussmann *et al.*, 1977; Jack and Levitt, 1978; Konnert and Hendrickson, 1980; Moss and Morffew, 1982; Hendrickson, 1985), that are followed by rebuilding the model structure with interactive computer graphics (Jones, 1982). During the final stages of refinement solvent molecules are usually included and alternative conformations for some atoms or residues in the protein may be introduced.

The aim of least-squares refinement is to minimize the difference between the observed ($|F_{obs}(hkl)|$) and calculated ($|F_{calc}(hkl)|$) structure factor amplitudes, which is usually expressed as a weighted sum of the residuals

$$\sum_{hkl} W_{hkl}[|F_{obs}(hkl)| - |F_{calc}(hkl)|]^2 \qquad (1)$$

where hkl are the reciprocal lattice points of the crystal. An indicator for the progress of the refinement is the R factor

$$R = \sum_{hkl} ||F_{obs}(hkl)| - |F_{calc}(hkl)|| \ / \ \sum_{hkl} |F_{obs}(hkl)| \qquad (2)$$

The least-squares refinement process is easily trapped in a local minimum and it does not correct residues that are misplaced by more than about 1Å so that human intervention becomes necessary. Least-squares refinement can be understood as a nonlinear optimization problem with the aim of finding a minimum close to the global minimum of a target function containing the residual sum (Eq. 1), and stereochemical and other interactions of the macromolecule. In the past few years progress has been reported in nonlinear optimization of a function of many parameters by the introduction of simulated annealing (Kirkpatrick *et al.*, 1983). Simulated annealing is now in widespread use in areas such as electronic circuit design and pattern recognition. The method consists of simulating the many-parameter system by a Monte Carlo algorithm (Metropolis *et al.*, 1953); initially, the temperature is kept very high and then the system is 'annealed' by reducing the temperature slowly. A physical analogy is provided by a glass (high energy state) and a crystal (low energy state) of the same substance. To get from the glassy state to the crystalline state, the substance can be melted and then cooled slowly. In other words, the temperature of the substance can be increased temporarily in order to search for the lower energy state.

A first attempt to introduce simulated annealing into crystallographic refinement has been reported (Brünger *et al.*, 1987). A major difference from the optimization problems discussed in Kirkpatrick *et al.* (1983) is that one is optimizing a single covalently-connected structure representing a macromolecule rather than a fluid-like system consisting of many identical subunits. A direct application of the Metropolis algorithm to macromolecules turns out to be inefficient if all degrees of freedom are included, as the covalent bonds of the system will lead to rejection of most steps taken by the algorithm. Instead, molecular dynamics (Karplus and McCammon, 1983) can be used to explore the conformational space of the molecule. The conformational space searched is confined to regions allowed by the diffraction data through introduction of an effective potential energy which includes the crystallographic residual. This effective potential energy is an extension of the function used in the refinement by energy minimization (Jack and Levitt, 1978). Refinement by molecular dynamics will be referred to as MD-refinement. It proceeds essentially in two stages, a heating stage and a cooling stage in analogy to the simulated annealing technique.

In (Brünger *et al.*, 1987) it was shown that MD-refinement has a radius of convergence that is much larger than that of conventional least-squares restrained refinement and that the method can reduce the need for manual corrections. In this lecture a representative application to the structure of a mutant of aspartate aminotransferase (MAAT) (Smith *et al.*, 1986) solved by multiple isomorphous replacement (MIR) is discussed. In this case MD-refinement yields an improved R factor, a small decrease in the average difference between MIR phases and calculated phases, and slightly improved stereochemistry.

2 Molecular Dynamics

Molecular dynamics simulations involve the simultaneous solution of the classical equations of motion for all atoms of a macromolecule with an empirical potential energy E_i describing internal stereochemical interactions (bond, bond angle, dihedral torsion angle, improper torsion angle) as well as nonbonded (Van der Waals and electrostatic) interactions (Karplus and McCammon, 1983). The empirical potential energy is given by

$$E_i = \sum_{bonds} k_b(r - r_0)^2 + \sum_{angles} k_\theta(\theta - \theta_0)^2 \tag{3}$$
$$+ \sum_{dihedrals} k_\phi cos(n\phi + d) + \sum_{impropers} k_\omega(\omega - \omega_0)^2$$
$$+ \sum_{atom-pairs} \left(ar^{-12} + br^{-6} + cr^{-1}\right)$$

with parameters inferred from experimental as well as theoretical investigations (Brooks *et al.*, 1983). The solution of the classical equations of motion is carried out numerically by the Verlet algorithm (Verlet, 1967).

In MD-refinement the empirical potential energy E_i is augmented by an effective potential energy E_x containing information about the diffraction data. The total potential energy E_{total} is then given by

$$E_{total} = E_i + E_x \qquad (4)$$

At the present state of empirical potential energy parameterization a molecular dynamics simulation without the effective potential energy E_x cannot reproduce the crystal structure to sufficient accuracy. It is the combination of the empirical potential energy and the effective potential energy describing the crystallographic residual that makes it possible to use molecular dynamics in refinement.

3 Effective Potential Energy

The effective potential energy E_x consists of three terms describing the structure factor amplitude difference, structure factor phase difference, and crystal symmetry nonbonding interactions.

$$E_x = E_x^A + E_x^P + E_x^N \qquad (5)$$

The difference between observed and calculated structure factor amplitudes is described by

$$E_x^A = W_A/N_A \sum_{hkl} W_{hkl}[|F_{obs}(hkl)| - k|F_{calc}(hkl)|]^2 \qquad (6)$$

where k is a scale factor, W_{hkl} are individual weights for each reflection hkl, W_A is an overall weight factor which relates E_x^A to the empirical potential energy E_i, and N_A is a factor given by $\sum W_{hkl}|F_{obs}(hkl)|^2$. The normalization factor N_A ensures that the choice of the overall weight factor W_A is approximately independent of the resolution range employed during refinement. The scale factor k is set to the value which makes the derivative of E_x^A with respect to k zero,

$$k = \sum_{hkl} W_{hkl}|F_{obs}(hkl)||F_{calc}(hkl)| \; / \; (\sum_{hkl} W_{hkl}|F_{calc}(hkl)|^2) \qquad (7)$$

which is a necessary condition to make E_x^A minimal. The effective potential energy E_x^A is the same as the function used in Jack and Levitt (1978), except for the treatment of the scale factor k. In Jack and Levitt k is treated as an independently minimized parameter whereas in this work k is eliminated by inserting Eq. (7) into Eq. (6). Therefore, the functional dependence of k on $|F_{calc}(hkl)|$ is included implicitly in the computation of the derivatives of E_x^A with respect to the atomic model parameters.

In cases where some phase information about the diffraction data is available (e.g., through MIR) and one wants to include this information in the refinement, the difference between observed phase information and the calculated phases is described by

$$E_x^P = W_P/N_P \sum_{hkl} W_{hkl} SQ\{\phi(F_{obs}(hkl)) - \phi(F_{calc}(hkl)), \text{acos}(FOM(hkl))\}$$

(8)

where W_P is an overall weight factor, N_P is a normalization factor set equal to the number of phase specifications occuring in the sum, $\phi(F_{obs}(hkl))$ is the most probable phase obtained from MIR, $\phi(F_{calc}(hkl))$ is the phase of the calculated structure factors, FOM(hkl) is the individual figure of merit for each measured phase, and $SQ\{x,y\}$ is a square-well function with harmonic 'walls' given by

$$SQ\{x,y\} = \begin{cases} (x-y)^2 & x > y \\ 0 & -y < x < y \\ (y+x)^2 & -y > x \end{cases}$$

(9)

This form of the effective potential energy E_x^P ensures that the calculated phases are restrained only within the experimental error limits approximated by the individual figure of merit for each reflection, i.e. as long as a calculated phase $\phi(F_{calc}(hkl))$ is within the range $(F_{obs}(hkl)) \pm \text{acos}(FOM(hkl))$ no restraint is applied to that particular phase.

Experience has shown that the influence of crystal packing interactions can be important in crystallographic refinement. In the absence of crystal packing information backbone or sidechain atoms of a molecule in the asymmetric unit can penetrate symmetry-related molecules. This can occur in contact regions between molecules where the electron density does not clearly define the molecular boundaries. A potential energy which describes nonbonding interactions between the molecule(s) located in the asymmetric unit and all symmetry related molecules surrounding the asymmetric unit is given by

$$E_x^N = \sum_S \sum_{i \geq j} NB \{ \min_{x,y,z} \mid r_i - S r_j + xa + yb + zc \mid \} \qquad (10)$$

where the first sum extends over all symmetry operators S of the crystal, the second sum extends over all pairs of atoms (i,j) with i$\geq$j for which the minimum image distance

$$\min_{x,y,z} \quad \mid r_i - S r_j + xa + yb + zc \mid \qquad (11)$$

is less than a specified cutoff r_{cut}, a, b, c are vectors that define the unit cell of the crystal, and *(x,y,z)* are fractional coordinates. For the application of the minimum image expression (Eq. 11) one has to make the reasonable assumption that the cutoff r_{cut} is small compared to the physical dimension of the unit cell for macromolecules. NB{r} has the form of a nonbonding interaction potential, i.e. it is a sum of Van der Waals and electrostatic interactions

$$NB\{r\} = ar^{-12} - br^{-6} + cr^{-1}, \qquad (12)$$

where a,b,c are appropriate constants and switching or shifting functions (Brooks *et al.*, 1983). Eq. (10) includes only intermolecular interactions between different molecules since the nonbonding interaction energy for atoms in the asymmetric unit is already included in the empirical potential energy (see Eq. (3)). As no weighting or normalization has been applied to E_x^N the nonbonding interaction energy between two atoms is independent of whether it is an intermolecular or an intramolecular interaction.

4   FFT Calculation of Structure Factors and Derivatives

The most central processor unit (CPU) time-consuming step in crystallographic refinement is the evaluation of the structure factors of the atomic model in Eqs. (6–8) and the first derivatives with respect to the atomic coordinates of the system. This also applies to MD-refinement since the CPU time needed to compute the structure factors through the direct summation method exceeds the CPU time for a single empirical potential energy calculation E_i (Eq. 3) by one to two orders of magnitude; e.g. in the case of MAAT the direct summation takes 102 secs whereas the evaluation of E_i takes only 2.8 secs on a CRAY-2.

Evaluation of the atomic electron density on a finite grid followed by fast-Fourier transformation (FFT) provides a way to greatly speed up the calculation (Cooley and Tukey, 1965; Ten Eyck, 1973; Agarwal, 1978; Isaacs, 1984). The expression for the first derivatives of an arbitrary function of structure factor amplitudes and phases such as given in Eqs. (6–8) can be derived by using a multidimensional version of the chain rule in complex space as was pointed out by Lunin and Urzhumtsev (1985). It is not necessary to compute second derivatives for MD-refinement as only first derivatives enter the classical equations of motion. The conjugent gradient minimization used in this work for conventional least-squares refinement also requires only the energy and its first derivatives (Fletcher and Reeves, 1964; Powell, 1977).

The FFT algorithms and programs that are presently being used in crystallographic refinement were completely reformulated in order to achieve vectorization on supercomputers that increases the speed by a factor of 10 in the case of MAAT; a detailed description will be published elsewere. The two main differences with earlier FFT implementations (Ten Eyck, 1973; Agarwal, 1978; Isaacs, 1984) concern the organization of the finite grid and the application of symmetry operators after the FFT has been carried out. In cases where the finite electron density grid is too large to fit into the physical memory of the computer the FFT is carried out on subgrids and the results are then accumulated. This was accomplished in analogy to the factorization of a one-dimensional Fourier transformation which is the basis for the FFT technique (Ten Eyck, 1973).

It was already pointed out (Isaacs, 1984) that the modelling of the electron density on the finite grid is the most expensive part of the computation. This applies to a much larger degree to vector machines where very efficient, multidimensional FFT routines are available (CRAY Research Inc., personal communication); e.g., for MAAT the electron density calculation takes 2.0 secs and the FFT takes 0.75 secs on a CRAY-2. Therefore, little would be gained at present by introducing space-group specific Fourier transformations. This has the additional advantage that the method is space-group general. To reduce the time spent in the calculation of the atomic electron density, it is computed only over the asymmetric unit of the crystal. However, the FFT is carried out over the complete unit cell, and the transposed symmetry operators of the crystal space group are then applied to the computed structure factors. This yields the identical result as if the electron density would have been computed over the complete unit cell followed by the FFT.

During MD-refinement the first derivatives are kept constant until any atom has moved by more than 0.2Å relative to the position at which the derivatives were last computed; at that point all derivatives are updated. Tests have shown that this does not affect the convergence properties and it speeds up the computation by almost an order of magnitude.

5 Annealing Schedule

A typical protocol for MD-refinement is described in this section. The example protein MAAT contains 396 amino acids and a pyridoximine phosphate co-factor; 8786 reflections from 10 to 2.8Å were available; the space-group is $C222_1$. The initial model was obtained from MIR to 3Å resolution; the MIR phase information was included in the refinement.

Table I shows the protocol that was used for MAAT; the refinement proceeded in four stages. During the first stage conjugent gradient minimization was employed to relieve any bad inter- or intramolecular contacts in the structure that could pose a problem when starting the molecular dynamics calculation. During that stage C^α -backbone atoms are harmonically restrained to their initial positions. This avoids a large drift of the structure, if any bad nonbonded contacts have to be relieved.

Table I: Protocol for MD-refinement of MAAT

stage	description
1	conjugent gradient minimization, 160 steps, C^α-restraints 20 Kcal/(mole Å^2), resolution 8.0–2.8Å, W^A=100,000 Kcal/mole, W^P=70,000 Kcal/(mole degrees), B=12Å^{-2}
2	molecular dynamics, 0.5 ps, T=3000 K, timestep=0.5 fs, velocities rescaled every 250 steps, structure factor gradient updated if any atom moved by more than 0.2Å, resolution 8.0–2.8Å, W^A=100,000 Kcal/mole, W^P=30,000 Kcal/(mole degrees), B=12Å^{-2}
3	molecular dynamics, 0.25 ps, T=300 K, timestep=0.5 fs, velocities rescaled every 50 steps, structure factor gradient updated if any atom moved by more than 0.2Å, resolution 8.0–2.8Å, W^A=100,000 Kcal/mole, W^P=30,000 Kcal/(mole degrees), B=12Å^{-2}
4	conjugent gradient minimization, 80 steps, resolution 8.0–2.8Å, W^A=100,000 Kcal/mole, W^P=0 Kcal/(mole degrees), B=12Å^{-2}

During the second stage, molecular dynamics was carried at 3000 K, for 0.5 ps; the initial velocities were assigned from a Maxwellian distribution at 3000 K. In the third stage, the system was cooled to temperatures near 300 K. The fourth stage consists of conjugent gradient minimization which further optimizes the R factor and stereochemistry of the system. This annealing schedule was developed by trial and error. It differs from the one described in (Kirkpatrick *et al.*, 1983) in the way the temperature is controlled during annealing; i.e. in Kirkpatrick the cooling rate is slowed down if the system is near a critical point whereas here the system is cooled at a constant rate. It is likely that the protocol in Table I could be improved. This will be subject of future investigations.

The major difference between MD-refinement and conventional least-squares refinement consists of the heating stage 2 and the annealing stage 3. If these stages were bypassed the protocol would be similar to the Jack-Levitt refinement (Jack and Levitt, 1978).

6   Choice of Parameters

Stereochemical and nonbonded parameters for the empirical potential energy (Eq. 3) were taken from the explicit polar hydrogen parameter set PARAM19 and TOPH19 of CHARMM (Brooks *et al.*) with two modifications. The two-fold dihedral angle term for peptide bonds was replaced by a one-fold term (force constant set to 100 Kcal/(mole degrees^2)) allowing only trans-peptide bonds. With the original two-fold term transitions of peptide bonds from trans to cis can occur at the high temperature during stage 2. In the case of proline peptide bonds, the two-fold term was retained with a reduced force constant of 5 Kcal/(mole degrees^2) to allow transitions between cis and trans prolines. The force constant of the improper torsion angle which specifies the tetrahedral geometry of C^α-carbon atoms was increased to 500 Kcal/(mole degrees^2). This was done in order to avoid transitions from L- to D-amino acids during the high temperature stage 2.

The weight factor W_A in Eq. (6) was chosen such that the gradient of E_x^A was comparable in magnitude to the gradient of the empirical potential energy E_i for the structure of MAAT obtained after a 0.5 psec molecular dynamics simulation with W_A and W_P set to zero, starting with the initial x-ray structure. It is necessary to use a molecular dynamics structure rather than the initial x-ray structure to compute the gradients as the initial structure could be strained and thus would artifically increase the gradient. Experience has shown that this procedure is relatively insensitive to the choice of resolution and initial structure. The comparison of the gradients suggested a factor W_A of approximately 100,000. W_A was kept constant throughout the refinement. Experience has shown that the exact choice of W_A is not critical for MD-refinement.

The same procedure was applied to choose W_P in Eq. (8), i.e., the gradient of E_x^P for the initial structure was made comparable in magnitude to the gradient of E_i. This suggested W_P to be approximately 15,000. However, the actual choice of W_P was larger than this estimate during stages 1, 2 and 3. W_P was set to zero during stage 4 in order to check whether the average phase difference decrease could be maintained even without the E_x^P energy term. Again, the exact choice of W_P does not seem to be critical.

The high temperature during stage 2 ensures that the system can overcome certain energy barriers provided by the potential energy (Eq. (4)). Tests have shown that the exact choice of the temperature is not critical. However, it appears to be necessary that the R factor during the high temperature stage stays well below 0.59, which would correspond to a random structure (Jensen, 1985). R factors between 0.30 and 0.50 during stage 2 seem to work equally well. Due to the increase in temperature, the R factor can actually increase during the first steps of this stage.

It has been found already that in a time period as short as 0.5 ps MD-refinement performs better than minimization. Longer simulations could be used to obtain an ensemble of structures by cooling the system at various intervals. In this way one could obtain alternative structures that match the diffraction data equally well.

The resolution range (8.0–2.8Å) was kept constant throughout the refinement. Alternatively, one could have subdivided stage 2, starting at low resolution and then increasing the resolution (Brünger *et al.*, 1987). This is useful in cases where high-resolution diffraction data are available. However, contrary to conventional least-squares refinement it appears that the radius of convergence of MD-refinement is insensitive to the choice of resolution; e.g., the MD-refinement of crambin described in (Brünger *et al.*, 1987) was accomplished at a constant 8.0–1.5Å resolution range throughout the refinement.

Temperature factors were kept constant during all stages and were set equal for all atoms. Combined B-factor and positional refinement could proceed after stage four. The individual weights w_{hkl} in Eqs. (6), (8) were set to one. However, the MD-refinement provides no restrictions on the choice of a different weighting scheme.

7 Results of the MAAT refinement

Table II shows the results of the MAAT refinement. Listed are the R factor, average phase difference $\Delta\phi$ between observed (i.e., most probable MIR) and computed phases, deviations of bonds and angles from ideality (Δ_{bond} and Δ_{angle}), rms deviations from the initial structure for backbone and all atoms, and the CPU times on a CRAY-2; the information is provided for the initial structure, and the structures after each stage of the MD-refinement.

Table II: Course of MAAT refinement

stage	R factor	$\Delta\phi$	Δ_{bond}	Δ_{angle}	rms		CPU
					back	all	
		[deg.]	[Å]	[deg.]	[Å]	[Å]	[s]
initial	0.42	59.3	0.037	4.8	–	–	–
1	0.35	49.3	0.033	6.1	0.36	0.55	1327
2	0.35	56.6	0.12	13.0	1.01	1.61	3704
3	0.27	50.1	0.042	6.0	1.01	1.52	1250
4	0.25	55.7	0.020	4.6	1.01	1.52	650
control(1+4)							
	0.28	57.9	0.023	4.8	0.64	0.90	2000

The ideal values of bond lengths and angles correspond to those used in version 19 of CHARMM (Brooks *et al.*, 1983). As a control the result of a pure minimization protocol that bypasses stages 2 and 3 is also shown. This control calculation is equivalent to conventional least-squares restrained refinement without model building. The minimization method (Powell, 1977) that was employed in this work performed slightly better than the method used in PROLSQ (Konnert and Hendrickson, 1980).

In comparison with the least-squares method, MD-refinement improved the R factor by 0.03, slightly reduced the average phase difference

(55.7 *vs* 57.9) and decreased the bond and angle deviations from ideality. This improvement in the structure was accomplished during the heating and annealing stages 2 and 3 when 48 backbone atoms moved by more than 2Å and 17 atoms moved by more than 5Å, whereas in the control calculation only one atom moved by more than 2Å. The rms difference between the MD-refined structure and the control structure is 0.73Å and 1.29Å for backbone and all atoms, respectively. Other protein crystal structures where the starting model was worse than in the MAAT case (e.g., R > 0.47) showed a much greater difference between MD and conventional refinements (Brünger *et al.*, 1987). Restrained least-squares refinement would hang up at about 35%, requiring extensive rebuilding. The MD-refinement proceeded automatically to R-factors in the middle 20s.

The R factor is large and the stereochemistry relatively bad during the heating stage 2. This can be explained by the fact that the kinetic energy of the atoms is large and consequently the atoms are not at their equilibrium positions. The large kinetic energy allows the system to overcome local energy minima. After cooling the system it has reached a lower R factor than was accomplished by minimization.

8   Computational Requirements

The CPU times in Table II show that the increase in computational time of MD-refinement compared to conventional refinement (as approximated by the control calculation) is 3.5 in the case of MAAT. The required CPU time on the CRAY-2 supercomputer is quite modest and suggests that the MD-refinement can be carried out on standard computers such as a VAX 11/780 or 8600.

All calculations in this work were carried out with the program X-PLOR that was specifically designed for crystallographic refinement. X-PLOR includes MD-refinement and conventional refinement with both the direct summation technique as well as the FFT technique. An optimized VAX/VMS version and a vectorized CRAY version will be made available for distribution.

9   Conclusion and Outlook

Simulated annealing has been sucessfully applied to crystallographic refinement by molecular dynamics. The application to MAAT has shown that MD-refinement can be carried out on large systems at a quite modest supercomputing expense.

With the increased availability of fast diffraction data collection devices and the success of crystallizing larger and larger macromolecular complexes, refinement clearly becomes a bottleneck for structure determination and the need for improving the efficiency of the refinement process arises. It is my hope that MD-refinement is a stepping stone towards fully automated crystallographic refinement.

Acknowledgement

I am grateful to M. Karplus, J. Kuriyan, G.A. Petsko, D. Ringe, and D.L. Smith for useful discussions and D.L. Smith for providing x-ray diffraction data of MAAT.

REFERENCES

Agarwal, R. C. (1978). *Acta Crystallogr.* **A43**, 791–809.

Brooks, B. R., Bruccoleri, R. E., Olafson, B. D., States, D. J., Swaminathan, S., and Karplus, M. (1983). *J. Comput. Chem.* **4**, 187–217.

Brünger, A. T., Kuriyan, J., and Karplus, M. (1987). *Science* **235**, 458–460.

Cooley, J. W. and Tukey, J. W., (1965). *Math. Comput.* **19**, 297–301.

Fletcher, R. and Reeves, C. M. (1964). *Comp. J.* **7**, 149–154.

Hendrickson, W.A. (1985). *Methods in Enzymology* **115**, 252–270.

Isaacs, N. (1984). In *Methods and applications in crystallographic computing* (Hall, S. R. and Ashida, T., eds), pp. 193–205. Clarendon, Oxford.

Jack, A. and Levitt, M. (1978). *Acta Crystallogr.* **A34**, 931–935.

Jensen, L. H. (1985). *Methods in Enzymology* **115**, 227–234.

Jones, T. A. (1982). In *Computational Crystallography* (Sayre, D., ed.), pp. 303–317. Clarendon, Oxford.

Karplus, M. and McCammon, J. A. (1983). *Annu. Rev. Biochem.* **52**, 263–300.

Kirkpatrick, S., Gelatt, C. D.,Jr., and Vecchi, M. P. (1983). *Science* **220**, 671–680.

Konnert, J. H. and Hendrickson, W. A. (1980). *Acta Crystallogr.* **A36**, 344–349.

Lunin, V. Y. and Urzhumtsev, A. G. (1985). *Acta Crystallogr.* **A41**, 327–333.

Metropolis, N., Rosenbluth, M., Rosenbluth, A., Teller, A., Teller, E., (1953). *J. Chem. Phys.* **21**, 1087–1092.

Moss, D. S. and Morffew, A. J. (1982). *Comput. Chem.*, **6**, 1–3.

Powell, M. J. D. (1977). *Mathematical Programming* **12**, 241–254.

Smith, D. L., Ringe, D., Finlayson, W. L., and Kirsch, J. F. (1986). *J. Mol. Biol.* **191**, 301–302.

Sussman, J. L., Holbrook, S. R., Church, G. M., and Kim, S. H. (1977). *Acta Crystallogr.* **A33**, 800–804.

Ten Eyck, L. F. (1973). *Acta Crystallogr.* **A29**, 183–191.

Verlet, L. (1967). *Phys. Rev.* **159**, 98–105.

10

SYSTEMATICS IN INTENSITY DATA AVOIDANCE, CORRECTION, DETECTION AND USE

H. D. Flack

1   Introduction

The objective of single-crystal diffraction experiments is to determine structural information in the form of parameter values with associated error estimates. One requires that the parameter values should be free of any bias or systematic error and that the error estimates be representative of the real accuracy of the experiment. A recent comparison (Taylor and Kennard, 1986) on crystal structures determined independently in different laboratories continues to show that we are still falling far short of this objective. Under the four sub-titles of avoidance, correction, detection and use, illustrations of how information in the measured intensities may be used to move towards the attainment of our declared goal.

2   Detection — Durbin-Watson d statistic

Upon convergence of a least-squares refinement, one has available, apart from the desired parameters, estimated standard deviations and correlation matrix, a sequence of weighted deviates containing a wealth of untapped information on the fit between the observed and calculated quantities. The commonly-used method of analyzing these, is to compare statistics calculated on the deviates with those obtained from a known probability distribution. For example, for a Gaussian or normal distribution the goodness of fit, S, takes a value of 1. The most commonly applied statistics in crystallography are R, wR and S. The weighted deviates may also be examined as a function of scattering angle, $|F|$, or some other variable of the data. In this type of analysis one expects to find a flat distribution. The normal probability plot, based on weighted deviates ordered according to their value, is also most helpful. Each of these different analyses can be used to bring to light deficiencies in the model or systematic errors in the data (these are in fact the same thing). However all of the aforementioned methods treat the deviates as independent from each other and do not show up any potential correlations. The Durbin-Watson d statistic (Durbin and Watson, 1950, 1951, 1971; Theil and Nagar, 1961) is well-documented and well-authenticated and provides a convenient manner of searching for correlations. It is insufficiently known and used by the crystallographic community at large.   The d statistic is defined as follows:

$$d = \sum (D_i - D_{i-1})^2 / D_i^2$$

where $D_i, i = 1, ..N$ are the weighted deviates. For a normal distribution d will take a value of 2. If successive deviates tend to have the same sign, d will fall towards zero and one talks of positive serial correlation. When successive signs tend to be opposite, d rises towards 4 and it is a case of negative serial correlation. Note particularly that the value of d will depend upon the order into which the data are sequenced. It is known for a least-squares calculation where there is significant serial correlation of the deviates that the parameter estimated standard deviations are prone to be grossly in error, being either too large or too small.

Rietveld analysis, full-profile powder diffraction fitting, provides a convenient example of the use of the Durbin-Watson d statistic. It has been realized for some time by comparison with single crystal data that estimated standard deviations from Rietveld analysis are suspiciously low. The possible causes and potential remedies have been hotly disputed. Values of d from Rietveld refinements are generally significantly smaller than 2 (Hill and Madsen, 1986; Hill and Flack, 1987) and hence in line with the underestimates of the standard deviations that have been noted above. Clearly the d statistic is providing a powerful pointer to an unacceptable situation.

The underlying cause of the Rietveld serial correlation lies in the inadequacies of the model used. Any modelling error in either the integrated intensity of a Bragg reflection or the shape and width of the profile will cause neighbouring deviates to have the same sign. This immediately suggests that one means of attack is to improve the model used in the refinement. This has not proved easy to do. Another is to admit that the current model is only an approximation but adequate for the known objectives of the experiment. We would then, working within the Bayesian viewpoint of statistics, attempt to quantify our knowledge of the poorness of the model by suitably modifying the weighting scheme. This leads to an awful computational problem of a non-diagonal weights matrix which has been solved in one special case by Rollett (1984). Other means of improving the value of the d statistic at small cost are to reduce the number of steps under each peak (e.g. by making each new step, the sum of the values of N old steps) or by decreasing the counting time of each step (which makes the quantum counting statistics the major source of uncertainty in the system).

3   Use - absolute structure

Model improvement was mentioned above and it can be applied with great success to the treatment of the chirality-polarity of non-centrosymmetric crystals. Due to the effect of anomalous dispersion, Friedel's Law is no longer valid as differences in intensity between reflection and anti-reflection occur. The major application of this effect in modern-day crystallography is in the determination of the absolute configuration of molecules. It is often forgotten, however, that the effect is not limited to chiral crystals but is present in all non-centrosymmetric crystals. Indeed, as shown by Ueki, Zalkin and Templeton (1966) and quantified by Cruickshank and

McDonald (1967) refinement of a non-centrosymmetric structure with a polar axis (i.e. floating origin) can lead to significant bias in atomic positional coordinates if only an asymmetric region of the LAUE group is measured in reciprocal space. This is known as a polar dispersion error. In this respect the danger is greater with absolute structure than in Rietveld analysis as biased parameters will result rather than biased estimated standard deviations.

Flack (1983) and Bernardinelli and Flack (1985) have elaborated on a suggestion by Rogers (1981) to resolve this problem. Any non-centrosymmetric crystal is taken to be an inversion twin. The corresponding structure amplitude equation may be written:

$$|F(h,k,l,x)|^2 = (1-x)|F(h,k,l)|^2 + x|F(-h,-k,-l)|^2,$$

where x is called the absolute-structure parameter. It takes a value of 0 when the model and crystal are in the same chirality-polarity and a value of 1 when they are inverted one with respect to another. A value of 0.5 for x indicates that the crystal is a real inversion twin with 50% of each component, rather like an equi-proportional oriented single-crystal mixture. It has been found that refinement of x is extremely stable, converging in a cycle or two. x manages to discriminate between a model and its inverse in cases where the use of Hamilton's R-factor ratio test (Hamilton, 1965) is inconclusive. As x is refined in the least squares, an estimated standard deviation is obtained. This is a specific error estimate on the absolute-structure determination and most helpful. According to our own and others experience (Müller, 1987) about 5% of non-centrosymmetric crystals are twinned by inversion.

4   Avoidance — data region

An important facet of data collection with major implications in data treatment and refinement is the choice of the region of reciprocal space to be measured. Is it best to measure one asymmetric region at scan speed z or rather two asymmetric regions at scan speed 2z for the same total measuring time? The choice of two asymmetric regions will necessarily incur higher overheads in circle movement on the diffractometer and calculation time in data treatment due to the doubling of the number of reflections.

In least-squares refinement it is known that omitting measurements does not introduce bias into the derived parameters as long as the criterion for rejection does not depend in any way on the measured values themselves (Prince and Nicholson, 1985). Since measurement of only one asymmetric region is more efficient in diffractometer and computer time, and no bias will be incurred, it would seem inescapable that only one asymmetric region of reciprocal space need be measured. This whole approach, however, is based on the proviso that the model used in the least-squares refinement is a true representation of the diffraction experiment. The possibility of model deficiencies or remaining systematic effects (errors) has been ignored or avoided. Consider the polar dispersion error mentioned above. Least-squares refinement of a model with the wrong polarity will lead to unbiased atomic positional coordinates if the data set contains all reflections in the asymmetric region of the crystal point group. On the other hand, the same model introduces a significant bias if only data from the asymmetric region of the LAUE group are used.

As the object of a diffraction experiment is to try and correctly model the physical state of the crystal it would seem most unwise to accept one's initial attempt without some serious verification. As an example, the assumed symmetry is a property of the model and may be in error. Measuring more than one asymmetric region will enable the validity of the assumed point group to be tested. The symmetry-equivalent reflections can also be used to obtain estimates for the standard deviations of the intensities which are more realistic than those obtained from quantum counting statistics alone. Anisotropic effects such as extinction, absorption, radiation damage etc. will have their estimation greatly facilitated by measurements of symmetry equivalent reflections and integrated intensity measurements at different values of the azimuthal angle.

In conclusion, along with others (Hamor, Steinfink and Willis, 1985), I make a strong recommendation for the measurement of more than one asymmetric region of reciprocal space together with some measurements at different azimuths. At first sight this recommendation might appear to be contrary to the results of our own study (Bernardinelli and Flack, 1987) of the effect of data region on the determination of absolute structure. This is not the case. In the latter study, the model of the compound was adequate, at least for the absolute-structure parameter, for the conditions of the Nicholson and Prince theorem to hold. Hence omitting reflections did not produce any significant effect on the estimate of the absolute-structure parameter. This was demonstrated by refinements on the complete data set. In hindsight the study should be taken to testify strongly in favour of the chosen model of determining absolute structure by means of x, rather than a proof of insensitivity to data region.

5   Correction — fixed or variable?

Correction for such effects as absorption, Lorentz-polarization, and thermal diffuse scattering are customarily applied as a linear transformation of the measured intensity with constant coefficients. The correction is thus assumed to be exact and any error in the numerical values used in their evaluation (such as crystal dimensions and linear absorption coefficient for absorption, polarization ratio for Lp correction) will become a systematic error in the 'corrected observations'. Furthermore the estimated standard deviations of the atomic parameters derived from least-squares refinement will not reflect the inaccuracy of the correction. Corrections of this sort are computationally economic as they only have to be calculated once for each measurement. Other corrections such as crystal decomposition, long-term variation of the radiation source and background are applied in the same way but it also customary to increase the individual estimated standard deviations of the 'corrected observations' as well. The covariances of the 'corrected observations' are ignored. Yet a third type of correction is that of extinction where there is no way of measuring the extinction parameters such as mosaic spread and domain size other than through the intensity measurements themselves. These parameters thus enter the least squares as refineable variables whose values will be estimated simultaneously with the atomic parameters. Computationally this is, of course, more expensive as the necessary correction and derivatives need to be calculated on each least-squares cycle. A distinct advantage however is that the e.s.d.s of the atomic parameters allow naturally for the inaccuracy of the correction.

If the goal of improved atomic parameters with improved error estimates is to be achieved, there are two possible lines of action. The first is to improve the quality or accuracy of the corrections to the point where they make a negligible contribution in comparison with errors from other sources. For absorption correction, for example, the limiting factor is the accuracy with which one can measure the dimensions of the crystal. Due to the exponential nature of the absorption correction there seems little hope that sufficient accuracy could ever be achieved.

A second line of attack is to 'refine' these corrections by including their defining parameters as part of the least squares. For absorption this would mean that the crystal dimensions and linear absorption coefficient would become variables. Several workers have found that changing their measured crystal dimensions within less than a few e.s.d.s greatly improves the absorption correction. There are a couple of little known computer programmes available (Rae and Dix, 1971; Rigoult, Tomas and Guidi-Morosini, 1979). For Lp corrections it has generally been found that the polarization ratios may be refined with CuK radiation but not with shorter wavelength radiations (Suortti, Kvick and Emge, 1986). However all of these attempts have one major shortcoming to my mind. The experimentally measured (or prior estimated) values of crystal dimensions, linear absorption coefficient and polarization ratio have not been allowed to enter into the least-squares refinements as observed values in the way of intensity measurements, or bond length and angle restraints.

REFERENCES

Bernardinelli, G. and Flack, H. D. (1985). *Acta Cryst.*, **A41**, 500–511.

Bernardinelli, G. and Flack, H. D. (1987). *Acta Cryst.*, **A43**, 75–78.

Cruickshank, D. W. J. and McDonald, W. S. (1967). *Acta Cryst.*, **23**, 9–11.

Durbin, J. and Watson, G. S. (1950). *Biometrika*, **37**, 409–428.

Durbin, J. and Watson, G. S. (1951). *Biometrika*, **38**, 159–178.

Durbin, J. and Watson, G. S. (1971). *Biometrika*, **58**, 1–19.

Flack, H. D. (1983). *Acta Cryst.*, **A39**, 876–881.

Hamilton, W. C. (1965). *Acta Cryst.*, **18**, 502–510.

Hamor, T. A., Steinfink, H. and Willis, B. T. M. (1985). *Acta Cryst.*, **C41**, 301–303.

Hill, R. J. and Flack, H. D. (1987). *J. Appl. Cryst.*, Submitted.

Hill, R. J. and Madsen, I. C. (1986). *J. Appl. Cryst.*, **19**, 10–18.

Müller, G. (1987). *Acta Cryst.*, In preparation.

Prince, E. and Nicholson, W. L. (1985). In *Structure and Statistics in Crystallography* (ed. A. J. C. Wilson) pp. 183–195. Adenine Press, Guilderland, NY.

Rae, A. D. and Dix, M. F. (1971). *Acta Cryst.*, **A27**, 628–634.

Rigoult, J., Tomas, A. and Guidi-Morosini, C. (1979). *Acta Cryst.*, **A35**, 587–590.

Rogers, D. (1981). *Acta Cryst.*, **A37**, 734–741.

Rollett, J. S. (1984). In *Methods and applications in crystallographic computing* (ed. S. R. Hall and T. Ashida) pp. 161–173. Clarendon Press, Oxford.

Suortti, P., Kvick, A. and Emge, T. J. (1986). *Acta Cryst.*, **A42**, 184–188.

Taylor, R. and Kennard, O. (1986). *Acta Cryst.*, **B42**, 112–120.

Theil, H. and Nagar, A. L. (1961). *J. Amer. Stat. Ass.*, **56**, 793–806.

Ueki, T., Zalkin, A. and Templeton, D. (1966). *Acta Cryst.*, **20**, 836–841.

11

ERROR ANALYSIS

J.S.Rollett

1   Motivation

A crystal structure analysis gives us molecular dimensions which are compared with those from other crystals. There may be comparison with dimensions from microwave or infra-red spectroscopy, gas electron diffraction or nuclear magnetic resonance methods. Crystal structure results can also be tested by reference to predictions from quantum chemical calculations. Inevitably there will be discrepancies and we need to decide whether:

1. The molecules are in differing environments which alter their dimensions, or

2. The differences are due to errors in the measurement techniques.

We discount, here, the small differences between results caused by the various modes of averaging of vibrational motion for different experiments. These are relatively well understood and generally smaller than the uncertainties in crystal structure dimensions.

If there is evidence of a chemical difference, then further work may be desirable to provide an explanation. If, however, the difference is purely due to technical errors, then it is of little interest except to those who wish to improve the technique. No theoretical chemist would be likely to wish to spend time on a discrepancy which arose from failure to correct for the effects of multiple reflections.

We need to provide estimates of error which cover appropriately all errors in the technique of crystal structure analysis that we cannot eliminate. If we underestimate the effects of these errors, then time will be wasted in search for physical causes of discrepancies where causes do not exist. On the other hand, if we overestimate the effects of technical errors we shall conceal physical causes of discrepancies, so that our work will be partly wasted.

We see that the course of future research, and its effectiveness, depend on reliable error estimates and their acceptance by molecular structure analysts in general. Unfortunately, we shall see that our present methods are not reliable. Even when duplicate X-ray crystal structure analyses are compared (see Taylor and Kennard(1986)) the discrepancies are generally greater than those to be expected from their error estimates. This paper discusses the methods available for producing error estimates, and makes suggestions for ensuring that the estimates are appropriate.

2 Least-squares optimization

2.1 Minimization functions

The great majority of crystal structures (except, possibly, for macromolecular structures) are refined by the method of least squares. This is one of a class of 'null-methods' which minimize some measure of the discrepancies between the observed values of quantities and the values predicted by a model of the structure. Write

I_i^o for the ith observed value, and

I_i^c for the ith calculated value

Then we could choose to minimize one of the following:

$$M_1 = \max_i |I_i^o - I_i^c| \tag{1}$$

$$M_2 = \sum_i |I_i^o - I_i^c| \tag{2}$$

$$M_3 = \sum_i (I_i^o - I_i^c)^2 \tag{3}$$

$$M_4 = \sum_i w_i (I_i^o - I_i^c)^2 \tag{4}$$

$$M_5 = -\log P(I_i^o, I_i^c, i = 1, 2, \ldots, n) \tag{5}$$

where

w_i is a weight for observation i

n is the number of observations

P is the probability of the set of I_i^o and I_i^c.

It is expensive and difficult to minimize M_1 and M_2, which do not have continuous first derivatives with respect to the parameters of the model. M_1 is also sensitive to even one large error of observation. For M_3 the algorithm is simpler, terminating in one step if the I_i^c are linear functions of the parameters. We need M_4, however, if we are to allow for varying precision of the observations.

If we wish to produce a 'maximum-likelihood' solution we should use M_5, but M_4 and M_5 are equivalent if each $I_i^o - I_i^c$ is normally distributed with standard deviation $1/\sqrt{w_i}$, and if the errors are uncorrelated. This is usually a good assumption, and so we can use M_4, which corresponds to the weighted least-squares method. In certain circumstances, particularly where a small number of observations have exceptional errors, we may need to consider M_5. We also need to minimize something other than a sum of squares of $I_i^o - I_i^c$ if these have correlated errors. Discussion of these cases will come later.

2.2 Finding a minimum

In structure analysis we can take I_i^o to be the measured intensity of the ith order of diffraction in the set of Bragg reflections that we use. It is common practice to use instead this intensity corrected for Lorentz and polarisation factors, so that these need not be calculated in I_i^c. In either case the function M_4 will have continuous derivatives with respect to the structure parameters, which we will call

$$x_j, \quad j = 1, 2, \ldots, m.$$

Provided that $m < n$, so that there are more observations than parameters, we have an overdetermined situation. The method of least squares can then be used to find error estimates for the x_j and functions of them.

Note that it has, in the past, been common to take the square root of the intensity (corrected for Lorentz and polarisation factors) as I_i^o . If this is done, then M_4 has a discontinuity in its first derivatives when an intensity passes through zero. That makes the minimum-seeking process less effective and the error estimates invalid. It is recommended, therefore, that the square root should not be taken (i.e. that minimization should be based on the intensity, or on $|F^2(hkl)|$, rather than $|F(hkl)|$, as the observed quantity.)

Given a minimization function M, with continuous derivatives, we can use the Newton method to find the minimum. Write

$$\mathbf{x}^T = (x_1, x_2, \ldots, x_m) \tag{6}$$

$$\Delta_i = I_i^o - I_i^c \tag{7}$$

$$\mathbf{x}_k = \mathbf{x} \text{ at step } k \text{ of the process} \tag{8}$$

$$\delta\mathbf{x}_k = \mathbf{x}_{k+1} - \mathbf{x}_k, \text{ the } k^{th} \text{ set of shifts.} \tag{9}$$

Then we have

$$\sum_{l=1}^{m} \frac{\partial^2 M}{\partial x_j \partial x_l} \delta x_l = -\frac{\partial M}{\partial x_j}, \quad j = 1, 2, \ldots, m \tag{10}$$

as the equations for the shifts, where the derivatives are evaluated at $\mathbf{x}_k$. If

$$M = \sum_i w_i \Delta_i^2 \tag{11}$$

then

$$\frac{\partial M}{\partial x_j} = \sum_i 2 w_i \Delta_i \frac{\partial \Delta_i}{\partial x_j} \tag{12}$$

and

$$\frac{\partial^2 M}{\partial x_j \partial x_l} = \sum_i 2w_i \left(\frac{\partial \Delta_i}{\partial x_j} \frac{\partial \Delta_i}{\partial x_l} + \Delta_i \frac{\partial^2 \Delta_i}{\partial x_j \partial x_l} \right). \tag{13}$$

We can expand $\partial M / \partial x_j$ as a Taylor series about $\mathbf{x_k}$ to show that, when the parameters are approaching the minimum, the distance from the minimum to $\mathbf{x_{k+1}}$ is proportional to the square of the distance to $\mathbf{x_k}$. This means that the Newton method has fast convergence at the end. Unfortunately however:

1. The $\partial^2 \Delta_i / \partial x_j \partial x_l$ are inconvenient to find.

2. The Newton method can diverge if the matrix of the $\partial^2 M / \partial x_j \partial x_l$ is not positive definite.

Because of this we use the Gauss-Newton approximation in which we drop the term $\sum_i \Delta_i (\partial^2 \Delta_i / \partial x_j \partial x_l)$ from $\partial^2 M / \partial x_j \partial x_l$. The approximation is then positive definite unless we choose the x_j so that they are linearly dependent. We can always get convergence to a local minimum of M, but unless $\Delta_i \to 0$ for all i as the minimum is approached, the distance from the minimum to $\mathbf{x_{k+1}}$ is proportional to the distance to $\mathbf{x_k}$. Convergence is therefore slower (if surer) than for the Newton method. (If the function M has relatively large third derivatives there is still a danger of divergence, but this can always be prevented by reducing the shifts.)

2.3 Effects of data errors on errors of results

For convenience, write

$$\mathbf{\Delta}^{\mathbf{T}} = (\Delta_1, \Delta_2, \dots, \Delta_n), \tag{14}$$

$$(J)_{ij} = \frac{\partial \Delta_i}{\partial x_j}. \tag{15}$$

Then the Gauss-Newton equations become

$$J^T W J \delta \mathbf{x} = -J^T W \Delta \tag{16}$$

where

$$W = \mathrm{diag}(w_1, w_2, \dots, w_n) \tag{17}$$

so

$$\delta \mathbf{x} = -(J^T W J)^{-1} J^T W \mathbf{\Delta}. \tag{18}$$

It follows that if we change $\mathbf{\Delta}$ by ϵ, then $\delta \mathbf{q}$ changes by $\delta \mathbf{q}$ where

$$\delta \mathbf{q} = -(J^T W J)^{-1} J^T W \epsilon. \tag{19}$$

Hence

$$\delta \mathbf{q} \delta \mathbf{q}^T = (J^T W J)^{-1} J^T W \, \epsilon \epsilon^T W J (J^T W J)^{-1}, \tag{20}$$

since $J^T W J$ is symmetric. If we take statistical expectations and write $V = E(\epsilon \epsilon^T)$, we get

$$E(\delta \mathbf{q} \delta \mathbf{q}^T) = (J^T W J)^{-1} J^T W V W J (J^T W J)^{-1}. \tag{21}$$

Now if we are able to choose $W = V^{-1}$ much of this expression cancels and we are left with

$$E(\delta \mathbf{q} \delta \mathbf{q}^T) = (J^T W J)^{-1}. \tag{22}$$

Note that $E(\delta \mathbf{q} \delta \mathbf{q}^T)$ is the matrix whose i, j element is the expectation value of the product of the changes in x_i and x_j due to the changes in the data. The equation says that this expectation value—the covariance of parameters x_i and x_j—is simply the i, j element of the inverse of the Gauss-Newton normal matrix used in the least squares analysis. We do need to have $W = V^{-1}$ to make this so. That is to say that we must have a set of weights which form the inverse of the variance matrix of the Δ_i. So long as the Δ_i do not have correlated errors the W matrix is diagonal and we use the conventional least squares method. If, however the Δ_i errors are correlated, we should use the 'generalised' least squares method, in which W is not diagonal.

Since the square-roots of the diagonals of the variance matrix of the x_j give standard deviations we can readily test the significance of discrepancies between alternative parameter values. If we want the standard deviation of a function of parameter values, such as a bond length, we can find the differential of this function as a linear combination of differentials of parameters. The square of the standard deviation of the bond length is then just the quadratic form of the variance matrix with the vector of coefficients of the linear combination. In symbols, consider a bond, which, for simplicity, is parallel to the x axis. Then its length l is

$$l = x_i - x_j, \tag{23}$$

where the bond is between atoms i and j. So

$$dl = 1.dx_i - 1.dx_j \tag{24}$$

gives the differential of l. To get the square of the standard deviation of l we find the quadratic form

$$\mathbf{v}^T (J^T W J)^{-1} \mathbf{v} \tag{25}$$

where

$$\mathbf{v}^T = (0, \ldots, 0, 1, 0, \ldots, 0, -1, 0, \ldots, 0) \tag{26}$$

and the 1 and -1 are in the positions corresponding to the parameters x_i and x_j.

2.4 Data with rogue observations

It is not uncommon for sets of data measured by diffractometer to contain a small proportion of 'rogue' observations which have much larger errors than the rest. This leads to the need for a procedure which can remove the effect of the rogue reflections automatically. This is called 'robust/resistant refinement' (see for example Prince and Nicholson (1983)).

The minimization function can be taken to be

$$M = \sum_i \rho(\sqrt{w_i}\Delta_i)(\sqrt{w_i}\Delta_i)^2 \tag{27}$$

where

$$\rho(\sqrt{w_i}\Delta_i) = \frac{w_i\Delta_i^2}{3}(1 + (\phi(\sqrt{w_i}\Delta_i))^{1/2} + \phi(\sqrt{w_i}\Delta_i)) \tag{28}$$

and

$$\begin{aligned}
\phi(\sqrt{w_i}\Delta_i) &= (1 - (w_i\Delta_i^2/a^2))^2, \quad |\sqrt{w_i}\Delta_i| \le a \\
&= 0, \quad\quad\quad\quad\quad\quad |\sqrt{w_i}\Delta_i| > a.
\end{aligned} \tag{29}$$

This gives normal weight to reflections with $|\sqrt{w_i}\Delta_i| \ll a$ and zero weight if $|\sqrt{w_i}\Delta_i| > a$. An obvious problem is that of choosing a. it is difficult to do this by using $\sum_i w_i \Delta_i^2$ because this sum is too much affected by the rogue $\sqrt{w_i}\Delta_i$, and a convenient method is to use $(6/n)\sum_i |\sqrt{w_i}\Delta_i|$. The sum of the $|\sqrt{w_i}\Delta_i|$ is less modified by the rogues than that of the $w_i\Delta_i^2$.

The use of this minimization function gives normal equations

$$\sum_i (\phi(\sqrt{w_i}\Delta_i)w_i \frac{\partial F_i}{\partial \mathbf{x}} \frac{\partial F_i}{\partial \mathbf{x}})\delta\mathbf{x} = \sum_i \phi(\sqrt{w_i}\Delta_i)w_i\Delta_i \frac{\partial F_i}{\partial \mathbf{x}} \tag{30}$$

so each of the usual normal equation terms is multiplied by a ϕ value. The error estimates for the results are obtained by inverting the matrix of these equations, as in the usual case.

2.5   Pseudosymmetric cases

So far we have considered only cases in which the minimization function M can be taken to be a quadratic function of $\mathbf{x}$ over a region around the minimum comparable in size with the errors. This is not always so. There are cases in which a structure lies close to a conformation of higher symmetry. In that case there is a 'mirror-image' structure which closely resembles the true one. If departures from Friedel's Law (that $I(hkl) = I(\overline{hkl})$) are neglected, then the true and mirror-image structures give equal minimum values of M. There is a low saddle point of M in between these two minima. The curvature of M varies rapidly in the region of each minimum, and the usual theory, based on the approximation of constant curvature, is no longer applicable. We cannot assess the significance of errors by considering deviations as multiples of standard deviations, and have to fall back on significance tests based directly on values of M for postulated conformations. That limits the testing of hypotheses about the structure to those in possession of the intensity data.

3   Assessment of data errors

We saw, in section 2.3, that provided that we could assess the variances of the $\Delta_i = I_i^o - I_i^c$ correctly we could readily derive the variance matrix of the x_j—the parameters of our structure model. That in turn enables us to get standard deviations for any desired functions of the x_j. All this depends on choosing the weights for the least squares refinement to correspond correctly to the variances of the Δ_i.

The central problem—and the most difficult part of the business of error estimation—is to allow for all the errors which affect the Δ_i. We will now consider this.

3.1   Classes of error

It is useful to try to enumerate the different kinds of error which can affect the Δ_i. The writer hopes that the list which follows is reasonably complete.

> Statistical fluctuations in quantum counts
> Experimental accidents (sticking shutters, crystals moving
> on their mounts, or falling off, and so on)
> Errors in correction for background radiation
> Errors in correction for thermal diffuse scattering
> Effects of multiple reflections
> Errors in absorption corrections
> Errors in extinction corrections
> Effects of damage to the crystal by the incident beam
> Effects of fluctuating power in the incident beam
> Failure of the counting train to record all quantae
> Incomplete allowance for anisotropic vibration
> Incomplete allowance for anharmonic vibration
> Errors in correcting for substitution of one atomic species
> by another
> Errors in allowing for disorder

Effects of incorrect symmetry assignment

Errors in allowing for anomalous dispersion

Deviations from the isolated atom approximation

This is a long list. Some of the types of error listed may not affect a particular structure. Equally, it may be possible to find errors omitted from this list. That makes it clear that it is not easy to be sure that all errors have been allowed for. On the assumption that this can be done, one can, in principle, construct the variance matrix for the Δ_i by adding together the variance matrices for all the different types of error concerned. This is an ideal which is hard to achieve, and in the next section we will discuss the characteristics of error types to see why that is so.

3.2   Characteristics of types of error

In principle, whenever we can detect a class of errors in the data we should try to eliminate it. For most types of error this can be done by taking sufficient care in our experimental procedures or in the analysis, and this is obviously desirable. We shall see, however, that our ability to remove errors is limited by experimental difficulties, theoretical problems, and by the properties of crystals themselves. We then have to come to terms with the errors we cannot remove.

Errors from counting statistics have a special place. They can never be removed entirely and can be reduced only at great expense (although the advent of very powerful synchrotron X-ray sources may make them so small that this is less important than it used to be). The statistics of these fluctuations are also well understood, in contrast to those of most other types of error. That has led some to suggest that counting statistical errors are the only type of errors for which standard deviations can be quoted with any confidence, and that no other type of error should be considered in estimating standard deviations.

The only sense in which this would be valuable would be that of providing an irreducible minimum below which the errors for a given set of data could not be put by any analytical technique. This minimum would seldom bear a close relation to the actual errors of a practical analysis.

Most experimental 'accidents' are sufficiently catastrophic to come to notice and force a restart of the experiment. Exceptions are sticking shutters, intermittent failures of counting trains and crystal orientations which allow reinforcement of a reflection by simultaneous occurrence of two other strong reflections with suitably related indices. These are likely to produce large errors in a small proportion of the data. There may be consequent difficulties in solving a structure and certainly will be trouble in obtaining a satisfactory refinement with the standard least squares procedure.

The best way of dealing with the refinement problem appears to be to identify the rogues with a robust/resistant procedure and then to remeasure them experimentally to find out and correct the causes of error. Even this approach may not deal effectively with cases in which the errors are marginal, so that unusually large errors escape detection by the procedure.

Correction for background scattering would not be a distinct problem if it were not for the effect of thermal diffuse scattering. This tends to peak underneath a Bragg reflection. The correction for it is complicated and difficult to carry out. It may well therefore produce an error which is poorly corrected and which tends to increase reflection intensities at large Bragg angles. That, in turn, is likely to reduce systematically the values of the diagonal elements of vibration tensors. An account of the effects of first and second order approximations to the thermal diffuse scattering upon the electron density is given by Criado, Conde and Marquez(1985).

This is an example of a situation in which one type of error can be compensated by another, at least partly. That has implications which we will consider when we discuss weighting schemes.

Errors in correcting absorption and extinction have a similar effect to those arising from thermal diffuse scattering. Absorption and extinction tend to reduce intensities at low Bragg angles more than those at high ones. Any error in correcting these effects can therefore, again, be partly compensated by errors in the vibration tensors. Jones (1984) gives an interesting review of the effects of failure to correct for absorption, and other errors, on the position and vibration parameters.

Radiation damage to crystals tends to disorder their structures. Individual intensities may rise or fall, but there is a general tendency for them to decrease, especially at large Bragg angles. It might be argued that this provides a measure of a genuine physical effect, but if different reflections are obtained after differing exposures, then no single state of the crystal is represented by the data as a whole. For crystals subject to rapid damage, area detectors, which allow rapid data collection, offer the best hope of containing this problem.

Disorder, and replacement of one atomic species by another chemically similar one, both alter the electron density distribution, usually in localised regions of the unit cell. it is reasonable to hope that refinement will not be considered as complete without study of difference maps which can reveal such problems. If care is taken, such errors should be reduced to negligible size without the need for excessive numbers of parameters.

Anisotropy and anharmonicity of vibration lead to the need for second and higher order tensors to represent the effects on atomic scattering power. It has been customary to use models in which such tensors are provided for each (non-hydrogen) atom. This demands large numbers of parameters, and frequently the quality and quantity of data available will not support accurate determination of so many parameters. The present state of refinement packages may then force the use of crude approximations which model the vibration badly.

There is a need for economical vibration models which can be applied without difficulty and delay, to overcome the problem. See Wu, Karplus and Hendrickson (1985) for an application of restraints to provide satisfactory models. Unless a good model is achieved there will be errors in which various reflections are correlated. This greatly increases the difficulty of making good error estimates for parameters.

Pseudosymmetry can cause misleading error estimates. An example will illustrate this. The structure of Hexachloroborazine (see Haasnoot *et al.* (1972), Gopinathan *et al.* (1974)) has a six-ring of alternating B and N atoms, sitting on a three-fold symmetry axis, with a Cl atom attached to each B or N atom. The molecule is puckered with Cl atoms alternating on each side of the mean plane. The original analysis gave alternating long and short B-N bond lengths. No way was found of explaining this theoretically, and so the structure was re-examined. By reversing the pucker, the bond-lengths were made indistinguishable and agreement was slightly improved.

Both structures gave very low R-factors (.0254 and .0253 for un-weighted R-factors and .0274 and .0271 for weighted R-factors) and both gave position coordinate e.s.ds. small compared with the differences between the two sets of positions. The Hamilton ratio test did, however discriminate between them. The ratio was 1.010, whereas the probability of exceeding a ratio of 1.006 was .005. It is still difficult to see how the original error could have been detected and corrected without the insight of a theoretical chemist. This, again, is an example of a situation in which a set of errors compensate for one another's effects to produce relatively good agreement between I_i^o and I_i^c.

4 Checks on assessment of errors of data

4.1 Statistical expectation

We have a set of discrepancies

$$\boldsymbol{\Delta}^T = \{\Delta_i = I_i^o - I_i^c,\ i = 1, 2, \ldots, n\} \tag{31}$$

and we estimate their variance matrix V. If we choose our weight matrix W so that

$$W = V^{-1} \tag{32}$$

then we expect the weighted sum

$$\chi^2 = \boldsymbol{\Delta}^T W \boldsymbol{\Delta} \tag{33}$$

to be distributed like χ^2 with $n - m$ degrees of freedom, where m is the number of parameters we use to adjust the I_i^c.

For large values of $n-m$, the value of $\chi^2/(n-m)$ has a low probability of departing far from unity. If we find it is far from unity, then there is strong evidence that W has been chosen incorrectly. Such a value must mean

1. that we have incorrectly assessed the size of some source of error, or

2. that we have neglected one or more types of error, or

3. that differing types of error are compensating for one another.

The possibility of compensation between errors implies that it is dangerous to use a low value of $\chi^2/(n - m)$ as an excuse for increasing the elements of W. A value of $\chi^2/(n - m)$ significantly above unity is a different matter. This cannot arise because errors that we have estimated correctly are compensating for one another. We must have neglected, or underestimated, something. Various options are available.

1. Look for some gross error, such as an incorrect solution to the phase problem. (We may be at a false minimum of the minimization function.)

2. Review the check list of error types to see whether something that we have left out could be contributing, or something could be worse than we believed. If so adjust the variance estimates accordingly.

3. If we are not able to identify the cause of the problem, adjust the variance estimates anyway, to produce a credible value of χ^2.

The last option carries the obvious danger that we have failed to understand the problem, and that there may be a type of error which produces effects on the parameters which are not distributed like the normal law of error. In that case the usual rules about the probability of a deviation exceeding a certain number of standard deviations may be misleading.

We therefore need tools for identifying the cause of unexpectedly large discrepancies, and we consider these in the next section.

4.2 Agreement analysis

Not only should χ^2 take a reasonably probable value, but also (for a diagonal system of weights) the weighted sum of squares of discrepancies should be nearly $(n - m)/n$ times the number of terms, for any large group of terms. Any big departures from this are therefore helpful hints concerning sources of error. We can analyse weighted discrepancies according to

1. groups of reflections with different intensity ranges.

2. groups in different ranges of Bragg angle.

3. groups with different values, or parities, of Miller indices.

4. groups with incident, or scattered, beam directions close to significant dimensions of the crystal specimen.

It is also useful just to list the indices of reflections with particularly large weighted discrepancies, and consider whether any regularities are apparent.

If an unusual pattern of error appears, then there will be a need to apply a weighting scheme with unusual rules. This is unlikely to be achieved unless the package in use permits a flexible approach. There is a real danger that nothing will be done about a problem if the investment in time and effort required exceeds that for the rest of the analysis.

5 External tests of error estimates

5.1 Evidence for biased estimates

If the error estimates for an X-ray crystal structure analysis represent at least those factors which can vary when the analysis is repeated, then we

would expect that duplicate analyses of the same structure would agree
well enough to make the discrepancies probable according to the estimates.

This has been studied by Taylor and Kennard (1986) who found
100 cases in which the same structures had been analysed by at least
two essentially independent workers. They noted that the average ratios
between discrepancies and their standard deviations varied from slightly
less than unity to large values. It is not likely that all the large values
have the same explanation. It is however clear that in most cases the error
estimates are biased so that they are smaller than they should be.

In the next section we consider why.

5.2   Reasons for biased estimates

One reason why error estimates may be low is the use of weights purely
based on counting statistics. If all other errors are ignored then the estimates must surely be low. Even if the estimates are scaled by $\chi^2/(n-m)$,
the non-uniformity of the $w_i \Delta_i^2$ implies that the results will not have minimum variance. The reduction from equation 21 to equation 22 is invalid
if, as in such a case, the W matrix differs from V^{-1}. Hence we can expect
the error estimates to be incorrect.

A further reason for bias in error estimates is likely to be that some
classes of error cause the Δ_i to be correlated. This has been studied
by Clarke and Rollett (see Rollett (1982)) who investigated a neutron
diffraction powder profile analysis for trifluor-acetic acid. They found that
when diagonal weights were replaced by weights which took account of the
correlation of the residuals, the standard deviation estimates increased.
For positional coordinates the average increase was by a factor of 1.4 but
individual standard deviations altered by more or less than this.

The values of some parameters, particularly vibrations, also changed.
This altered a situation in which the error estimates excluded real amplitudes of vibration to one in which the results were consistent with real
amplitudes. Clearly, the error estimates based on diagonal weights were
inconsistent with any physically reasonable structure model, and allowance

for correlation removed the difficulty.

It is quite difficult computationally to extend the procedure for allowing for correlated errors from the one-dimensional powder diffraction case to the three-dimensional single-crystal case. The writer does not know of any check on the effects of such a procedure for a single crystal analysis, but the substantial factor by which the e.s.ds. increased suggests that it might explain a large part of the discrepancy noted by Taylor and Kennard (1986).

6   Conclusions

It may be helpful to summarise the conclusions that the writer derives from the discussion of error estimates in this paper:

1. That there are so many sources of error which affect materially the results of a crystal structure analysis that a check-list of error types would be of value. This paper presents one in section 3.1.

2. That no crystal structure analysis can be considered to have reliable error estimates unless it has been shown that the weights lead to weighted residuals of reasonably uniform and appropriate size.

3. That present methods of error estimation do not allow for compensation of errors. Cross checking against independent evidence is highly desirable. (See the discussion at the end of section 3.2.)

4. That the duplication of effort represented by the independent determination of the same structure has been valuable. Further duplicate determinations in suitably chosen cases could throw light on problems in error estimation.

5. That the errors of the parameters may not be normally distributed in case that they are dominated by a small number of large errors in residuals, or that residuals are systematically related to part of the unit cell. Checks on difference electron density are needed to rule out the latter possibility.

REFERENCES

Criado, A., Conde, A., and Marquez, R. (1985). *Acta Crystallographica*, **A41**, 491–494.

Gopinathan, M. S., Whitehead, M. A., Coulson, C. A., Carruthers, J. R., and Rollett, J. S. (1974). *Acta Crystallographica*, **B30**, 731–737.

Haasnoot, J. G., Verschoor, G. C., Romers, C., and Groeneveld, W. L. (1972). *Acta Crystallographica*, **B28**, 2070–2073.

Jones, P. G. (1984). *Chemical Society Reviews*, **13**, 157–172.

Prince, E., and Nicholson, W. L. (1983). *Acta Crystallographica*, **A39**, 407–410.

Rollett, J. S. (1982). In *Computational Crystallography*, (ed. D. Sayre), Clarendon Press, Oxford.

Taylor, R., and Kennard, O. (1986). *Acta Crystallographica*, **B42**, 112–120.

Wu, H., Karplus, M., and Hendrickson, W. A. (1985). *Acta Crystallographica*, **B41**, 191–201.

Fibre diffraction

12

X-RAY FIBER DIFFRACTION

R.P. Millane

1. INTRODUCTION

X-ray fiber diffraction analysis refers to a collection of
techniques that are used to determine the structures of molecules
that do not form regular three-dimensional crystals, but can be
oriented to form fibers in which the molecules or assemblies are
approximately parallel. The numbers and types of such materials
is large, including simple polysaccharides (cellulose), complex
linear (gellan) and branched (xanthan) polysaccharides, nucleic
acids, fibrous macromolecular assemblies (tobacco mosaic virus,
bacteriophages, microtubles) and some membrane structures that
form two-dimensional periodic arrays. Since these molecules do
not form extensive regular crystals they are not suitable for
conventional crystallographic analysis. However, diffraction
studies have played an important role in determinations of the
structures of DNA and a wide range of other linear biopolymers
because specimens can be prepared in which the long axes of the
molecules are approximately parallel. They sometimes further
organize laterally into very small regions of three-dimensional
crystallinity. However, the orientations of the crystallites
about their long axes are random. Diffraction patterns from these
fibrous specimens often contain sufficient information to make
structure determination possible. However, the number of measured

diffraction data is usually less than the number of independent atoms in the structural unit of the molecule, and they rarely extend beyond 3 Å resolution. The diffraction data must therefore be supplemented by other information.

Fiber diffraction techniques fall into three classes, depending on the type of molecule under investigation:

(1) Techniques for which the number of possible configurations of the molecule is so small that each can be modelled and refined against the x-ray data. This method can be applied to polymers of relatively simple sequence such as polysaccharides and nucleic acids.

(2) Techniques which more closely mimic those of traditional macromolecular crystallography. These make use of isomorphous replacement and molecular replacement to phase fiber diffraction data to ultimately produce a three-dimensional electron density map. These methods can be used to determine the structures of complex molecules such as fibrous viruses that have many degrees of freedom.

(3) Techniques that use low resolution modelling to obtain information on large macromolecular assemblies.

It is not possible to cover all aspects of fiber diffraction in a short review such as this, so the salient computational aspects of fiber diffraction analysis are covered here. The review is therefore comprehensive but not detailed. This should allow readers to determine if their particular system may be susceptible to fiber diffraction analysis, what techniques should be used and critical references to the literature. The theory of diffraction by helical molecules, data collection, and the various methods of structure determination are covered in the following sections.

2. DIFFRACTION BY HELICAL STRUCTURES

Fibrous molecules always display some helical symmetry that can be determined by examining the systematic absences on the meridian (axis parallel to the fiber) on the diffraction pattern.

Consider an isolated molecule that has a structural repeating unit, which we call a *residue*, that is repeated by helical symmetry along the molecular axis. On defining a cylindrical polar coordinate system (r,ϕ,z), the points (r,ϕ,z) and $(r,\phi+2\pi mv/u, z+mPv/u)$ are equivalent where m is any integer, P is the helix *pitch* and u and v are integers that define the helix symmetry. This is referred to as a u_v helix that has u residues in v turns. The crystallographic repeat is $c = vP$. If $v = 1$ the helix is called *integral* and $c = P$. The diffraction pattern from such a molecule is non-zero on *layer planes* in reciprocal space, spaced by $1/c$. Denoting a cylindrical polar coordinate system in reciprocal space by (R,ψ,Z), the Fourier transform $F_\ell(R,\psi)$, where ℓ indexes the layer planes, of the molecule is given by (Sherwood, 1976)

$$F_\ell(R,\psi) = \sum_j \sum_n f_j \, J_n(2\pi Rr_j)$$

$$\times \exp\{i[n(\psi+\pi/2) - n\phi_j + 2\pi\ell z_j/c]\} \tag{1}$$

where J_n is the nth order Bessel function of the first kind, (r_j,ϕ_j,z_j) are the coordinates of the jth atom with scattering factor f_j, and the sum over j is over the atoms in one residue. The sum over n is over all integers satisfying the *helix selection rule* (Cochran, Crick and Vand, 1952)

$$\ell = um + vn \tag{2}$$

where m and n are integers (positive and negative).

In a *non-crystalline* fiber specimen, the molecules are approximately parallel but are randomly rotated relative to each other about their long axes and are not organized laterally. The intensity $I_\ell(R)$ on the diffraction pattern then depends on R and ℓ

only and is equal to the intensity of the molecular transform averaged over ψ, so that

$$I_\ell(R) = (1/2\pi) \int_0^{2\pi} |F_\ell(R,\psi)|^2 \, d\psi. \tag{3}$$

Making use of eqn (1) allows eqn (3) to be put in the form (Franklin and Klug, 1955)

$$I_\ell(R) = \sum_n G_{n\ell}(R) \, G_{n\ell}^*(R) \tag{4}$$

where n satisfies the selection rule and (Klug, Crick and Wyckoff, 1958)

$$G_{n\ell}(R) = \sum_j f_j \, J_n(2\pi R r_j) \, exp[i(-n\phi_j + 2\pi\ell z_j/c)] \tag{5}$$

Equations (4) and (5) are used to compute continuous amplitudes (i.e. the equivalent of structure amplitudes) for non-crystalline specimens. Because a Bessel function is small for values of its argument less than its order, only a finite number of Bessel orders need be included in the calculation. The maximum order required is determined by the outer radius of the molecule (maximum r_j) and the maximum resolution required (maximum R). For higher order helical symmetry (larger u/v), fewer Bessel orders are required.

   In a *polycrystalline* specimen, the molecules are also parallel but they also organize laterally into microcrystallites, each of which is a very small regular three-dimensional crystal. The crystallites however are randomly rotated relative to each other so that the diffraction pattern again depends only on R and ℓ, and is the cylindrical average of the intensity diffracted by a single crystal. The diffraction pattern therefore contains Bragg reflections with structure factors that can be calculated using the Bessel form

$$F_{hk\ell} = \sum_p exp[i2\pi(hu_p + kv_p + \ell w_p)]$$

$$\times \sum_n G_{n\ell}(R_{hk}) \, exp(-in\mu_p) \tag{6}$$

where the sum over p is over all the molecules or equivalent positions in the unit cell with fractional coordinates (u_p, v_p, w_p) and relative rotation μ_p about the molecular axis, and R_{hk} is the cylindrical radius of the reciprocal lattice point. Alternatively, the structure factors can be calculated using the usual triginomeric form

$$F_{hk\ell} = \sum_j f_j \, exp[i2\pi(hx_j + ky_j + \ell z_j)] \tag{7}$$

where in this case the sum is over all the atoms j in the unit cell with fractional coordinates (x_j, y_j, z_j). Computation using the Bessel form is more efficient for high symmetry helices (because of the selection rule) whereas the triginometric form is faster for low symmetry helices. Because of the random crystallite orientations, reflections that have the same, or nearly the same, R_{hk} will be coincident or inseparable on the diffraction pattern. The recorded data are therefore amplitudes H_j given by

$$|H_j|^2 = \sum |F_{hk\ell}|^2 \tag{8}$$

where the sum is over all (h, k, ℓ) for which the reflections overlap in spot j.

There is a continuous spectrum of types of disorder betwee‘

the polycrystalline and oriented types described above present for different fibrous specimens. Various types of disorder give rise to both Bragg and continuous diffraction in the same or different regions of a diffraction pattern (Arnott, 1980). The relationship of these different diffraction components to the structure must be considered when using such patterns for structural analysis. Two common types of disorder are *statistical crystallinity* and *screw disorder* (Arnott, 1980) and both of these have been incorporated into structure determinations (Arnott *et al*, 1974; Arnott *et al*, 1986; Park *et al*, 1987).

3. DATA COLLECTION

Since diffraction data from a fibrous specimen are two-dimensional, they can be collected with a single exposure of a stationary specimen. Traditionally, the positions of reflections on the film were measured manually and intensities measured using a radial microdensitometer trace through each spot (Arnott and Hukins, 1973; Fraser and MacRae, 1973). Many laboratories now have access to a computer controlled microdensitometer that allows two-dimensional digitizing of the whole diffraction pattern followed by computer-aided determination of the intensities (Fraser *et al*, 1976; Makowski, 1978; Fraser, Suzuki and MacRae 1984; Millane and Arnott, 1985a). The digitized pattern is first analyzed to determine its center and orientation to the raster coordinate system, the film-to-fiber distance and the fiber tilt to the x-ray beam (Fraser *et al*, 1976). Background can be removed using either an exposure with the specimen removed, by interpolating between sample points between reflections (Millane and Arnott, 1985b) or both. It is sometimes convenient to map the digitized pattern from film space to reciprocal space (Fraser *et al*, 1976). Subsequent processing depends on whether one is dealing with a diffraction pattern containing Bragg reflections or continuous intensities so these are described separately.

3.1 Measuring Bragg Intensities

By examining ratios of the R values of reflections close to the meridian, the type of unit cell and the approximate cell constants can be determined. These can then be refined by minimizing the sum of the deviations between the calculated and measured reciprocal space coordinates (Arnott and Hukins, 1973).

If the unit cell is small and/or the space group symmetry is high, the reflections are well separated and the intensities can be measured either by summing the optical densities over the region of the spot (Fraser *et al*, 1976; Millane and Arnott, 1985a) or by profile fitting (Fraser, Suzuki and MacRae, 1984). If the unit cell is large, the symmetry is low or the specimen is not well ordered, then adjacent reflections may overlap. The total intensity of such composite reflections can be measured in the same way as described above. More x-ray data can be obtained if the intensities of individual overlapping reflections are measured. Some success has been achieved with using profile fitting to resolve overlapping reflections (Fraser, Suzuki and MacRae, 1984) but the applications have not yet been extensive enough to consider these methods routine. Geometrical corrections are described by Fraser *et al*, (1976) and Millane and Arnott (1985a).

3.2 Measuring Continuous Intensities

X-ray data from continuous diffraction patterns consist of closely spaced samples of the intensity along each layer line. Imperfect parallelism of the molecules (disorientation) causes individual points on the layer lines to be smeared along arcs. Consequently, layer lines overlap with their neighbours beyond some value of R. To collect data to the highest possible resolution, the contributions from the individual layer lines must be separated. The measured layer line intensity $D(\rho,\sigma)$, where ρ is the length of the reciprocal space vector and σ is the angle it makes with the z-axis (Millane and Arnott, 1985a), can be well approximated by a one-dimensional convolution along circles of constant ρ in reciprocal space (Makowski, 1978), so that

$$D(\rho,\sigma) = \sum_i I_i(\rho,\sigma_i) \, g(\sigma-\sigma_i) \qquad\qquad (9)$$

where $I_i(\rho,\sigma_i)$ is the intensity at the center of the ith layer line and $g(\sigma)$ is the angular profile of the layer lines that is related to the distribution of molecule orientations. The profile $g(\sigma)$ is almost independent of position in reciprocal space except within about $2\sigma_0$ (where σ_0 is the half-width of the profile) of the meridian. The profile can be measured close to the center of the pattern where there is no overlap. By making measurements of the intensity D at a number of values of σ for a particular value of ρ, eqn (9) can be solved by linear least squares for the individual layer line intensities I_i. Repeating this for a range of closely spaced values of ρ followed by interpolation gives values for the intensities I_i along the individual layer lines. This procedure was first described and implemented by Makowski (1978) in film space (i.e. on (r,ϕ) rather than (ρ,σ)). The procedure is slightly more accurate when applied in reciprocal space since $g(\sigma)$ is more invariant than $g(\phi)$ (Makowski, 1978; Namba and Stubbs, 1985) although this can be partially accounted for by applying a systematic variation to $g(\phi)$ with position and solving eqn (9) as a profile fitting problem, i.e. replacing $g(\phi-\phi_i)$ by $g(\phi,\phi_i)$ in eqn (9) (Millane and Arnott, 1985a). The maximum spacing ΔR between the intensity samples along the layer lines that retains all the information in the continuous intensity is $1/2d$ where d is the maximum diameter of the molecule (Makowski, 1981). The effects of noise can be reduced by applying the deconvolution for a smaller ΔR and smoothing the resulting values. Residual background can be estimated as part of the deconvolution procedure (Makowski, 1978) but there is a danger of lowering the reliability of the results if more than a single unknown parameter is used to describe the background at each radius. Close to the meridian the profile is strongly dependent on position but since there is little overlap there, the intensity can be measured by summing along an arc (Millane and Arnott, 1985a; Namba and Stubbs, 1985).

The layer lines can only be separated reliably out to a maximum value of R on each layer line which is approximately where the profiles cross at about 0.75 of their peak values (Makowski, 1978; Millane and Arnott, 1985a). The higher layer lines can be reliably measured to a higher resolution (larger ρ) than the lower ones because their angular separation is greater (Millane and Arnott, 1985a).

4. STRUCTURE DETERMINATION BY ATOMIC MODELLING

The structures of polymers such as polysaccharides and nucleic acids can often be determined by using an atomic model of the molecule. The primary (chemical) structure of the polymer must be known in order to use these methods. The principle is to refine each of the possible molecular and crystal models against the x-ray data and stereochemical restraints to produce the best model of each kind. The optimized models can be compared using various figures of merit. In favorable cases, one of the resulting models will be sufficiently superior to the remainder for it to unequivocally represent the correct structure.

The primary steps are as follows.
(1) Construction of all types of stereochemically acceptable model with the correct pitch and symmetry. Examples of different types of models are, conformation angles in different domains, left and right-handed helices, single and double helices, or parallel and antiparallel double helices. If the specimen is polycrystalline, then the possible packings must also be considered. These may include different juxtapositions of molecules within a particular space group, or different space groups. The space group cannot

always be assigned unequivocally - for example the space groups $P3_121$, $P3_221$, $P6_2$ and $P6_4$ cannot be distinguished from the positions of reflections on a fiber diffraction pattern.

(2) Refinement of each of the models against the x-ray data and stereochemical restraints.

(3) Adjudication among the resulting models.

Implementation of the first two steps are embodied in computer programs that define and refine polymer models. The two most widely used and versatile programs are the *linked-atom least-squares* system (LALS) (Smith and Arnott, 1978) and the *variable virtual bond* method (Zugenmaier and Sarko, 1980). Although the way in which structures are defined and refined in these two systems are different, the principles, applications and outcomes are the same. The LALS method has been used more extensively and is the one described here.

The LALS concept was first applied to helical polymers in the 1960s (Arnott and Wonacott, 1966) and the programs extended and generalized since then (Smith and Arnott, 1978). This system has been used to refine a wide range of polynucleotides, polysaccharides, polyesters and polypeptides - refer to the references in Arnott (1980).

The overall scheme for defining and refining polymer structures is outlined in Fig. 1. The bond lengths and angles in polymers have very nearly the same values as in the corresponding monomers which are often susceptible to conventional crystallographic analysis so that the bond lengths and angles are known rather precisely. This reduces the solution of the polymer structure to that of determining conformation angles about single bonds: one parameter per atom rather than three. Moreover, the conformations of some structural entities such as rigid rings (e.g. sugar rings, planar conjugated bases) may also be assumed to persist from monomer to polymer. It is therefore possible to prepare a *linked-atom* description of a molecule: that is, one in which interatomic relationships are described in terms of bond

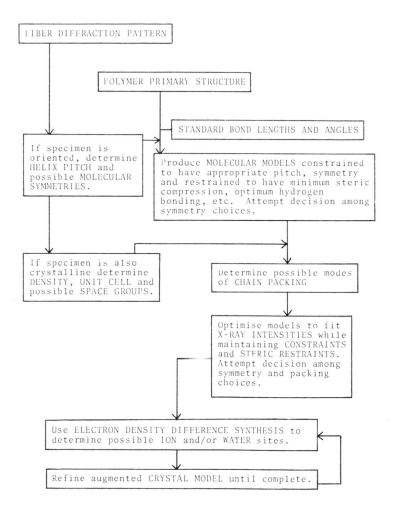

FIG. 1. A scheme for defining and refining polymer structures using x-ray fiber diffraction data and stereochemical restraints (from Arnott, 1980).

lengths, bond angles and conformation angles. A further source of stereochemical data is the requirement that a model exhibit no over-short nonbonded interatomic distances.

In the LALS system, the quantity Ω given by

$$\Omega = \sum_m \omega_m \, \Delta F_m{}^2 + \sum_m k_m \, \Delta d_m{}^2 + \sum_m \lambda_m \, G_m \qquad (10)$$

$$= \quad X \quad + \quad C \quad + \quad L$$

is minimized by varying a set of chosen parameters consisting of conformation angles, packing parameters, and x-ray scale and attenuation factors. The term X involves the differences ΔF_m between the model and experimental x-ray amplitudes - Bragg and/or continuous. The term C involves restraints that can be used to ensure that over-short non-bonded interatomic distances are driven beyond acceptable minimum values, that conformations are within desired domains, that hydrogen bond and coordination geometries are close to the expected configurations as well as a variety of other relationships (Smith and Arnott, 1978). The ω_m and k_m are weights that are inversely proportional to the estimated variances of the data. The term L involves constraints which are relationships that are to be satisfied exactly ($G_m = 0$) and the λ_m are Lagrange multipliers. Constraints are used, for example, to ensure connectivity from one helix pitch to the next and to ensure that chemical ring systems are closed.

Structure determination usually involves first using eqn (10) with the terms C and L only to establish the stereochemical acceptability of each of the possible molecular models and packing arrangements. It is worth emphasizing that it is usually advantageous if the specimen is polycrystalline even though the continuous diffraction contains, in principle, more information than the Bragg reflections (since the latter are sampled). This is because the molecule in a non-crystalline specimen must be refined in steric isolation whereas for a polycrystalline specimen it is refined while packed in the crystal lattice. The extra information provided by the intermolecular contacts can often help eliminate incorrect models. This can be particularly significant if the molecule has flexible sidechains. The models surviving are then confronted with the x-ray data by further refinement with X included in eqn (10). The ratio $(\Omega_P/\Omega_Q)^{\frac{1}{2}}$ can be used in Hamilton's test (Hamilton, 1965) to decide between models P and Q. Also, the ratio $(X_P/X_Q)^{\frac{1}{2}}$ can be used to indicate whether the x-ray amplitudes are sensitive to the differences between the two models. These tests can be used to decide if one model is clearly superior to the others. In the final stages of refinement, bond angles can be allowed to vary in a "stiffly elastic" fashion from their mean values if there are sufficient data to justify the increase in the number of degrees of freedom.

If sufficient x-ray data are available, it is sometimes possible to locate additional ordered molecules such as counterions and water molecules by difference Fourier synthesis using phases derived from the polymer model. Since the signal to noise ratio in these syntheses is usually low, they must be interpreted with caution. The assignment of counterions or water molecules to peaks in the difference synthesis must be supported by sensible interactions with the rest of the structure. The positions of molecules located in this way may then be co-refined with the polymer conformation while hydrogen bonds and coordination geometries are optimized. The resulting structure can then be used to compute improved phases to search for additional molecules. For a good description of the application of difference Fourier synthesis in this manner, the reader is referred to Mitra *et al* (1983). Further current examples of the application of atomic model building to structure determination may be found in Arnott *et al* (1986) and Park *et al* (1987).

5. STRUCTURE DETERMINATION BY OTHER METHODS

Although molecular model building has been used to determine the majority of macromolecular structures from fiber diffraction data, more complex structures have been solved using quite different techniques. These fall into two different classes and are outlined in the next two subsections.

5.1 Phasing fiber diffraction data

In conventional macromolecular crystallography multiple x-ray data sets (from isomorphous derivatives for example) are used to phase the measured structure amplitudes allowing an electron density map to be calculated. Analagous techniques can be applied to fiber diffraction data although, because of the cylindrical averaging of the diffraction pattern, they are more complicated.

The three-dimensional electron density function $\rho(r,\phi,z)$ in a cylindrical coordinate system can be put in the form (Klug, Crick and Wyckoff, 1958)

$$\rho(r,\phi,z) = (1/c) \sum_{\ell} \sum_{n} g_{n\ell}(r) \ exp[i(n\phi - 2\pi\ell z/c)] \qquad (11)$$

where

$$g_{n\ell}(r) = \int_{0}^{\infty} G_{n\ell}(R) \ J_{n}(2\pi Rr) \ 2\pi RdR \qquad (12)$$

Hence the electron density can be calculated if the complex functions $G_{n\ell}(R)$ are determined from the measured data $I_{\ell}(R)$ given by eqn (4). The phase problem in fiber diffraction therefore involves separating the individual $G_{n\ell}(R)$ terms, for different n from $I_{\ell}(R)$ as well as determining the phase of each $G_{n\ell}(R)$. Stubbs and Diamond (1975) show that if there are *N* Bessel terms contributing to a layer line then fiber diffraction data from a minimum of *2N* derivatives can be used to separate and phase the $G_{n\ell}(R)$ terms.

A phenomenon known as *layer line splitting* can also be used to help phase fiber diffraction data (Stubbs and Makowski, 1982). Layer line splitting occurs when the molecule has almost, but not exactly, an integral number (u) of residues in v turns. This is a result of the true axial repeat distance of the structure being much greater than vP. The diffraction associated with some of the G_n terms is offset very slightly from the expected position on a layer line. Careful digital processing of the diffraction pattern can reveal this *splitting* and provide additional information to separate the Bessel terms. If N derivatives (including the native) are available and M give accurate measurements of splitting, then, in principle, $(N+M)/2$ Bessel orders can be separated and phased. The Hendrickson-Konnert refinement program has been modified to allow refinement of structures built into the resulting electron density maps (Stubbs, Namba and Makowski, 1986).

The techniques described above have been used to determine the structure of tobacco mosaic virus to 3.6 Å resolution. A detailed description of this work appears in Stubbs and Diamond (1975), Stubbs and Makowski (1982), Namba and Stubbs (1985) and Namba and Stubbs (1987).

5.2 Low resolution model building

The advantage of model building is that the phases do not need to be determined experimentally. For larger molecules however, too many possible atomic models would need to be considered. Structural information on larger molecules can sometimes be obtained using low resolution model building if appropriate independent information is available. The amount of information at low resolution (how low depends on the symmetry and diameter of the structure) in the continuous intensity is higher (because it is not sampled) than from a crystalline diffraction pattern, despite the angular averaging (Makowski *et al*, 1980). With some independent information, the phase, and hence the low resolution structure, may be determinate. This low resolution structure may then indicate how an approximate atomic model can be built.

The maximum diameter of the structure can often be estimated and incorporated into a refinement procedure (similar to solvent levelling) to refine phases from an initial model and find an electron density distribution of the required diameter that is consistent with the measured intensities (Makowski, Casper and Marvin, 1980; Makowski, 1982). This does not generally lead to a unique solution however - different starting models (or phases) may lead to different consistent electron densities. The ambiguity may be resolved if other structural information is available and is incorporated into the refinement procedure. At each iteration the electron density is modified to conform to the known features to calculate a new set of phases for the Bessel terms that are then combined with the measured amplitudes to calculate a new electron density. In favourable cases this procedure may converge to a unique electron density map that can be usefully interpreted.

An example of the application of such a procedure is the solution of the structure of the filamentous bacteriophage Pf1 to 7 Å resolution (Makowski, Casper and Marvin, 1980). Independent measurements showed that a large proportion of the structure is α-helical. At each cycle of refinement therefore, the electron density was searched for rod-like features that might represent α-helical segments. These features were replaced by rods of the appropriate diameter and electron density to form the new electron density map from which new phases were calculated. In subsequent cycles of refinement, adjustment of the rod positions was alternated with imposition of the particle diameter until convergence was attained. Many of a variety of starting models converged to the same final model that had the best x-ray agreement of all models tested. Other final models had either poor x-ray agreement or insufficient α-helical features. The solution therefore appears to be unique and some correlation of the resulting map with the amino acid sequence was possible.

6. CONCLUSIONS

Procedures are available for performing detailed analyses of the structures of a wide range of fibrous structures using their x-ray diffraction data. Unique solutions can often be obtained. When utilizing model building, alternative models must be scrupulously examined to ensure a reliable solution.

I am grateful to my colleagues, Drs. Struther Arnott and R. Chandrasekaran, from whom I learned many of the idiosyncrasies of fiber diffraction analysis. This work was supported by the U.S. National Science Foundation.

REFERENCES

Arnott, S. and Wonacott, A.J. (1966). *Polymer*, **7**, 157-166.

Arnott, S. and Hukins, D.W.L. (1973). *J. Molecular Biology*, **81**, 93-105.

Arnott, S., Scott, W.E., Rees, D.A. and McNab, C.G.A. (1974). *J. Molecular Biology*, **90**, 253-267.

Arnott, S. (1980). In *Fiber Diffraction Methods* (ed. A.D. French and K.H. Gardner) pp. 1-30. ACS Symposium Series Vol. 131, American Chemical Society, Washington, DC.

Arnott, S., Chandrasekaran, R., Millane, R.P. and Park, H.S. (1986). *J. Molecular Biology*, **188**, 631-640.

Cochran, W., Crick, F.H.C. and Vand, V. (1952). *Acta Crystallographica*, **5**, 581-586.

Franklin, R.E. and Klug, A. (1955). *Acta Crystallographica*, **8**, 777-780.

Fraser, R.D.B. and MacRae, T.P. (1973). *Conformation in Fibrous Proteins*, Academic Press, New York.

Fraser, R.D.B., MacRae, T.P., Miller, A. and Rowlands, R.J. (1976). *J. Applied Crystallography*, **9**, 81-94.

Fraser, R.D.B., Suzuki, E. and MacRae, T.P. (1984). In *Structure of Crystalline Polymers* (ed. I.H. Hall) pp. 1-37. Elsevier, New

York.

Hamilton, W.C. (1965). *Acta Crystallographica*, **18**, 502-510.

Klug, A., Crick, F.H.C. and Wyckoff, H.W. (1958). *Acta Crystallographica*, **11**, 199-213.

Makowski, L. (1978). *J. Applied Crystallography*, **11**, 273-283.

Makowski, L., Casper, D.L.D. and Marvin, D.A. (1980). *J. Molecular Biology*, **140**, 149-181.

Makowski, L. (1981). *J. Applied Crystallography*, **14**, 160-168.

Millane, R.P. and Arnott, S. (1985a). *J. Macromolecular Science - Physics*, **B24**, 193-227.

Millane, R.P. and Arnott, S. (1985b). *J. Applied Crystallography*, **18**, 419-423.

Mitra, A.K., Arnott, S., Atkins, E.D.T. and Isaac, D.H. (1983). *J. Molecular Biology*, **169**, 873-901.

Namba, K. and Stubbs, G. (1985). *Acta Crystallographica*, **A41**, 252-262.

Namba, K. and Stubbs, G. (1987). *Acta Crystallographica*, **A43**, 64-69.

Park, H.S., Arnott, S., Chandrasekaran, R., Millane, R.P. and Campagnari, F. (1987). *J. Molecular Biology*, in press.

Sherwood, D. (1976). *Crystals, X-rays and Proteins*, Longman, London.

Smith, P.J.C. and Arnott, S. (1978). *Acta Crystallographica*, **A34**, 3-11.

Stubbs, G.J. and Diamond, R. (1975). *Acta Crystallographica*, **A31**, 709-718.

Stubbs, G.J. and Makowski, L. (1982). *Acta Crystallographica*, **A38**, 417-425.

Stubbs, G., Namba, K. and Makowski, L. (1986). *Biophysical J.*, **49**, 58-60.

Zugenmaier, P. and Sarko, A. (1980). In *Fiber Diffraction Methods* (ed. A.D. French and K.H. Gardner) pp. 225-237. ACS Symposium Series vol. 131, American Chemical Society, Washington, DC.

Electron diffraction

13

CALCULATION OF ELECTRON WAVEFUNCTIONS FOR ELECTRON DIFFRACTION

P.G. Self

The scattering power of atoms for fast electrons is of the order of 10^4 times greater than for x-rays. Consequently, it is not possible to use a kinematical approximation when calculating diffracted electron amplitudes; a full dynamical treatment must be used. The formulation of dynamical diffraction for electrons is considerably simpler than the formulation for x-rays because, for electrons, it is not necessary to consider polarization. However, because of the strong interaction and the relative flatness of the Ewald sphere, the formulation for the dynamical diffraction of electrons must take into account a large number of diffracted beams. Thus, in most cases, electron diffraction wavefunctions must be calculated by numerical methods.

Two major approaches have been developed for the solution of the dynamical electron diffraction problem. The first approach was developed by Bethe (1928) and is based on the direct solution of Schrödinger's equation. The second is the physical-optics approach developed by Cowley and Moodie (1957,1958). From these two methods many other formulations have been developed e.g. the Howie-Whelan equations (Howie and Whelan, 1961), the scattering-matrix method (Sturkey, 1962) and the real-space method (Van Dyck and Coene, 1984). However, the original two approaches still remain the most suited to numerical solution. This article outlines the methodology of these two approaches.

1.   THE EIGENVALUE (BLOCH WAVE) FORMULATION

In this formulation Bloch's theorem is used to solve the time-independent Schrödinger equation for an electron wavefunction $\psi(\mathbf{r})$. If the crystal potential is $V(\mathbf{r})$ and the accelerating voltage of the microscope is E, Schrödinger's equation can be written in the form

$$\nabla^2\psi(\mathbf{r}) + \frac{8\pi^2 me}{h^2} [V(\mathbf{r}) + E] \,\psi(\mathbf{r}) = 0 \tag{1}$$

In this equation m is the relativistic mass of the electron, e is the electron charge and h is Planck's constant. Equation (1) is simplified by making the substitutions $K = \dfrac{8\pi^2 meE}{h^2}$ and $U(\mathbf{r}) = \dfrac{8\pi^2 meV(\mathbf{r})}{h^2}$ so that

$$\nabla^2\psi(\mathbf{r}) + [\, U(\mathbf{r}) + K \,] \,\psi(\mathbf{r}) = 0 \tag{2}$$

In this equation, K represents the magnitude of the wavevector of the electron *in vacuo*.

The term $U(\mathbf{r})$ in equation (2) is determined by the crystal potential and therefore is a function that is periodic in three dimensions. This term can be expressed as a Fourier series such that

$$U(\mathbf{r}) = \sum_{\mathbf{g}} U_{\mathbf{g}} \exp(2\pi i\, \mathbf{g}{\cdot}\mathbf{r}). \tag{3}$$

where the summation extends over all reciprocal lattice vectors $\mathbf{g}$. The Fourier coefficients $U_{\mathbf{g}}$ are related to the electron structure factors $F_{\mathbf{g}}$ by the formula $U_{\mathbf{g}} = \dfrac{F_{\mathbf{g}}}{\pi V_c} \dfrac{m}{m_e}$ where m_e is the rest mass of the electron and V_c is the volume of the unit cell. The electron structure factors are given by the standard formula

$$\sum_{j} {}^{e}f_j (s_{\mathbf{g}}) \; \exp (-B_j\, s_{\mathbf{g}}^2) \; \exp (2\pi i\, \mathbf{g}{\cdot}\mathbf{r}_j) \tag{4}$$

where the index j represents a sum over all atoms in the unit cell with the fractional coordinates of the jth atom being $\mathbf{r}_j$.

The electron scattering factor for the jth atom ${}^{e}f_j (s_{\mathbf{g}})$ may be obtained either from experimental determinations or from tabulated values for free atoms (International Tables for X-ray Crystallography Vol.IV, and Doyle and Turner, 1968). If electron scattering factors are unavailable for any particular atom they can be deduced from the corresponding x-ray scattering factor using the Mott formula (Mott, 1930). The parameter $s_{\mathbf{g}}$ is specified as $s_{\mathbf{g}} = (\sin\theta_{\mathbf{g}})/\lambda$, where $\theta_{\mathbf{g}}$

is the Bragg angle for the **g**th reflection. Thus by Bragg's law, $s_g = 1/(2d_g)$ where d_g is the interplanar spacing of the diffracting planes. Finally, B_j is the Debye-Waller factor for the jth atom. Again, the values of B for various atoms can be obtained from experimental determinations or from tabulated values such as those given in Vol.III of the International Tables for X-ray Crystallography. A vector notation is adopted in (4) to indicate that the equation is applicable to the case of anisotropic scattering factors and Debye-Waller factors.

Bloch's theorem states that, as $U(\mathbf{r})$ is periodic, the solution to (2) must also be periodic and have the same period as $U(\mathbf{r})$. Therefore, the particular solutions, $C(\mathbf{r})$, of (2) can be written as a Fourier series with the same form as (3), i.e.

$$C(\mathbf{r}) = \sum_{\mathbf{g}} C_{\mathbf{g}} \exp [2\pi i\, (\mathbf{k}+\mathbf{g}){\cdot}\mathbf{r}]. \tag{5}$$

These particular solutions are called Bloch waves and have a wavevector **k**. The Fourier coefficients $C_{\mathbf{g}}$ are termed the Bloch wave coefficients and the values of these coefficients can be found by substituting (5) into (2) to give a set of simultaneous equations. The number of simultaneous equations will equal the number of reciprocal lattice vectors **g** included in the summation (5). Strictly, this summation is over all reciprocal lattice vectors but for numerical solution the summation must obviously be restricted to a manageable subset of vectors. The set of simultaneous equations can be written in the matrix form

$$\begin{bmatrix} K^2{-}(\mathbf{k}{+}\mathbf{h})^2 & U_{\mathbf{g}{-}\mathbf{h}} & \cdots & U_{\mathbf{l}{-}\mathbf{h}} \\ U_{\mathbf{h}{-}\mathbf{g}} & K^2{-}(\mathbf{k}{+}\mathbf{g})^2 & \cdots & U_{\mathbf{l}{-}\mathbf{g}} \\ \bullet & \bullet & & \bullet \\ \bullet & \bullet & & \bullet \\ \bullet & \bullet & & \bullet \\ U_{\mathbf{h}{-}\mathbf{l}} & U_{\mathbf{g}{-}\mathbf{l}} & \cdots & K^2{-}(\mathbf{k}{+}\mathbf{l})^2 \end{bmatrix} \begin{bmatrix} C_{\mathbf{h}} \\ C_{\mathbf{g}} \\ \bullet \\ \bullet \\ \bullet \\ C_{\mathbf{l}} \end{bmatrix} = 0 \tag{6}$$

where K is the wavenumber for an electron in the mean inner potential of the crystal, i.e. $K^2 = (1/\lambda)^2 + U_0$.

If there are N reciprocal lattice points (i.e. N diffracted beams) to be included in the calculation, there are necessarily N equations in the set (6) for the N Bloch wave coefficients. These equations have non-trivial solutions if and only if the determinant of the N by N matrix is zero. The characteristic equation of the matrix is a vector polynomial of order N, the roots of which define N concentric surfaces collectively known as the dispersion surface. Each point on the dispersion surface defines one value of the vector **k** and a corresponding set of Bloch wave coefficients.

The particular values of **k** and the coefficients C for a given set of operating conditions are determined by the boundary conditions. Conservation of an incident electron's momentum parallel to the crystal surface implies that the end point of the vector **k** must lie on a line perpendicular to the crystal surface. It is impractical to find the solutions of the characteristic vector equation of the matrix in (6) and then construct the surface normal to specify **k**. It is more practical to apply the boundary conditions before attempting to solve (6).

The most widely used boundary condition is that the surface normal is perpendicular to the Laue zone of interest. Then, if only zero layer reflections are considered it is possible to separate the wavevector into components parallel and perpendicular to the zero layer. Thus **k** can be written as $\mathbf{k} = k_\perp\mathbf{n}+\mathbf{u_o}$ where **n** is a unit vector parallel to the surface normal and u_o is the vector from the centre of the Laue circle to the origin of reciprocal space (Fig.1a). Using this expansion $(\mathbf{k}+\mathbf{g})^2 = k_\perp^2+(\mathbf{g}+\mathbf{u_o})^2$. The vector $\mathbf{u_g} = \mathbf{g}+\mathbf{u_o}$ is a two dimensional vector and the term $\mathbf{u_g}^2$ can be readily calculated from geometrical considerations (Fig.1b). Thus (6) can be rewritten in the form

$$\mathbb{M} \begin{bmatrix} C_\mathbf{h} \\ C_\mathbf{g} \\ \bullet \\ \bullet \\ \bullet \\ C_\mathbf{l} \end{bmatrix} = (k_\perp^2-K^2) \begin{bmatrix} C_\mathbf{h} \\ C_\mathbf{g} \\ \bullet \\ \bullet \\ \bullet \\ C_\mathbf{l} \end{bmatrix} \tag{7}$$

where

$$\mathbb{M} = \begin{bmatrix} -\mathbf{u_h}^2 & U_{\mathbf{g-h}} & \bullet\bullet\bullet & U_{\mathbf{l-h}} \\ U_{\mathbf{g-h}} & -\mathbf{u_g}^2 & \bullet\bullet\bullet & U_{\mathbf{l-g}} \\ \bullet & \bullet & & \bullet \\ \bullet & \bullet & & \bullet \\ \bullet & \bullet & & \bullet \\ U_{\mathbf{h-l}} & U_{\mathbf{g-l}} & \bullet\bullet\bullet & -\mathbf{u_l}^2 \end{bmatrix}$$

Thus the Bloch wave coefficients form an eigenvector of the matrix $\mathbb{M}$ and the wavenumbers are specified by the eigenvalues.

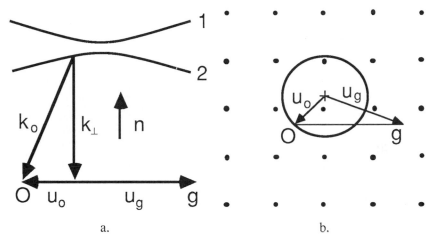

a.   b.

FIG. 1 Geometry of the wavevector k and it's components: (a) geometry of $k_\perp$, (b) geometry of $\mathbf{u_g}$.

As $\mathbb{M}$ is a N by N matrix it will have N eigensolutions, i.e. there will be N Bloch waves. (As the eigenvalues of $\mathbb{M}$ equal $k_\perp^2 - K^2$, $k_\perp$ can take both positive and negative square roots giving a total of 2N solutions; however, negative values of $k_\perp$ represent backscattering which is neglected for fast electrons.) The diagonalization of $\mathbb{M}$ can be carried out on a computer using one of several standard routines available to programmers. As the crystal potential is real the factors U must satisfy the condition $U_g = U_g^*$. Thus $\mathbb{M}$ is Hermitian which simplifies the diagonalization procedure. The procedure is simplified even further when the crystal has a centre of symmetry because then the factors U can be expressed as purely real numbers so that $\mathbb{M}$ becomes a real symmetric matrix.

The complete solution to the Schrödinger equation is the linear combination of the particular solutions. That is

$$\psi(\mathbf{r}) = \sum_j \varepsilon_j \sum_g C_g^{(j)} \exp[2\pi i\,(\mathbf{k}^{(j)}+\mathbf{g})\cdot\mathbf{r}]. \tag{8}$$

The terms ε_j are the excitation coefficients of the Bloch waves. These coefficients are determined by matching the incident wavefunction to the crystal wavefunction at the top surface of the crystal. If the electron wavefunction is written in vector form so that each element is the amplitude and phase of a specific diffracting component then the wavefunction at a depth H in the crystal may be expressed as

$$\psi(r) = \mathbb{C} \begin{bmatrix} \exp(2\pi i k_\perp^{(1)} H) & & & O \\ & \exp(2\pi i k_\perp^{(2)} H) & & \\ & & \cdot & \\ & & & \cdot \\ O & & & \exp(2\pi i k_\perp^{(N)} H) \end{bmatrix} \mathbb{E} \qquad (9)$$

$$\text{where } \psi(r) = \begin{bmatrix} \psi_h \\ \psi_g \\ \cdot \\ \cdot \\ \cdot \\ \psi_l \end{bmatrix} \quad \mathbb{C} = \begin{bmatrix} c_h^{(1)} & c_h^{(2)} & \cdots & c_h^{(N)} \\ c_g^{(1)} & c_g^{(2)} & \cdots & c_g^{(N)} \\ \cdot & \cdot & & \cdot \\ \cdot & \cdot & \cdots & \cdot \\ \cdot & \cdot & & \cdot \\ c_l^{(1)} & c_l^{(2)} & \cdots & c_l^{(N)} \end{bmatrix} \text{ and } \mathbb{E} = \begin{bmatrix} \varepsilon_1 \\ \varepsilon_2 \\ \cdot \\ \cdot \\ \cdot \\ \varepsilon_N \end{bmatrix}$$

At the top surface of the crystal (H=0) (9) becomes $\psi(0) = \mathbb{C} \cdot \mathbb{E}$. Thus if the incident wavefunction at the crystal surface is ϕ then $\psi(0) = \phi$ so that $\mathbb{E} = \mathbb{C}^{-1} \cdot \phi$. Necessarily ϕ must be in a form that is consistent with the calculation performed to find the Bloch waves, i.e. ϕ should contain only intensity in the direction of the N beams included in the calculation.

In many cases it is not necessary to invert the matrix $\mathbb{C}$ to find the excitation coefficients. For centrosymmetric crystals, $\mathbb{M}$ is real symmetric and so its eigenvectors form an N-dimensional orthogonal set and therefore $\mathbb{C}^{-1} = \tilde{\mathbb{C}}$. Consequently, if ϕ contains intensity only in the zeroth order beam $\varepsilon_j = c_0^{(j)}$.

Fig.2 shows results from the eigenvalue calculation of the amplitude and phase of beams in the (111) systematic row of silicon. (Only the results for the (000), (111) and (222) beams are displayed.) For Fig.2(a) only 3 beams were included in the calculation while for Fig.2(b) 9 beams were included. It can be seen that, in this case, the eigenvalue formulation provides a very good evaluation of the amplitude and phase of the strongly diffracting reflections even when only 3 beams are included in the calculation.

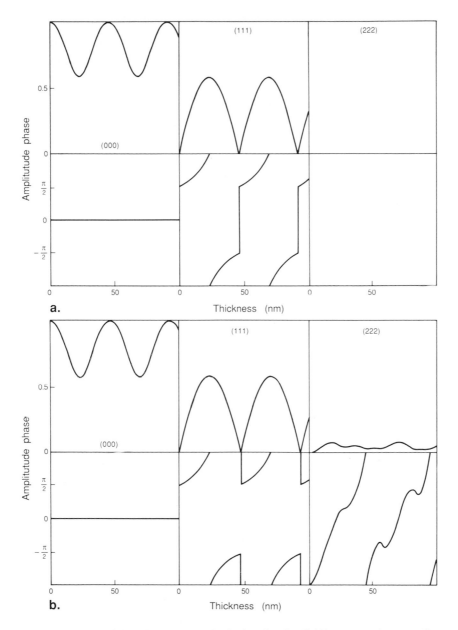

FIG.2 Results of the eigenvalue calculation for the (111) systematic row of silicon. Accelerating voltage = 200 KV, and incident beam direction perpendicular to g_{111}. (a) 3 beam calculation, (b) 9 beam calculation.

The above formulation does not include absorption effects. These effects can be included by adding a complex component to the structure factor so that U_g becomes $U_g - iU_g'$ where $U_g'/U_g \approx 0.1$. Suitable values for U_g' are given by Humphreys and Hirsch (1968) and Radi (1970). With the inclusion of absorption the matrix M becomes complex and so for complete specification of its eigensolutions it is necessary to use an eigenvector program for complex matrices. This can considerably increase computation time. Alternatively it is possible to use a perturbation theory (Hirsch *et al.*, 1977, p.217). Using this theory it can be shown that $k_\perp^{(j)}$ should be replaced by $k_\perp^{(j)} + iq^{(j)}$ where

$$q^{(j)} = (1/2K) \sum_g \sum_h U_{g-h}' C_h^{(j)} C_g^{(j)*}. \tag{10}$$

This equation provides an extremely good approximation to the full complex solution for values of U_g'/U_g up to 0.2, which is well above the value required in all situations.

Two other factors that can be important in the calculation of diffraction amplitudes are the inclusion of out-of-zone (upper and lower Laue zones) and the allowance for incident beams that cut the crystal surface at non-normal angles. In both cases the wave vectors of the Bloch waves do not all have the same component in the plane of the zeroth Laue zone. Hence it is not possible to use the formulation given above. In these special situations exact solutions can be found but the computation involves the solution of a 2N by 2N set of equations (Metherell, 1975). A more widely used approach in both of these cases is to use an approximation that makes the diagonal terms of M linear in terms of some parameter thus eliminating the cross-terms introduced by $(k^{(j)} + g)^2$. One such approximation is the small angle approximation, which states

$$K^2 - (k+g)^2 \approx 2K(\zeta_g - \gamma\cos\theta_g),$$

where ζ_g and θ_g are the excitation error and the Bragg angle of the gth reciprocal lattice point respectively. The excitation error is discussed in more detail in section 2. The parameter γ is the unknown quantity defining the dispersion surface.

The above is necessarily only the briefest of outlines of the eigenvalue formulation, for a fuller discussion of this formulation see Metherell (1975).

2.   THE MULTISLICE FORMULATION

The basis of the multislice method is to divide the specimen into a number of thin slices perpendicular to the incident beam direction. In each slice the effects of wave propagation and the effects of the specimen potential (transmission) are treated separately. By using Huygens' principle the real space wavefunction after the nth slice, $\psi_n(xy)$, is given by

$$\psi_n(xy) = [\psi_{n-1}(xy) * p_{n-1}(xy)] \bullet q_n(xy) \qquad (11)$$

where $q_n(xy)$ is the transmission function of the nth slice and $p_{n-1}(xy)$ is the free-space propagator for the distance between the (n-1)th and nth slices. The symbol * represents convolution and xy represent the real-space coordinates in planes perpendicular to the incident beam direction. The calculation is initiated by setting the zeroth-slice wavefunction to the incident wavefunction, ϕ.

The multislice iteration is best performed in reciprocal space where the wavefunction exists only at the discrete points of the reciprocal lattice. Also, as the Fourier components of the wavefunction fall off with increasing order, it is a good approximation to consider the reciprocal space wavefunction as bandwidth limited. The form of the reciprocal space multislice equation is obtained by Fourier transforming (11) to give

$$\Psi_n(hk) = [\Psi_{n-1}(hk) \bullet P_{n-1}(hk)] * Q_n(hk) \qquad (12)$$

where hk represents the reciprocal space coordinates corresponding to xy.

The distance between slices will necessarily be small and so the free-space propagator will best be described by the Fresnel formulation of diffraction. Thus, for a slice thickness of Δz, the reciprocal space propagation function is

$$P(hk) = \exp[-2\pi i \, \zeta(hk)\Delta z] \qquad (13)$$

where $\zeta(hk)$ is the excitation error of the hk reciprocal lattice point as shown in Fig.3. The excitation error is defined as negative when the corresponding reciprocal lattice point lies outside the Ewald sphere. Using geometric arguments it can be shown

$$\zeta(hk) = \{(1/\lambda)^2 - [(h-u)^2 a^{*2} + (k-v)^2 b^{*2} + 2(h-u)(k-v)a^*b^*\cos\beta^*]\}^{1/2}$$
$$- \{(1/\lambda)^2 - [u^2 a^{*2} + v^2 b^{*2} + 2uva^*b^*\cos\beta^*]\}^{1/2}$$

where a^* and b^* are the lengths of the reciprocal lattice basis vectors, β^* the angle between these vectors and (u,v) the centre of the Laue circle. By introducing a term dependent on the centre of the Laue circle in the expression for $\zeta(hk)$, allowance is made for changing the direction of the incident illumination.

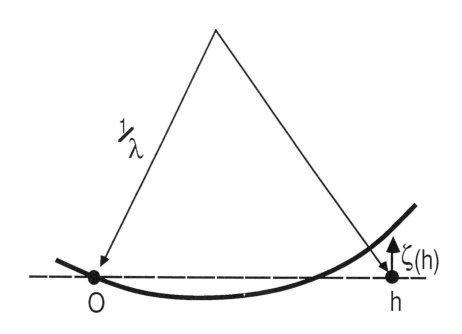

FIG.3 Geometry defining the excitation error $\zeta(h)$ in reciprocal space.

The transmission function in (11) is taken to be the phase grating as developed by Cowley and Moodie (1957). The phase grating models the amplitude and phase changes of the wavefunction of an electron as it passes through the electrostatic potential of the crystal. The phase grating for the nth slice is

$$q_n(xy) = \exp\left[-i\sigma\, \phi_n(xy)\, \Delta z_n\right] \qquad (14)$$

where Δz_n is the slice thickness of the nth slice and $\phi_n(xy)$ is the nth slice projected potential per unit length for a projection direction parallel to the incident beam direction. The interaction constant is defined as

$$\sigma = \frac{2\pi m e \lambda}{h^2}$$

where the parameters are as defined in section 1. Of course, the slice thickness must be small enough that the phase grating approximation gives an accurate representation of diffracted amplitudes and phases. For most cases a slice thickness of 0.1 to 0.3 nm is sufficient.

The projected potential per unit length for the nth slice is found by integrating the crystal potential over the distance Δz_n. In general, for a slice of thickness Δz centred on z_0 the result of this integration is

$$\phi(xy) = \sum_{hkl} V(hkl) \frac{\sin(\pi l \Delta z/c)}{(\pi l \Delta z/c)} \exp(2\pi i l z_0/c) \; \exp[2\pi i(hx/a+ky/b)] \quad (15)$$

where a and b represent the real space repeat distances of the crystal in the plane perpendicular to the beam direction and c represents the real space repeat distance of the crystal parallel to the beam. The values of the Fourier coefficients of the crystal potential, V(hkl), can be found from the structure factors using the relationship

$$V(hkl) = \frac{h^2 F(hkl)}{2\pi m_e e V_c}$$

where the parameters are as defined in section 1.

When evaluating the multislice iteration there is considerable computational advantage if the slices can be defined such that $Q_n(hk)$ and $P_n(hk)$ are unchanged from one slice to the next. For most crystals it is possible to divide the crystal into slices of equal thickness with equal potential in each slice. The projected potential per unit length can then be expressed as

$$\phi_p(xy) = \sum_{hk} V(hk0) \exp[-2\pi i(hx/a+ky/b)]. \quad (16)$$

This equation contains no information about scattering to diffraction points out of the zeroth Laue layer. Thus if out of zone effects are strong for the system under consideration (e.g. Gold, see Lynch, 1971) then (15) must be used.

As in the eigenvalue method, absorption is modelled by adding an imaginary component to the crystal potential so that the phase grating becomes

$$q(xy) = \exp\{-1\sigma[\phi(xy)-i\phi_a(xy)]\Delta z\}. \quad (17)$$

The effect of the absorption potential is to remove some electrons from the calculation. In reality these electrons are not lost; they simply do not contribute to the elastic wavefunction and instead form a diffuse background. The values of Q(hk) are calculated by Fourier transforming (14) or (17).

The conventional method of carrying out the multislice iteration is to use the reciprocal space sum

$$\Psi_n(hk) = \sum_{h'k'} \Psi_{n-1}(h'k')P(h'k')Q(h-h',k-k'). \quad (18)$$

If N beams are included in the calculation then the above sum involves $2N^2$ complex multiplications and N^2 complex additions. The time required for this procedure can be prohibitive, considering that it must be repeated for every slice until the required crystal thickness is reached. An alternative method of carrying out a convolution is to make use of the formulation

$$G * P = \mathcal{F}[\mathcal{F}^{-1}(G) \bullet \mathcal{F}^{-1}(P)].$$

That is, Fourier transform (represented by $\mathcal{F}$) to real space, multiply the two real space functions and then return to reciprocal space thereby completing the convolution. Using this method the iteration becomes

$$\Psi_n(hk) = \mathcal{F}\{\mathcal{F}^{-1}[\Psi_{n-1}(hk)P(hk)] \bullet \mathcal{F}^{-1}[Q(hk)]\}. \qquad (19)$$

If the Fourier transforms are calculated by using the conventional Fourier summation the iteration (19) is slightly slower than using the direct summation (18). However, by using the fast Fourier transform (FFT) algorithm the use of (19) becomes much quicker than using (18) (Ishizuka and Uyeda, 1977).

Calculating a convolution in reciprocal space by use of a Fourier transform involves sampling of the real space wavefunction at discrete points. Thus, as a consequence of the sampling theorem for Fourier transforms, the reciprocal space wavefunction will appear periodic. The term for this false periodicity in reciprocal space is aliasing and in order to avoid aliasing in the FFT multislice formulation, the Fourier coefficients of the wavefunction must be embedded into an array such that the diffracted beams extend to at most 2/3 of the array size in any direction. The array elements outside this limit must be set to zero after each iteration.

As $\mathcal{F}^{-1}[Q(hk)]=q(xy)$ it is not necessary to find $Q(hk)$ to evaluate (19). However to avoid aliasing effects it is necessary to use $q(xy)$ such that it's Fourier coefficients are truncated in the way specified above. The FFT should be used in the calculation of the values of $Q(hk)$ and in any other parts of the multislice calculation that require Fourier transforms such as the calculation of the projected potential.

The computation time required to evaluate a N point FFT is proportional to $N\log_2 N$ while the time take to evaluate the eigenstates of an N by N matrix is proportional to N^2. Thus, the multislice method has become the preferred method of computation where large numbers of beams are involved such as zone-axis calculations.

3.   HIGH-RESOLUTION IMAGING

Having calculated the wavefunction at the exit surface of the crystal, electron images are calculated by propagating the wavefunction to the image plane. As for the free-space propagator used in the multislice iteration, the form of the propagator is given by the Fresnel formulation of diffraction. The propagator must include lens effects such as defocus and spherical aberration. In general, the real-space wavefunction at the image plane can be written as

$$\psi_{Im}(xy) = \mathcal{F}^{-1}\{\Psi_e(hk)\exp[i\chi(hk)]A(hk)S(hk)\} \qquad (20)$$

where $\Psi_e(hk)$ is the reciprocal-space wave function at the exit surface of the crystal, $\chi(hk)$ is the phase factor introduced by the lens system, $A(hk)$ is the aperture function and $S(hk)$ models the effects of astigmatism. The electron intensity at the image plane is given by the modulus of $\psi_{Im}(xy)$ squared.

For a defocus value of Δ and a spherical aberration coefficient of C_s

$$\chi(hk) = \pi\lambda\Delta U(hk)^2 + \frac{1}{2}\pi\, C_s\, \lambda^3 U(hk)^4. \qquad (21)$$

In this equation the generalised vector notation is adopted for $U(hk)$, the reciprocal space coordinate of the point (hk) relative to the optic axis of the imaging (objective) lens. The magnitude of this vector is $|U(hk)| = [\sin\theta(hk)]/\lambda$ where $\theta(hk)$ is the angle between the (hk) reflection and the optic axis.

The effect of the lens aperture is to truncate the number of beams contributing to the wavefunction at the image plane. The aperture function may be written as

$$A(hk) = \begin{cases} 1 \text{ if hk lies inside the aperture} \\ 0 \text{ if hk lies outside the aperture} \end{cases}$$

Astigmatism in an image is caused by the objective lens being of unequal strength in different directions, so that beams travelling at equal angles to the optic axis but in different directions do not experience the same phase change. Thus astigmatism introduces an extra phase factor to the wavefunction at the image plane so that

$$S(hk) = \exp[\frac{1}{2}\pi i\lambda C_A U(hk)^2\cos(2\phi)]$$

where C_A is the defocus difference between the maximum and minimum astigmatism-induced defocus values and ϕ is the angle between $U(hk)$ and the direction of maximum defocus change. It is usually assumed that astigmatism is fully corrected in all high-resolution TEM images (i.e. $C_A = 0$).

Because (20) deals only with the coherent wavefunction it neglects two very important incoherent effects in image formation; namely chromatic aberration and beam divergence (O'Keefe and Sanders, 1975). Chromatic aberrations are caused by instabilities in the objective lens power supply and by the spread of energy in the incident electron beam. Beam divergence is the result of using a converged incident beam in order to increase brightness.

The effect of both incident-beam energy spread and lens instability can be modelled by a change of objective lens defocus. Hence, to form the chromatically aberrated image, the images calculated using (20) must be averaged over a range of defocus values. That is

$$I_D(xy) = \int E(d,\Delta)\, I(xy)\, dd$$

where $E(d,\Delta)$ is the distribution of defocus values (d) around the average value (Δ) and $I(xy)$ is calculated using (20). Given that the instabilities are random processes

$$E(d,\Delta) = [1/(D(2\pi)^{1/2})]\, \exp[-(d-\Delta)^2/(2D^2)].$$

The standard deviation in defocus, D, is given by the expression

$$D = C_c\, [(\Delta E/E)^2 + 4(\Delta J/J)^2]^{1/2}$$

where ΔE is the mean energy spread in the incident beam energy, E, and ΔJ is the variation in the objective lens current, J. The coefficient of chromatic aberration is C_c. In practice, $I(xy)$ varies fairly slowly with defocus and so $I_D(xy)$ can be evaluated by averaging the values of $E(d,\Delta)I(xy)$ calculated at defocus intervals of roughly 5 nm over the defocus range $\Delta-2D$ to $\Delta+2D$.

Beam divergence means that the overall exit surface wavefunction is made up of a set of wavefunctions corresponding to slightly different specimen tilts and that images are formed with beams covering a range of angles relative to the optic axis of the objective lens. The total image, $I_{D\alpha}(xy)$, is therefore the average of the set of images formed for all angles in the incident beam. Thus

$$I_{D\alpha}(xy) = \int B(s)I_D(xy)\, d^2s = \iint B(s)\, E(d,\Delta)\, I(xy)\, dd\, d^2s$$

where $B(s)$ is the angular distribution of intensity in the incident illumination. Usually, the incident illumination fills the condenser lens aperture uniformly and so

$$B(s) = \begin{cases} \lambda^2/(\alpha^2\pi) & \text{for } |s| \leq \alpha/\lambda \\ 0 & \text{for } |s| > \alpha/\lambda \end{cases}$$

where α is the divergence half angle.

Examples of calculated high-resolution images are shown in Figure 4. These images show the effects of instrumental parameters such as spherical and chromatic aberration on high-resolution images. A fuller discussion of electron imaging calculations is given by Self and O'Keefe (1987).

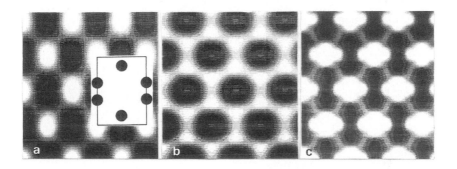

FIG.4 Computed images of the [110] zone of Si for three values of microscope resolution. In all figures the accelerating voltage is 100 KV and the crystal thickness is 20 nm. Images (a) and (b) are for C_S=0.7 mm, 7.5 nm spread in focus, 0.5 mrad beam divergence and defocus of -90 nm while image (c) is for C_S=0.1 mm, no spread in focus, no beam divergence and defocus of -50 nm. Images (b) and (c) have no aperture and image (a) is for an aperture including the (000) and {111} beams. The insert shows the position of the Si atoms (from Self *et al.*, 1985).

4.   SIGN CONVENTIONS

In order to obtain consistent results for the calculation of electron wavefunctions it is necessary to choose a convention for the form of a forward propagating wave. Either $\exp(-2\pi i \mathbf{k} \cdot \mathbf{r})$ or $\exp(+2\pi i \mathbf{k} \cdot \mathbf{r})$ may be chosen. The consequences of the choice are shown in Table 1.

TABLE 1

Wave conventions

Free space wave	$\exp[-2\pi i\,(\mathbf{k}\cdot\mathbf{r}-\omega t)]$	$\exp[+2\pi i\,(\mathbf{k}\cdot\mathbf{r}-\omega t)]$
Fourier transforming from real to reciprocal space	$\int a(x)\,\exp(+2\pi ihx)dx$	$\int a(x)\,\exp(-2\pi ihx)dx$
Fourier transforming from reciprocal to real space	$\int A(h)\exp(-2\pi ihx)dh$	$\int A(h)\exp(+2\pi ihx)dh$
Structure factors	$\sum_j f_j(s_j)\,\exp\,(+2\pi i\mathbf{g}\cdot\mathbf{r}_j)$	$\sum_j f_j(s_j)\,\exp(-2\pi i\mathbf{g}\cdot\mathbf{r}_j)$
Transmission function	$\exp(-i\sigma\phi_p(x)\Delta z)$	$\exp(+i\sigma\phi_p(x)\Delta z)$
Phenomological absorption	$\phi_p(x)-i\phi_a(x)$	$\phi_p(x)+i\phi_a(x)$
Propagation function	$\exp(-2\pi i\zeta(h)\Delta z)$	$\exp(+2\pi i\zeta(h)\Delta z)$
Propagation to image plane	$\Psi'(h)=\Psi(h)\exp[-i\chi(h)]$	$\Psi'(h)=\Psi(h)\exp[+i\chi(h)]$

$$\chi(h)=2\pi\eta\,(h)\,[\Delta f-\lambda C_s\eta(h)]$$
$$=-\pi\lambda u^2\,(h)\,[\Delta f+\lambda^2 C_s u^2\,(h)\,/2]$$

$\zeta(h)$: Excitation error relative to incident beam direction (defined as negative for reflections lying outside the Ewald sphere)

$\eta(h)$: Excitation error relative to the optic axis

Other symbols as defined in text.

5.   DEFECT STRUCTURES AND PERIODIC CONTINUATION

The above formulations deal only with periodic structures because of the
requirement that the reciprocal space wavefunction exists at discrete points only.
For non-periodic objects, such as defects in crystalline materials or amorphous
specimens, both the real and reciprocal space form of the electron wavefunction
are continuous. To calculate diffraction patterns and images from non-periodic
objects, the electron wavefunction must be approximated by a discrete valued
function. This is achieved by defining a supercell that is representative of the
non-periodic object and then laying the supercells together on a regular lattice.
The introduction of periodicity into the system by this method is called periodic
continuation. In the calculation of the wavefunction, reciprocal space is sampled
at positions corresponding to the Bragg reflections for the supercell.

Periodic continuation can be applied in many systems. Point defects can
be modelled by placing perfect crystal above and below the defect region.
Surfaces parallel to the beam direction can be modelled by surrounding a sliver of
crystal by free space. Planar defects are also easily modelled but there is often a
bulk displacement of the crystal on either side of the planar fault and a returning
fault must be included to allow a supercell to be defined. Some care is required
in the choice of a returning defect as this defect may dominate the images and
diffraction patterns of the system. Fig. 5 shows computed images for a planar
defect (I_2 stacking fault) in CdSe with a returning defect at the edge of the
supercell. In Fig. 5a there is poor matching at the edges of the supercell so that
there is two heavy atoms lying very close together. The calculated images for
this returning defect show an extended modulation around the returning defect
that interfers with the region of image around the prime defect. A more suitable
returning defect can be formed by extending the supercell a few atomic widths
(Fig. 5b); the images of this system are more satisfactory for the study of the
prime defect. Thus returning defects should be chosen to have as little difference
as possible from the surrounding structure.

In systems where a returning defect is required to form a supercell, the
calculated diffracted amplitudes are made up of a contribution from both defects
in the system and therefore are not representative of the diffraction amplitude of
either of the defects. Wilson and Spargo (1982) show that by adding, with
appropriate phase factors, the diffracted amplitude results from two calculations

for two supercells of different size, the diffraction pattern for a single defect can be extracted from the mixed pattern. Furthermore, in diffraction patterns of defects calculated by periodic continuation, the calculated intensity of the Bragg reflections that arise from the matrix surrounding defects cannot be regarded as absolute because increasing the size of a supercell will increase the amount of perfect crystal and thus increase the Bragg scattering.

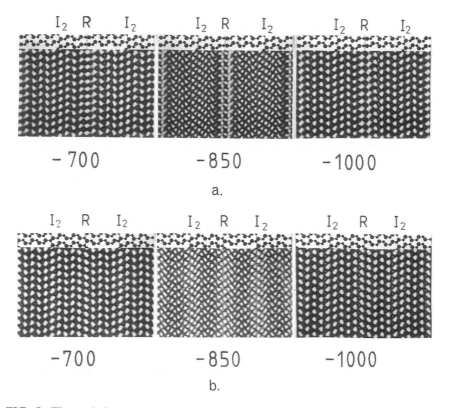

FIG. 5 Through focal series of images of an I_2 stacking fault (marked I_2) in CdSe using two different periodically continued cells. (a) The supercell is such that two atoms in the returning defect lie unrealistically close together. (b) The returning defect is also an I_2 stacking fault. The defocus values are marked in Å. The specimen thickness is 200 Å and the zone is [110]. The instrumental parameters are C_s=0.7 mm, accelerating voltage = 100 KV, chromatic aberration spread in defocus = 100 Å and beam divergence half angle = 1 mrad. (Courtesy R.W. Glaisher, PhD thesis, University of Melbourne).

ACKNOWLEDGEMENTS

The author would like to thank Ms D.L. Smith for the careful typesetting and Mr D.A. Wright and Mr J.A. Coppi for the preparation of the figures.

REFERENCES

Bethe, H.A. (1928). *Annals der Physik* (Leipzig), **87**, 55.

Cowley, J.M. and Moodie, A.F. (1957). *Acta Crystallographica*, **10**, 609.

Cowley, J.M. and Moodie, A.F. (1958). *Proceedings of the Physical Society*, **71**, 533.

Doyle, P.A. and Turner, P.S. (1968). *Acta Crystallographica*, **A24**, 390.

Hirsch, P., Howie, A., Nicholson, R.B., Pashley, D.W. and Whelan, M.J. (1977). *Electron Microscopy of Thin Crystals*, Kreiger, Huntington, NY.

Howie, A. and Whelan, M.J. (1961). *Proceedings of the Physical Society*, **A263**, 217.

Humphreys, C.J. and Hirsch, P.B. (1968). *Philosophical Magazine*, **18**, 115.

International Tables for X-ray Crystallography, Kynoch, Birmingham, UK.

Ishizuka, K. and Uyeda, N. (1977), *Acta Crystallographica*, **A33**, 740.

Lynch, D.F. (1971). *Acta Crystallographica*, **A27**, 399.

Metherell, A.J.F. (1975). In *Electron Microscopy in Materials Science Part II* (ed. U. Valdre and E. Ruedl). Commission of the European Communities, Luxembourg.

Mott, N.F. (1930). *Proceedings of the Royal Society* (London), **A127**, 658.

O'Keefe, M.A. and Sanders, J.V. (1975). *Acta Crystallographica*, **A31**, 307.

Radi, G. (1970). *Acta Crystallographica*, **A26**, 41.

Self, P.G., Glaisher, R.W. and Spargo, A.E.C. (1985). *Ultramicroscopy*, **18**, 49.

Self, P.G. and O'Keefe, M.A. (1987). In *High Resolution Transmission Electron Microscopy* (ed. P.R. Buseck, J.M. Cowley and L. Eyring). Oxford University Press. In Press.

Sturkey, L. (1962). *Proceedings of the Physical Society*, **80**, 321.

Van Dyck, D. and Coene, W. (1984). *Ultramicroscopy*, **15**, 29.

Wilson, A.R. and Spargo, A.E.C. (1982). *Philosophical Magazine*, **A46**, 435.

Accurate electron density analysis

14

ELECTROSTATIC PROPERTIES FROM BRAGG DIFFRACTION DATA

B.M. Craven

The first stage in studying electrostatic properties of molecules in crystals is to determine the detailed charge density distribution from diffraction data. Other electrostatic properties, such as the molecular dipole moment or the electrostatic potential, can then be derived. For success, there are requirements involving the experimental procedures and methodology which take more work and computing time than a conventional structure determination. However, there are ample rewards in the additional physical and chemical information which can be extracted from the observed structure amplitudes.

Experimental considerations can only be outlined here. The diffraction data collection requires high standards of accuracy with care to avoid systematic errors, such as extinction, absorption, temperature fluctuations, and crystal mis-centering on the diffractometer. Such errors may have effects on the structure amplitudes which rival and obscure the small effects due to chemical bonding (seldom more than 5-10%). It is desirable to collect high resolution data $(\sin\theta/\lambda > 1 \text{ Å}^{-1})$ at a reduced crystal temperature, so that the electronic charge density can be more easily deconvoluted from the atomic thermal vibrations in the crystal. Also, it is desirable to collect neutron as well as X-ray data, so as to obtain nuclear positional and thermal parameters unbiased by the effects of local asphericity in the charge density. This is especially important for H atoms. Crystal structures chosen for study should be made as simple as possible by avoiding unnecessarily heavy atoms and non-essential functional groups. These contribute to the total number of electrons in the unit cell without contributing to the charge density features that are of interest.

In what follows, we assume that a suitable set of X-ray structure amplitudes are available. We describe a procedure for charge density analysis based on Stewart's rigid pseudoatom model (1976) and outline how it can be implemented with computer programs written at the University of Pittsburgh (Craven, Weber & He, 1987).[*]

The molecular charge density in the crystal is partitioned as the sum of rigid pseudoatom densities where each pseudoatom is centered on an atomic nucleus. The pseudoatom is said to be rigid because its electron density is assumed to accompany the nucleus during thermal vibrations. Each pseudoatom has an invariant core, usually consisting of a neutral spherical Hartree-Fock atomic density just as in a conventional structure analysis. The X-ray scattering factors for such cores are well known (International Tables, Vol. IV). In addition, the pseudoatom has a series of nuclear-centered electron density deformation terms, each with an adjustable weight (electron population parameter). These terms are intended to model the effects of the molecular and crystal environment, whereby the pseudoatom may become aspherical and carry a net charge.

---

[*]Alternative procedures are described by Spackman & Stewart (1984), Hansen & Coppens (1978). A related procedure is described by Hirshfeld (1971).

1. PSEUDOATOM DEFORMATION TERMS IN CRYSTAL SPACE

Charge density deformation terms are each assumed to have the form

$$\Delta\rho(r,\theta,\phi) = C_{n\ell m}R_n(r)Y_{\ell m}(\theta,\phi) \tag{1}$$

when expressed in polar coordinates with respect to the Cartesian crystal axial directions shown in Fig. 1, and with the origin at the pseudoatom nucleus. In our procedure, $R_n(r)$ is a radial function of Slater form, $Y_{\ell m}(\theta,\phi)$ is an angular function selected as one of the surface spherical harmonics shown in Fig. 2, and $C_{n\ell m}$ is the weight or electron population coefficient, which is to be determinated from the experimental data. The subscripts n, ℓ, m are simple integers (see below).

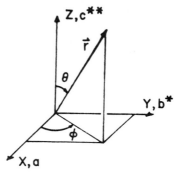

FIG. 1. Crystal Cartesian reference axes.

The surface harmonics can be written as "even" or "odd" functions

$$Y^e_{\ell m}(\theta,\phi) = \cos(m\phi)P^m_\ell(\cos\theta) \text{ or } Y^o_{\ell m}(\theta,\phi) = \sin(m\phi)P^m_\ell(\cos\theta)$$

where $P^m_\ell(\cos\theta)$ are unnormalized associated Legendre functions [see p.1264 of Morse & Feshbach (1953)]. They may be considered as having values defined on the surface of a unit sphere (r = 1), with nodal lines (Y = 0) which are great or small circles. Nodal circles are shown as heavy lines in the stereographic projections in Fig. 2. The value of ℓ determines the total number of nodal circles, both of latitude or longitude, while m, which cannot exceed ℓ, is the number of nodal circles of longitude. When $\ell = 0$, the angular function (Y^e_{00}) is said to be a monopole, for which the surface of the

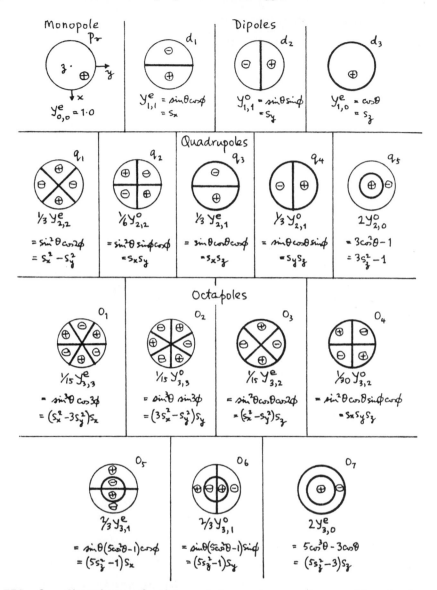

FIG. 2. Sketches of the nodal surfaces (heavy lines) for surface spherical harmonics shown in stereographic projection. Expressions are given for these angular functions in terms of polar coordinates (θ, ϕ) and also the direction cosines of a unit vector $\vec{s}$. Reference axes are shown at top left.

sphere consists of one (positive) region. A positive population parameter will thus confer an excess electron density on the pseudoatom, while a negative value confers a deficiency. When $\ell = 1$ there are 3 dipole functions (Y_{10}^e, Y_{11}^e, Y_{11}^o) where the surface consists of hemispheres with opposite signs. A positive population for Y_{11}^o (Fig. 2) corresponds to a charge density deficiency over the left hemisphere and an excess over the right, without changing the net charge of the pseudoatom. Similarly, for $\ell = 2$ there are five quadrupole functions and for $\ell = 3$ there are seven octapoles. At each level of ℓ, these multipole functions form a complete orthogonal set. We have found that expansions up to the octapole level give adequate descriptions of the deformation charge density in most organic molecules (Swaminathan, et al., 1984). At this level, the total deformation density for each pseudoatom is a linear combination of 16 terms, requiring 16 adjustable population parameters.

Each multipole function is associated with a single one-electron radial function

$$R_n(r) = \frac{\alpha^{n+3}}{4\pi(n+2)!}\, r^n \exp(-\alpha r)$$

which is also a function of ℓ to the extent that $n \geqslant \ell$ in order for the charge distribution to give rise to a Coulombic potential. The variable α, which governs contraction or dilation of the deformation term (see Fig. 3) is usually given the same value for pseudoatoms of the same type. Often we assign standard values such as $\alpha = 6.42, 7.37, 8.50, 4.69$ Å^{-1} for C, N, O and H respectively (Hehre, Stewart & Pople, 1967).

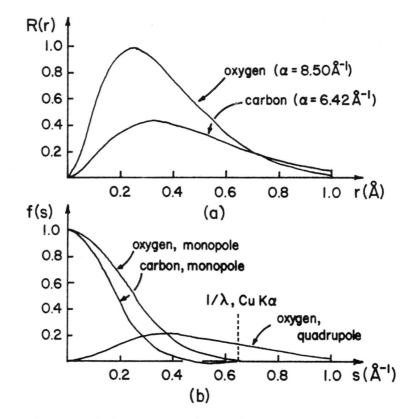

FIG. 3. Radial functions with n = 2.

(a) In crystal space, $R = \dfrac{\alpha^5}{96\pi} r^2 \exp(-\alpha r)$, (b) In reciprocal

space, $f = (1-c^2)/(1+c^2)^4$ for a monopole term; $f = 2c^2/(1+c^2)^4$

for a quadrupole $(c = 4\pi s/\alpha$ and $s = \sin\theta/\lambda)$.

2.   THE PSEUDOATOM X-RAY SCATTERING FACTOR IN RECIPROCAL SPACE

The Fourier transform of the expression in (1) is

$$f_{n\ell m}(\alpha,\vec{Q}) = \iiint \Delta\rho_{n\ell m}(r,\theta,\phi)\exp(i\vec{Q}\cdot\vec{r})r^2\sin\theta\,d\theta\,d\phi\,dr$$

where $|\vec{Q}| = 4\pi\sin\theta/\lambda$ is the magnitude of the scattering
vector. As shown on p. 1464 in Morse & Feshbach (1953), this
can be written

$$f_{n\ell m}(\alpha,\vec{Q}) = 4\pi i^{\ell} Y_{\ell m}(u,v) \int_{o}^{\infty} r^2 j_{\ell}(Qr) R_{n\ell}(r) dr \qquad (2)$$

where $\vec{Q}$ is expressed in polar coordinates ($|Q|,u,v$) and j_{ℓ} is the ℓth order spherical Bessel function. Because of the restrictions on ℓ and n, the integral in (2) always leads to a simple polynomial in $c = (Q/\alpha)$ (see Epstein & Stewart, 1977). Thus with $n = 2$, $\ell = m = 0$, so that $Y_{00}(u,v) = 1.0$, we have $f_{200} = (1-c^2)(1+c^2)^{-4}$, as in Fig. 3. See the Appendix in Epstein, Ruble & Craven (1982) for further details, including a tabulation of the explicit radial and angular functions used in the POP procedure.

Note the advantage in using the multipole angular functions, that a given deformation term involves one surface harmonic which is the same both in crystal space (eqtn.1) and reciprocal space (eqtn.2).

3. LEAST SQUARES REFINEMENT WITH THE PSEUDOATOM MODEL

The program POP (Craven, Weber & He, 1987) carries out full matrix or block diagonal least squares refinement in which the variables may include an overall scale factor, an isotropic extinction parameter, radial parameters (α) for each pseudoatom type, the usual positional and thermal parameters for each pseudoatom and the electron population parameters for the deformation terms. The calculated structure factor is given by

$$\vec{F}_{calc}(\vec{h}) = y \sum_{q}^{Nat} \sum_{j}^{Nsym} [s_q(f^o+\Delta f'+i\Delta f'')_q/K + \sum_{k=1}^{Nm} c_{qk}f_{qjk}]$$
$$\times \exp[2\pi i\vec{h}\cdot\vec{r}_{qj}] \times T_{qj}(h) \qquad (3)$$

where y = isotropic extinction factor (Becker & Coppens, 1974)

$\quad s_q$ = site factor for q^{th} atom

f_q^o, f_q', f_q'' = spherical core scattering factors for the qth atom (Int. Tables, Vol. IV; p.71)

K = scale factor to give F_{calc} on the scale of $|F_{obs}|$

c_{qk} = population parameter for the kth deformation term on the qth atom

f_{qjk} = scattering factor for the kth deformation term on the qth atom in the orientation of the jth symmetry operation

r_{qj} = position vector for the qth atom transformed by the jth symmetry operation

$T_{qj}(h)$ = temperature factor for the qth atom in the orientation of the jth symmetry operation which may include terms up to 4th order in the Gram-Charlier expansion (Int. Tables, Vol. IV, p.316)

Note: (a) For the purpose of least squares refinement, the site and scale factors apply only to the invariant core component of the atomic scattering factor.

(b) For large structures, a block diagonal refinement with considerable flexibility is available. Thus the scale factor, anisotropic thermal parameters (U_{ij}) and monopole and quadrupole population parameters for all atoms might be in one block, since these parameters may be strongly correlated. Similarly, positional parameters, dipole and octapole parameters and third order thermal parameters might be in a second block.

(c) It is desirable to have fixed values for r_{qj} and $T_{qj}(h)$ determined from neutron diffraction.

(d) It is advisable to complete a conventional refinement before including c_{qk} as variables. The R-value at this stage is typically 0.05 or more.

4. AUXILLIARY PROGRAMS IN THE 'POP' PROCEDURE

POPROT is for transforming electron population parameters from the crystal Cartesian axes to local atomic or molecular axial systems. The axial rotations follow the procedure given by Cromer, Larson & Stewart (1976). Population parameters are also normalized according to Hansen & Coppens (1978) so that all deformations are on a common scale.

ORTHAX, FORSEX and SIGRO are for the Fourier synthesis of the electron density, the residual density or the e.s.d. in the electron density for a general section or at selected points in the unit cell. This program is important for showing whether the pseudoatom model accounts for all the observed electron density.

CHARGE uses pseudoatom charge parameters in order to map the total electron density, the deformation density or its e.s.d. in a general section in the unit cell. These maps give electron densities for pseudoatoms at rest, that is, after deconvolution from the atomic thermal vibrations.

MOM calculates pseudoatom and molecular dipole and quadrupole moments with e.s.d.'s, following the procedure of Stewart (1972).

CRYPOT is for mapping the electrostatic potential in a general section through the crystal structure. The calculations make use of a Fourier synthesis in which the coefficients are the structure factors weighted by $(\sin\theta/\lambda)^{-2}$, as described by Stewart (1982).

MOLPOT makes use of the pseudoatom charge parameters to map the electrostatic potential in sections through molecules or groups of molecules isolated from the crystal structure. The procedure is that of Stewart (unpublished).

This work has been supported by NIH Grants GM-22548 and HL-20350.

REFERENCES

BECKER, P. J. & COPPENS, P. (1974). Acta Cryst. A30, 129-147.

CRAVEN, B. M., WEBER, H. P. & HE, X. M. (1987). "The POP Refinement Procedure," Techn. Report, Department of Crystallography, University of Pittsburgh.

CROMER, D. T., LARSON, A. C. & STEWART, R. F. (1976). J. Chem. Phys. 65, 336-349.

EPSTEIN, J., RUBLE, J. R. & CRAVEN, B. M. (1982). Acta Cryst. B38, 140-149.

EPSTEIN, J. & STEWART, R. F. (1977). J. Chem. Phys. 66, 4057-4064.

HANSEN, N. K. & COPPENS, P. (1978). Acta Cryst. A34, 909-921.

HEHRE, W. J., STEWART, R. F. & POPLE, J. A. (1969). J. Chem. Phys. 51, 2657-2664.

HIRSHFELD, F. L. (1971). Acta Cryst. B27, 769-781.

MORSE, P. M. & FESHBACH, H. (1953). "Methods of Theoretical Physics," New York : McGraw-Hill.

SPACKMAN, M. A. & STEWART, R. F. (1984). "Methods and Applications in Crystallographic Computing," Editors: S. R. Hall & T. Ashida, pp. 302-320, Oxford : Clarendon Press.

STEWART, R. F. (1972). J. Chem. Phys. 57, 1664-1668.

STEWART, R. F. (1976). Acta Cryst. A32, 565-574.

STEWART, R. F. (1982). God. Jugoslav. cent. kristalogr. 17, 1-24.

SWAMINATHAN, S. CRAVEN, B. M., SPACKMAN, M. A. & STEWART, R. F. (1984). Acta Cryst. B40, 398-404.

Computer graphics

15

ANATOMY OF
A MOLECULAR GRAPHICS PROGRAM:
DOCK

Robert K. Stodola, William P. Wood, Jr., Frank J. Manion,
Helen M. Berman, and Norman Badler

1   Abstract

An interactive computer graphics program which enables scientists to display and manipulate molecules easily on an Evans
and Sutherland PS–300 family graphics workstation is described.
Molecules are represented on a vector display workstation and
can be rotated and moved independently. Electron density
contours, atomic surface representations, and arbitrary drawings can also be displayed. This paper consists of a guided
tour through the molecular graphics program, DOCK, which
has been used at the Fox Chase Cancer Center since 1978. Emphasis is placed on the techniques used to write the program
and the considerations which led to the use of those techniques.

A number of programs for interactive molecular graphics have been written
and distributed. Their goals, software architecture, and the hardware they
use vary widely. For anyone designing and implementing such a program,
however, there are some common considerations. This paper will address
some of these using the program DOCK as a model. We will describe the
framework used to meet our molecular graphics needs, the considerations
given to the design of the user interface, the structure of the program, the
design of a graphics package, and a few hints at programming the Evans
and Sutherland PS–300.

Our initial forays into the field of computer graphics involved writing programs to solve specific problems(Berman, Carrell, Stodola, Andrews, Bernstein, Bernstein, Koetzle, Meyer, and Morimoto 1974; Berman, Neidle, and Stodola 1978). As the scope and nature of these studies grew, it quickly became apparent that new, large graphics programs could not be written each time the research problem changed. We therefore decided to design a more general molecular graphics program which could address all of the molecular graphics problems envisioned by the crystallographers. After extensive planning, we were able to convert all the requirements into a model suitable for implementation on the existing graphics systems. Few details were given concerning specific functions to be carried out by the program. The important result was the establishment of a framework for the organization of molecular graphics problems.

DOCK emerged from this effort, and has since been the principal graphics program in use at the Fox Chase Cancer Center. Though many new features have been incorporated into DOCK, and its user interface has been changed extensively over the years, the original framework is essentially intact through nearly a decade of expanding research needs. Its longevity is largely a result of this thorough analysis of the problems to be addressed. Though the elaborate superstructure required a large initial investment of effort, it has given us a very flexible tool for molecular graphics.

2   DOCK View of a Molecular Graphics Problem

DOCK views a molecular graphics problem as a hierarchy of graphical structures. For convenience in DOCK these are given the names of the chemical terms they typically represent, such as 'atoms' and 'molecules'. It is important to remember, however, that DOCK knows very little about chemistry, and the application of these terms to a graphics problem is largely under the control of the user. To differentiate the graphical structures from their chemical counterparts, capital letters will be used in this paper to denote the graphical entities. The needs of a particular problem, for example, may cause the user to treat several molecules as a single MOLECULE entity within DOCK. On the other hand, to treat parts of a molecule independently, the user may describe the molecule as a number of separate MOLECULEs. A BOND may be drawn between two distant ATOMs for the purposes of illustration, and so on.

The DOCK hierarchy is shown in Table 1. The entities are listed along with some of their principal properties. The description is most easily understood by beginning at the bottom.

TABLE 1

DOCK Hierarchy

Entity	Principal Characteristics
WORKSPACE	Split bond rendition, list of SYMMETRIES, CONTOUR basket descriptions
SYMMETRY	List of MOLECULEs
MOLECULE	SYMMETRY relation, name, cell parameters, list of residues, list of RIGID UNITs
RIGID UNIT	List of BONDs, list of ATOMs, ATOMs defining TORSION bond, rotation sense, dominating RIGID UNIT, next RIGID UNIT, list of RIGID subUNITs
BOND	ATOMs defining the BOND, line mode, color, and bond radius
ATOM	Coordinates, radius, name, element, class, color, label color, temperature factor, and structure factor code

At the bottom is the **ATOM**. An atom has, among other things, a chemical element type associated with it, a color, and a radius. (Again, the correspondence of these properties with those of the 'atom' they represent is controlled by the user).

A **BOND** connects two ATOMs together.

A collection of ATOMs and BONDs makes up a **RIGID UNIT**. Within a RIGID UNIT, ATOMs and BONDs are fixed relative to each other.

RIGID UNITs may be connected to other RIGID UNITs by TOR-SION BONDs. One RIGID UNIT may move relative to the ones it is connected to only by rotating about the connecting TORSION BONDs. One or more connected RIGID UNITs make a **MOLECULE**.

One or more MOLECULEs can be combined into a **SYMME-TRY GROUP**. Each MOLECULE in a SYMMETRY GROUP maintains a specified symmetric relationship with all other MOLECULEs in the group. Thus, when one MOLECULE moves, all others move in a symmetric fashion.

Finally, multiple SYMMETRY GROUPs can be combined in a single **WORKSPACE**. SYMMETRY GROUPs can move about in the WORKSPACE independently of one another.

3   The User Interface

This superstructure may sound a bit complicated for the user who wishes only to display a single molecule and vary a single torsional angle, or to show a couple of molecules and move them about each other. However, the user interface to DOCK was designed to make the simple problems simple, and to allow more complex problems to be built up from the basic elements in a stepwise fashion.

3.1   Menus

DOCK is controlled by a hierarchical set of menus built around the structures described above. When beginning a new DOCK session, the user is presented with the menu shown below:

```
DOCK:
    WORKSPACE
    SCRIPT
    CONTOUR
    SYMMETRY
    MOLECULE
    SURFACE
    REFINE
```

TORSION
BOND
ATOM
STOP
LATER

Selecting one of these options invokes another menu with new options related to these structures. Many of the submenus have equivalent or analogous operations which apply to different levels in the DOCK hierarchy. For example, both the ATOM menu and the MOLECULE menu have the function 'Identify'. In both cases, DOCK will identify the thing you point to. In the ATOM menu, the ATOM is identified (and labelled), and in the MOLECULE menu, the MOLECULE is identified. Likewise, the MOLECULE, TORSION, BOND, and ATOM menus all contain a 'Color' selection. In each case, the purpose is to re-color a particular entity at the level associated with the menu. Certain functions in the submenus may require further clarification, and third and fourth level menus are used in this case. At no time is the user faced with a large number of options.

Each of the submenus has a long and a short version. The short menus contain all the functions normally required for simple applications and by the less sophisticated user, while the long menus contain all functions available at that level. The long DOCK menus are shown in Table 2.

The user interactions with the program were designed with the following goals in mind:

1. All operations are selected by menus.

2. Menus are used to specify one of a small, fixed set of options.

3. All menus are relatively short.

4. Text information is read from the keyboard. Prompts are provided to guide the user and to indicate what the default responses are.

5. Entity selection (pointing at an atom, for example) is made either via tablet/joystick type devices or via the keyboard, at the option

of the user. The user need not point exactly to an entity. In most cases the nearest entity will be chosen.

6. Each of the basic functions of the graphics workstation (i.e. menus on/off, clipping on/off, long/short menus, etc.) is controlled by a single function button. The buttons controlling these basic functions are not used for other functions. This allows the user to learn one set of controls which remains consistent.

TABLE 2

DOCK Menus

Workspace	Script	Contour	Symmetry	Molecule
Clear	Read	Contour file	Create	Read
Attach	Execute	Add contour	Attach	Write
Center	Build	Delete contours	Relation	Write frac
RCenter	Exit	Change levels	Reset	Attach
Plot		New Colors	Exit	Reset
Local plot		New Draw planes		Merge frag
Checkpoint		Status		Delete
Restore		Exit		Color
Lsq				Clone
Split on/off				Plot
Reset				Identify
Zoom				Plane
Intensity				Center
AutoCkp On/Off				Rcenter
AutoCkp Time				Exit
Exit				

Surface	Refine	Torsion	Bond	Atom
Sphere all	Unfix all	Create	Create	Add
Unsphere all	Fix mol	Uncreate	Delete	Names
Sphere mol	Unfix mol	Angle?	Length	Distance
Unsphere mol	Fix rigid	Setangle	Bond angle	Bond angle
Sphere rigid	Unfix rigid	Attach	Torsion angle	Torsion angle
Unsphere rigid	Fix atoms	Sense	Clear	Identify
Sphere atoms	Unfix atoms	Reset	Range	Delete
Unsphere atoms	Adjust wts.	Reverse	Radius	Color
Change template	Refine	Pucker	Type	Radius
Exit	Add criteria	Color	Change color	Type
	Exit	Exit	Exit	Center
				Rcenter
				Exit

Due to the potentially large number of independently controlled entities in DOCK (a simple display of two molecules would normally use six control dials for each molecule and six for the workspace, for a total of eighteen), the PS–300 version of DOCK allows the user to select eleven dial sets with a total of 88 dials. This is done with the shifted function keys F1–F11. The user is encouraged to develop his or her own conventions for using the dials based on the needs of the problem at hand, rather than to remember a program–determined set of controls.

3.2   An example

To illustrate the power of DOCK's user interface, the following steps would be used to create a pair of symmetrically related dinucleotides with symmetric rotation about a bond in each of the nucleic acid groups:

1. The user enters the MOLECULE submenu and selects the 'Read' option. First the name of the file containing the coordinates and a single command specifying the connection scheme of the dinucleotide are typed in. DOCK draws the MOLECULE.

2. The user enters the TORSION submenu and selects the 'Create' option. The user points to the two BONDs about which rotation is desired. Two dials are specified (via a menu), one to control each TORSION BOND rotation.

3. The user enters the SYMMETRY submenu and selects the 'Create' option. The user must then type in the cell parameters and two symmetry relations ('x, y, z' and '$-x, y, -z$', for instance), and specify that the second molecule is to be copied from the first (the TORSION BONDs and their dial attachments are automatically copied in this operation).

4. The user selects the SYMMETRY submenu's 'Attach' option and attaches six dials to control symmetric rotation and translation of the related molecules.

5. The user enters the WORKSPACE submenu and selects the 'Attach' option. This attaches six dials to control rotation and translation of the entire picture and two additional dials for zoom and intensity control.

In an actual replay of this session, an experienced user accomplished the setup in under five minutes, with 16 menu selections made, eight buttons pressed, seven short lines of text typed, and four atoms pointed to.

A full list of DOCK features is given in the Appendix.

4   The DOCK Screen

DOCK divides the screen into four areas (see Figure 1).

1. A small header bar across the top showing the time, program name, memory usage, and so on.

2. A vertical menu area on the right side of the screen.

3. A graphics window in the central part of the screen. This contains the picture of the WORKSPACE. A single function button is used to select single, stereo, or quarter views (quarter view shows the workspace from four orientations — down each axis and from a diagonal — see Figure 2). Another function button will enlarge the graphics window to fill the screen, causing the other three areas to disappear.

4. A six line terminal area at the bottom of the screen. A function button allows the user to shrink the graphics window and expand the terminal area to a standard 24 line VT100 compatible terminal (Figure 3).

This partitioning of the screen into areas was felt to be more effective and less confusing than superimposing text and menus on the graphics window.

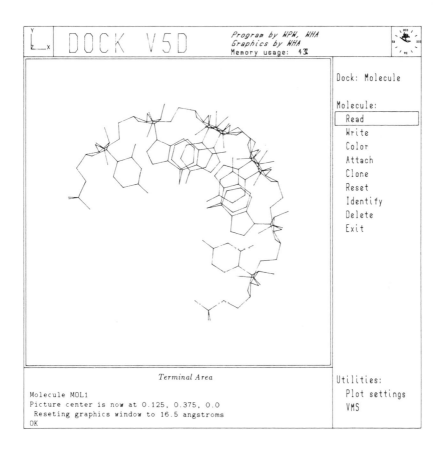

Figure 1: The standard DOCK screen

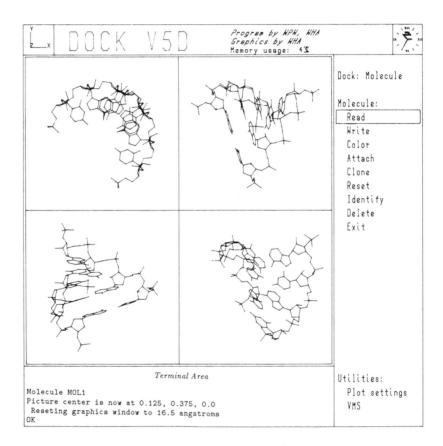

Figure 2: Quarter view mode — four orientations

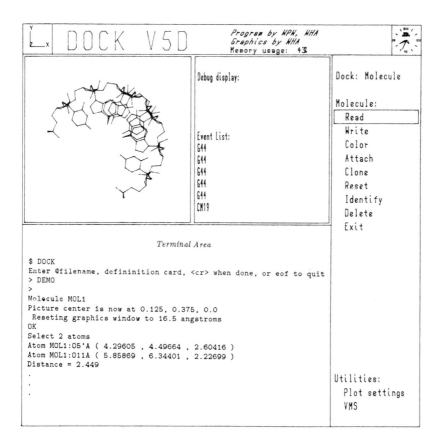

Figure 3: Big (24 line) terminal mode

5   Overview of a Graphics Program

Any interactive graphics program will find itself caught between a person and a graphics machine:

This will be conceptually true even if all or part of the program is resident on the graphics machine. We found it useful to separate the program into two large pieces — the application and the graphics package:

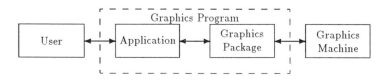

This has several advantages:

1. The presence of a well–defined interface between the application and the graphics package makes it easy to implement the program on displays with dissimilar architectures.

2. A general purpose graphics package is useful for other programs.

3. Changes in the graphic display arrangement or the user interface have little or no impact on the applications.

4. The application can be modified and rearranged with little or no concern for graphics programming.

The current implementation of DOCK is on a Digital Equipment Corporation VAX computer and an Evans and Sutherland PS–300 graphics computer. Where the program is spread across two machines, it is desirable for the graphics package to be resident in both machines and span the link between them. Not only does this remove the need to control a communications path from the application, but it also allows the interface between the application and the graphics package to be made through subroutine calls.

6 DOCK Internals

For a number of practical reasons involving the hardware upon which DOCK originated and the user community it was designed to serve, DOCK was written in FORTRAN, some translated from RATFIV, a structured version of FORTRAN created by Wood by combining the RATFOR (Kernighan 1975) compiler with the *Software Tools* macro processor (Kernighan and Plauger 1976). The organization of the data within DOCK, however, will be most easily understood by those familiar with C, PASCAL, or other like language. All DOCK structures are described in record structures dynamically allocated within one large array. The structures are linked together with pointers into this array, and offsets are used to indicate the various elements of a structure. For example, the structure describing RIGID UNITs (TORSION) contains a pointer to the first BOND structure. This structure, in turn, points to the next BOND data structure, as well as to the structures describing the ATOMs connected by that bond.

An example of a code fragment which might write out the BOND information is:

```
BND = MEM(BONDL+TOR)
TYPE 100,'Bond connecting ',MEM(ANAME+MEM(ATOM1+BND)),
*          ' with ',MEM(ANAME+MEM(ATOM2+BND))
TYPE 110,'Bond type is ',MEM(BTYP+BND)
etc.
```

MEM(BONDL+TOR) is the pointer in the TORSION structure (based at MEM(TOR)) to the BOND structure. MEM(ATOM1+BND) is the pointer to the ATOM structure at one end of the BOND, so MEM(ANAME+MEM(ATOM1+BND)) is the pointer to the ATOM name in that

ATOM structure. A real array (RMEM) is equivalenced to the integer array (MEM), to allow easy referencing of real numbers within the data structures.

A map of the major data structures within DOCK indicating their relationships, is shown in Figure 4.

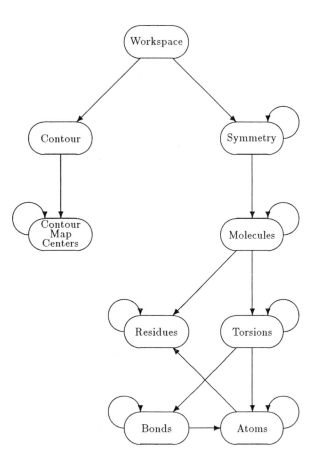

Arrows represent references to other data structures

Figure 4: The DOCK data structure hierarchy

7   The Graphics Package

Though DOCK and its graphics package were designed in tandem, they were viewed as separate projects, and a great deal of effort was made to keep the graphics package general. Not only did we want it to be useful for other computer graphics programs, but we tried to make it flexible enough to be responsive to unforeseen needs of DOCK itself as the program matured. There were two major issues to be addressed.

First, which specific functions should be part of the application and which belong in the graphics package? We used two guiding principles: 1) things which would be useful to other programs belonged in the graphics package; 2) things which required information about the application belonged in the application. Things which were placed in the graphics package under the first rule were workstation control functions, hard copy, graphic object building and control, device interaction with objects, line rotation of objects, a generalized menu interface, a generalized pointing interface, and all communications to the graphics workstation. Things omitted from the graphics package under the second rule included all specifications of graphic object contents, menu contents and organization, what devices are to control what objects, how to interpret pointing information, etc. Thus, while the graphics package maintains menus and invokes them on request, it knows nothing of DOCK's multilevel or long/short menu scheme; while the graphics package allows for a hierarchical tree of graphic objects, it does not understand symmetric movement of objects. These higher level operations which would require some integration of the DOCK framework into the graphics package are left to the application. While this made DOCK more difficult to implement initially, it has served to make DOCK very easy to modify and restructure. Much of the original DOCK code has survived over ten years, two changes in operating systems, and a migration from a Vector General Series 3 machine to the very different Evans and Sutherland system.

The second issue is which functions within the graphics package should be handled in the graphics machine, and which in the host computer. These decisions are far less clear cut, and depend much more heavily on the capabilities of the graphics machine. Our success in addressing this issue resulted from having, initially, to deal with rather slow serial interfaces between the graphics machine and the host. This helped us formulate two guiding principles: 1) let the graphics machine do what it does best; 2) minimize communication between the graphics machine and the host. We maintained hard copy screen plotting in the host in order to be independent of the hardware vendor's choice of graphic output devices. There is necessarily much redundancy between the PS–300 and the VAX database. For example, all vectors drawn are stored on the host by the graphics package, as well as being present in the PS–300, and regeneratable from the application. When the user presses the hard copy button, the PS–300 part of the package sends a message indicating the request and the current display mode (i.e. stereo, frame on or off, etc.). The VAX updates its copies of the object transformation matrix, and can then draw the entire picture without further communication with the graphics display or any assistance from the application. This is particularly important with a complex picture on the screen and a slow communications interface.

8   Programming the Evans and Sutherland PS–300

Programming the PS–300 series graphics computers requires unusual techniques. This is due to the use of function networks as a programming language. In the remainder of the text, ESFN will refer to Evans and Sutherland's implementation of function networks for the PS–300 family of graphics computers, and all programming examples will be in ESFN or FORTRAN–77.

The basic component of ESFN is the function. The function is a 'black box' with a defined set of inputs and a defined set of outputs. For example, the ADD function is used to add two values:

In this function, the input types may be integers, floating point scalars, or any of the allowable vector types. The input types must, however, be compatible — you cannot, for instance, add a single integer to a vector of floating point numbers. ESFN has many defined functions, and there are provisions for writing your own functions. The programmer can create 'instances' of these functions and connect them together in a network. A function network to add three numbers together would look like:

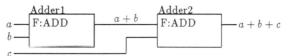

'*a*', '*b*', and '*c*' are sent as messages to the inputs of Adder1 and Adder2. Each ADD function will remain inactive until a message appears at both inputs. The order in which '*a*', '*b*', and '*c*' are presented to the functions is irrelevant to the outcome. 'Adder1' will not produce an output until both '*a*' and '*b*' are present, and 'Adder2' will not function until both inputs ('*c*' and the output of 'Adder1') are present.

Functions may also have 'continuous' inputs, as, for example, the function ADDC:

Once an initial value has been sent to '*b*', each value sent to '*a*' will produce an output. New values sent to '*b*' do NOT produce outputs, but merely replace the former value of '*b*'.

The basic difference between ESFN and 'standard' programming languages such as PASCAL or FORTRAN lies in the non-sequential nature of ESFN. Most computer programming languages proceed sequentially. Consider the following FORTRAN program segment, which might have been taken from DOCK:

```
c(1) = cx + wsxc
c(2) = cy + wsyc
c(3) = cz + wszc
call vecmul (c,object_transform,actual_coordinates)
write (mollun,'(3g12.5)',err=100) actual_coordinates
go to 999
```

Even though these six lines contain three control redirection statements (one subroutine call, one conditional transfer if a write error occurs, and an unconditional control transfer by the GO TO statement), one can reasonably predict the results by reading and interpreting each statement in turn from top to bottom. Imagine the results if the computer could randomly choose the order in which these statements are executed. Consider them executed in the following order:

```
c(2) = cy + wsyc
call vecmul (c,object_transform,actual_coordinates)
c(3) = cz + wszc
write (mollun,'(3g12.5)',err=100) actual_coordinates
go to 999
c(1) = cx + wsxc
```

This is precisely the difficulty in dealing with ESFN. Several functions may be waiting to execute at any point, and there is no guarantee as to the order of execution. Consider a possible network which is intended to calculate the following:

$$n + \sum_{i=1}^{n} x_i$$

The proposed network is:

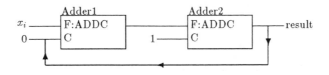

The way the network is intended to work is as follows:

1. Initially, a one is sent to the continuous input of Adder2 (denoted
   $\langle 2 \rangle$Adder2) and a zero will be sent to the continuous input of Adder1
   ($\langle 2 \rangle$Adder1).

2. A new x_i value is sent to $\langle 1 \rangle$Adder1, and it is added to $\langle 2 \rangle$Adder1.

3. This sum (Adder1$\langle 1 \rangle$) is sent to $\langle 1 \rangle$Adder2 where one is added, and
   this sum is sent to result and also stored in $\langle 2 \rangle$Adder1 for the next
   value.

   In practice, this network will not work correctly. This is because
of the non-sequential nature of function networks. If two x_i values are
queued to the first input of Adder1 ($\langle 1 \rangle$ Adder1), it is quite possible that
both will be processed through Adder1 before the first result is processed
by Adder2. In this manner, the first value will end up being ignored, and
the network result will be unpredictable.

   While there are many ways to correct (and improve) this network,
we will use it to show the common synchronization technique. A corrected
version of this network might be:

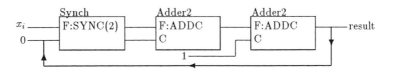

   The SYNC function, which can have any number of inputs and outputs, passes through values from each input to the corresponding output,
but only when all inputs are present. In this fashion, one may impose some
order on function networks. DOCK's graphics package makes extensive
use of the SYNC function to impose sequentiality on the network.

9   Some final remarks on program development

Professional programmers who are asked to deal with programs written by scientists often express the opinion that scientists should not write programs. Scientists who are forced to deal with programmers often express the opinion that programmers operate in a vacuum and question their priorities. These expressions arise from a fundamental difference in their points of view: the scientist is focussed on the problem and generally views the program as a means to an end, while the programmer tends to view the program as an end in itself.

The lesson to be learned from the DOCK development project is that both viewpoints have some merit. Our success in producing a usable program with sufficient flexibility to evolve to meet research needs resulted from a careful blend of these attitudes. Scientists were forced to explain their problems to a computer scientist. This helped to refine the scientists' definition of what was needed, and a model was produced to address these needs. No code was written until everyone was convinced of the completeness of the model and that its implementation was practical. The program was written by professional programmers who were concerned with building in flexibility and producing an artful and maintainable product. The program and its user interface were constantly tested in the environment of active research programs, and a cooperative attitude toward the development project was maintained among the scientists and the programmers. The careful initial definition of the program and the mutual respect of the parties involved were principally responsible for DOCK's success.

Acknowledgements

This work was supported by NIH grants GM21589, CA06927, and RR05539, and by an appropriation from the Commonwealth of Pennsylvania. We are also grateful for support from Smith Kline &French. Our thanks also to Leslie Lessinger whose critical and helpful comments on the program and this manuscript were invaluable.

REFERENCES

Berman, H. M., Carrell, H. L., Stodola, R. K., Andrews, L. C., Bernstein, F. C., Bernstein, H. J., Koetzle, T. F., Meyer E. F., and Morimoto, C. N. (1974). *X International Congress of Crystallography*, Abstract 23.1–7.

Berman, H. M., Neidle, S., and Stodola, R. K. (1978). *Proc. Natl. Acad. Sci.*, U.S.A. **75**, 828–832.

Kernighan, B. W. (1975). in *Software — Practice and Experience*, October, 1975.

Kernighan, B. W., and Plauger P. J. (1976) in *Software Tools*, Chapter 8. Addison-Wesley Publishing Company, Reading, Massachusetts.

Appendix: DOCK Features

High level functions:

— Non-linear least squares minimization to optimize user specified geometric goals. Allows any or all of user specified SYMMETRY, MOLECULE rotations and translations, and TORSION BOND rotations to vary.

— Add ATOM according to a specified geometry.

— Change pucker of five-membered ring.

— Merge two MOLECULES according to specified geometry.

Graphic feature controls:

— Either split BOND representation using ATOM colors or solid BOND color.

— Color change of ATOMs, BONDs, RIGID UNITs, MOLECULEs, or CONTOUR baskets.

— Add or delete ATOMs, BONDs, MOLECULEs, or CONTOUR baskets.

— Add or delete ATOM names from display.

— Modify ATOM or BOND display format or radius.

— Specify and change CONTOUR basket color and contour levels.

— Add or remove dotted surface representation of atoms.

— Split a RIGID UNIT in two across a TORSION BOND, or merge two adjacent RIGID UNITs into one.

— Create SYMMETRY GROUPs for one or more MOLECULES in the WORKSPACE.

— Clear BONDs from a MOLECULE.

— Copy (clone) MOLECULE (including all TORSION BOND definitions, controls, colors, etc.).

Hard copy options:

— Write Cartesian or fractional coordinates.

— Screen copy on QMS laser printers or HP plotters.

— Write coordinates for VIEW program for ball and stick display.

Graphic spatial controls:

— User assignment of 88 dials (11 logical sets of 8 physical dials).

— Assignment of control of entire WORKSPACE: rotation, translation, zoom, and intensity.

— Assignment of control of SYMMETRY GROUPs: rotation and translation of the entire SYMMETRY GROUP within the WORK-SPACE; point or center rotation and translation of MOLECULEs within the SYMMETRY GROUP.

— Assignment of control of MOLECULEs: rotation and translation.

— Assignment of control of TORSION BOND rotation.

— Reset TORSION BOND, MOLECULE, SYMMETRY, or WORK-SPACE to their original orientation.

— Orient MOLECULE for a planar presentation.

— Specify rotational center of a MOLECULE or the entire WORK-SPACE about an ATOM or the WORKSPACE center.

— Script control (for filming) of MOLECULEs, TORSION BONDs, and the WORKSPACE.

— Modify symmetry relations within a SYMMETRY GROUP.

— Specify new TORSION BOND angle.

— Invert direction of TORSION BOND rotation (move in opposite direction of dial rotation), or reverse the TORSION BOND (change which rigid unit moves when you rotate about the BOND).

Interrogation:

— Distance defined by two ATOMs, or length of a BOND.

— BOND angle, or the angle between three ATOMs.

— TORSION angle across a BOND, about a TORSION BOND, or defined by four ATOMs.

— Identify ATOMs or MOLECULEs.

— Determine status of contour map files.

Session Controls:

— Clear WORKSPACE.

— Checkpoint WORKSPACE (save all session information: display contents, device attachments, etc.).
— Restore WORKSPACE (restore session to the checkpointed state).
— Turn automatic checkpointing on or off, and specify the time between automatic checkpoints.

16

MOLECULAR GRAPHICS: PRACTICAL APPLICATIONS

H. L. Carrell

Introduction

The use of computers for graphics in small molecule crystallography had its beginning in the 1960's with the introduction of **ORTEP** (Johnson 1965) which allowed one to plot the results of a structure determination in an elegant and very informative manner. The graphics afforded by **ORTEP** were therefore useful primarily for presenting one's final results. The development of vector displays led to the introduction of a number of interactive graphics program systems, many of which were cited by Huml (1985) in a survey of graphics program packages. At that time, the interactive graphics packages available were aimed primarily toward macromolecular applications or the interaction of macromolecules and small molecules. Missing from the listings of graphics program packages at that time were the programs **VIEW** (Carrell 1976) and **DOCK** (Stodola, Manion, Berman, Wood, and Badler 1987), both of which were developed at the Fox Chase Cancer Center in Philadelphia, PA (USA); these two programs allow one to view molecules, small or large, manipulate molecules, study crystal packing, compare molecules and dock two or more molecules together. It is these two programs which I shall discuss in the course of this tutorial.

First, we should address the question 'What do we want to use computer graphics for?' Several applications include: 1) the rapid examination of the results of a structure determination, 2) an easy way to present the results in publication form, 3) the utilization of graphics to aid in structure solution, 4) visualization of a number of molecules simultaneously to determine how they compare or perhaps how they might interact, or 5) perhaps do model building. Any one of these reasons is sufficient for developing programs; the two programs to be discussed here together cover a wide range of graphics applications.

The program **VIEW** was written by this author to fulfill the need to visualize results of crystal structure determinations and in the process the program was found to be quite useful in the solution of crystal structures. On the other hand, the much more powerful program **DOCK** was developed later and is still under development; it can be used for a much wider variety of applications. First, the program **VIEW** will be discussed then we will look at the more powerful program **DOCK**.

1　View

1.1　Program description

The program **VIEW** was initially written in 1976 in **FORTRAN** for the purpose of displaying the results of crystal structure determinations in ball and stick form with elimination of hidden lines. The initial version of **VIEW** was merely another plotting program written because we could not manage to make the hidden line routines in **ORTEP** fit into the memory of our PDP–11/40. The program can be run without having a display screen available, thus allowing one to produce ball and stick drawings on a plotter without previewing the results on a screen; there is considerable flexibility for orientation of the molecule via commands. If a graphics display is available, the process of orienting the molecule or molecules becomes much easier. Since the Vector General Series 3 (VG3) interactive graphics terminal was available for our use through the **CRYSNET** project (Berman, Carrell, Stodola, Andrews, Bernstein, Koetzle, Meyer, and Morimoto 1974), it was only natural to utilize the VG3 as way to orient a particular drawing. The program is driven by instruction keywords or commands and any line of input which does not begin with a recognized instruction keyword is ignored. In addition to non-graphics terminals, **VIEW** currently may be run on the Evans and Sutherland PS–300 series of graphics terminals as well as a TEKTRONIX 4107. Since the use of a graphics display is done within a single subroutine of the program, one should be able to modify the display routine so it can be used with almost any type of terminal display having graphics capability.

1.2 Operating environment

The program **VIEW** may be executed at any terminal connected to a VAX/VMS system. The SHOW command requires the presence of the PS–300 in the case of **VIEW** or the Tektronix 4107 in the case of **TEKVIEW** so that if a user is able to make do without a graphics display, the program may be used to produce a disk file of instructions which may then be plotted, using the program **VUEPLOT**, on the appropriate device which in our laboratory is either a QMS laser printer or a Hewlett Packard pen plotter. Many of the commands require that arguments be present, however none of these arguments are prompted for, therefore the user must become somewhat familiar with the various commands. Missing arguments on a command line may cause unpredictable results. Although **VIEW** is not a graphically interactive program such as **DOCK**, it affords the opportunity to do some graphics in a rapid and simple manner. Because the program has been written as several modules, the implementation of new commands is not a complicated matter so that future additions to **VIEW** should not be difficult to accomplish.

When the E&S PS–300 is used for display, the dials are used for rotations, translations, depth cueing and scaling. When the user has displayed the molecule on the graphics screen, a small menu is available which allows the user to display selected atom names, to save the coordinates of the current display, to alter some of the plotting parameters, to generate a plot file or to return to the main calling routine. The graphics display may only be used for orientation, other changes in the contents of the picture must be done through the command mode of the program.

When a graphics terminal such as the TEKTRONIX 4107 is used, the on screen rotation of the structure must be carried out by redrawing the molecule for each rotation angle. In our case, the terminal keypad is polled and the picture is rotated about the appropriate reference direction in steps of 1, 5, or 30 degrees depending on the key pushed. The columns of keys on the keypad are used to control rotation about the axes, for example the 1, 4, and 7 control the rotation step size about the horizontal or X axis. Similarly, the 2, 5, and 8 control the Y axis (vertical) rotation and the 3, 6 and 9 control the rotation about Z, i.e., rotation in the plane of the display screen. The 0 (zero) key toggles continuous/stepwise rotation and the '.' (decimal point) key toggles the direction of rotation. A small menu is also available with the same options as available for the PS–300 version,however, the menu selections are made by striking the appropriate key.

1.3　Commands in VIEW

The program is run in VMS with the command **VIEW** whereupon the user is greeted with a display of all the current commands is and the prompt symbol '>' is displayed. The commands available to the user are shown in Table 1 as they appear on the terminal screen. The user then enters commands via the terminal or an external script file using the FILE command or by way of a VAX indirect command file using '@' preceding the file name.

Table 1

Initial VIEW display of commands

CELL	ATOM	CLRA	CLRB	CURR	RNGE	GEOM	DIST	ANGL
TRSN	AXI*	AXIS	BOND	PLAN	SHOW	PLOT	SCAL	SIZE
ROTR	LIST	SYMM	BILD	FILE	HELP	HALT	QUIT	STOP
END	DISP	DELE	SAVE	TITL	PERS	LINK	BNDP	

The commands may be grouped into those which control input and picture composition, orientation, plotting functions and miscellaneous. Those commands concerned with input and picture composition are FILE, CELL, ATOM, RNGE, DELE, LINK, BNDP, SYMM, BILD, and PERS. The commands associated with orientation are AXIS, AXI*, BOND, PLAN, SHOW, and ROTR. The commands which control the final plot are PLOT, SCAL, and SIZE. The commands in the miscellaneous category are CURR, GEOM, DIST, ANGL, TRSN, LIST, TITL, SAVE, CLRA, CLRB, HALT, QUIT, STOP, END, and HELP. The single most important command is the HELP command since the user may then obtain on screen documentation of all the commands. By the use of the available commands, the user may input the crystal cell parameters and atomic coordinates, generate symmetry related atoms, orient the atoms as desired, obtain listings of the molecular geometry, save the results for later use, delete unwanted atoms, and finally generate a file of plot instructions in order to obtain a plot of the results. A summary of commands currently available to the user of **VIEW** is found in the Appendix.

1.4 Example

The following example of a session running **VIEW** at a PS–300 graphics terminal is offered. A sample of an input file for a session is shown in Table 2 and for purposes of this tutorial has been assigned the filename of SAMPLE.PIC. After the user has initiated the program by typing **VIEW** on the terminal and the welcome messages have been come up, the prompt character '> ' appears on the left side of the terminal screen or page. Whenever this prompt appears, the program is awaiting further instructions or commands. The session might look as follows:

>FILE SAMPLE.PIC   causes **VIEW** to process commands fro m the file SAMPLE.PIC.

>PLAN 1,2,3,4,5,6   calculates plane of atoms 1-6 and orients the molecule so that the plane lies in the plane of

TABLE 2

Input file example for VIEW and DOCK

```
NAME M1
TITL OXIRANE I
CELL 11.413,10.487,7.246,90.,103.03,90.
ATOM    C1      0.00654 0.34633 0.34614   0.000   1.0000   2 0.25   1
ATOM    C2     -0.11453 0.27946 0.32272   0.000   1.0000   2 0.25   1
ATOM    C3     -0.17324 0.25771 0.12000   0.000   1.0000   2 0.25   1
ATOM    C4     -0.09497 0.23889-0.00929   0.000   1.0000   2 0.25   1
ATOM    C5      0.02546 0.24372 0.03896   0.000   1.0000   2 0.25   1
ATOM    C6      0.09941 0.21053-0.10197   0.000   1.0000   2 0.25   1
ATOM    C7      0.19637 0.30900-0.10867   0.000   1.0000   2 0.25   1
ATOM    C8      0.27289 0.33472 0.08892   0.000   1.0000   2 0.25   1
ATOM    C9      0.19170 0.38148 0.21877   0.000   1.0000   2 0.25   1
ATOM    C10     0.09262 0.28224 0.23772   0.000   1.0000   2 0.25   1
ATOM    C11     0.26323 0.43097 0.41302   0.000   1.0000   2 0.25   1
ATOM    C12     0.36344 0.52657 0.39813   0.000   1.0000   2 0.25   1
ATOM    C13     0.44450 0.46946 0.28049   0.000   1.0000   2 0.25   1
ATOM    C14     0.36727 0.43656 0.08367   0.000   1.0000   2 0.25   1
ATOM    C15     0.45847 0.41849-0.04103   0.000   1.0000   2 0.25   1
ATOM    C16     0.54723 0.52768 0.02702   0.000   1.0000   2 0.25   1
ATOM    C17     0.63722 0.55828 0.22729   0.000   1.0000   2 0.25   1
ATOM    C18     0.51830 0.35590 0.38092   0.000   1.0000   2 0.25   1
ATOM    C19     0.15128 0.16543 0.34643   0.000   1.0000   2 0.25   1
ATOM    C19'    0.13969 0.03615 0.28908   0.000   1.0000   2 0.25   1
ATOM    O19     0.07692 0.07785 0.41479   0.000   1.0000   1 0.25   2
ATOM    O3     -0.28208 0.25057 0.06525   0.000   1.0000   1 0.25   2
ATOM    O17     0.59485 0.63846 0.32859   0.000   1.0000   1 0.25   2
ATOM    H1      0.04427 0.34855 0.48346   5.176   1.0000   3 0.10   1
ATOM    H1'    -0.00606 0.43700 0.29440   5.288   1.0000   3 0.10   1
ATOM    H2     -0.10663 0.18877 0.38438   7.389   1.0000   3 0.10   1
ATOM    H2'    -0.16371 0.33437 0.38142   9.074   1.0000   3 0.10   1
ATOM    H4     -0.12916 0.22080-0.13333   5.158   1.0000   3 0.10   1
ATOM    H6      0.13856 0.12234-0.06561   6.230   1.0000   3 0.10   1
ATOM    H6'     0.04288 0.19537-0.21795   6.739   1.0000   3 0.10   1
ATOM    H7      0.24896 0.27616-0.19630   4.198   1.0000   3 0.10   1
ATOM    H7'     0.16173 0.38863-0.16451   3.935   1.0000   3 0.10   1
ATOM    H8      0.31104 0.25598 0.13951   2.919   1.0000   3 0.10   1
ATOM    H9      0.15206 0.45596 0.15709   1.839   1.0000   3 0.10   1
   .       .         .        .       .        .        .        .     .
   .       .         .        .       .        .        .        .     .
   .       .         .        .       .        .        .        .     .
ATOM    H20'    0.20873-0.02656 0.33952   8.098   1.0000   3 0.10   1
RNGE1,23,1,23,1.,1.6,.08,1,
RNGE1,23,24,49,.1,1.2,.02,1,
```

the viewing surface.

>CURR   informs user about number of atoms and bonds and current PLOT scale factor.

>GEOM   calculates the bond lengths, bond angles and torsion angles for all bonded atoms and writes the result to the file VIEWPRT.DAT.

>SHOW   draws the molecule on the graphics screen allowing the user to adjust the orientation or just to view the handiwork. User returns to command mode by selecting EXIT from the menu.

>PLOT   writes a file of plot instructions for the plotting program. The plot orientation is the last one seen on the screen.

>ROTR 1,90.   rotate the molecule by 90 degrees about the horizontal axis.

>PLOT   write a file of plot instructions for the current orientation.

>END   submit any plot jobs to batch and exit the program.

The above sample session allows the user to process all the commands contained in the file SAMPLE.PIC which includes the calculation of the bonding. The molecule is then oriented looking onto the least squares plane of atoms 1-6. The SHOW command allows the user to view the results on the graphics screen and to adjust the orientation using the rotation devices. The resulting orientation is then recorded for plotting then turned by 90 degrees and the new orientation recorded for plotting. Upon leaving **VIEW**, the plot instructions for both plots are submitted to the batch queue for processing and plotting by the program **VUEPLOT**. If a non-graphics terminal such as a DEC VT220 were used, the user would refrain from issuing the SHOW command.

2 Dock

The program **DOCK** as described in 'Anatomy of a Molecular Graphics Package: DOCK' by R.K. Stodola, *et al.*, is an interactive computer graphics program which enables the user to display and to manipulate molecules on a graphics display with relative ease. The program allows the user to independently manipulate a large number of molecules. These manipulations may include torsional rotations about specified bonds in the molecules as well as rotations and translations of each molecule. In addition, global rotation, translation and scaling is available. **DOCK** also contains a provision for displaying electron density maps in the form of 3-dimensional baskets for one or more levels of constant electron density. One of the more powerful features of **DOCK** is that the user may introduce crystallographic symmetry operations, and in addition to examining the packing of molecules, the user can attach the symmetry operations to dials and investigate the consequences of rotation or translation of a molecule. This allows the user to attempt to solve crystal structures graphically by examining the packing of the molecules in the crystal. **DOCK** also contains a non-linear least squares procedure which is extremely useful for overlaying similar molecules or for producing models of intermolecular interactions given reasonable restraints on the system.

DOCK is a menu driven program and any input that must be entered by the user is generally prompted for. The most important step in using a program such as **DOCK** is to be able to get started. The user must prepare coordinate files in advance unless he/she feels comfortable typing in atomic parameters at run time. Before dealing with the actual running of the program, it might be useful to describe the preparation which must be done in order to get started and what the user will encounter at the graphics terminal.

TABLE 3

DOCK function buttons

Key	Unshifted	Control	Shifted
F1	Frame ON/OFF	LEDs ON/OFF	Dial set 1
F2	Clip ON/OFF	Control enable/disable	Dial set 2
F3	Plot device select	Do local plot	Dial set 3
F4	Dial enable/inhibit	Reserved	Dial set 4
F5	Display format	Reserved	Dial set 5
F6	Terminal 6/24 lines	Reserved	Dial set 6
F7	Axes ON/OFF	Control reset	Dial set 7
F8	Reserved	Analog/digital clock	Dial set 8
F9	Big/small menus	Unused	Dial set 9
F10	Contours On/OFF	Unused	Dial set 10
F11	Level 0/1	Unused	Dial set 11
F12	Exit/quit/done	Unused	Interrupt

The unshifted functions are labeled on the LEDs above the function keys.

2.1 The graphics keyboard

The PS–300 keyboard is currently used as the terminal input keyboard for **DOCK**. In addition to the standard terminal keyboard functions, the numeric keypad is used to provide menu control; twelve function keys and their associated LED's along the top of the keyboard are used by **DOCK** for function selection. Table 3 contains a summary of the function keys used by **DOCK**. The column labeled 'Control' refers to the function when the labeled key is pressed while holding the 'ctrl' button on the keyboard. The selection of the currently operative dial set is done by using the shifted function keys and the dial set number is displayed on the LED display of the control dials. Thus in device level 0 or 1 as selected by button F11, the user has available eleven sets of eight dials (88 devices) which may be used to control objects drawn on the graphics screen. Level 0 is normally used, however the two levels are independent and the very ambitious user may wish to attach objects to dials in both levels making it possible to have as many as 176 devices available although only 88 of these may be active at the same time. It is the user's responsibility to remember which functions have been attached to which dials and dial sets. Attachment to dials is usually done interactively via menus, but may be done by way of a command in a MOLECULE definition file during a MOLECULE READ by the ATCH (attach command) with a list of the devices for the attachment of rotation and translation.

2.2   Data Preparation

DOCK views the molecular graphics problem as a hierarchy of graphical constructs with the ATOM as the basic unit, a collection of ATOMs and BONDs comprise a MOLECULE, and a single MOLECULE or many MOLECULEs may be combined into a single WORKSPACE. In general, it is most convenient to use one file to define the contents of each MOLECULE that is to comprise the WORKSPACE. Thus, the preparation of the file containing the ATOMs is crucial. In the current version, an ATOM may defined by one of three formats, namely **CRYSNET**, **PDB** or **DOCK** format. The **CRYSNET** format derives from the atomic coordinates format used in the **CRYSNET** project and is the same format as used in **VIEW**. **PDB** format derives from the format used by the Protein Data Bank (Bernstein, Koetzle, Williams, Meyer, Brice, Rodgers, Kennard, Shimanouchi, and Tasumi 1977) while the **DOCK** format was derived specifically as a means of inputting all the information about an ATOM which **DOCK** needs. The format which is desired by a user may be specified by a command when the user first enters **DOCK**. Regardless of which format for atomic coordinates the user prefers, the input record must begin with the word 'ATOM' in columns 1-4 and in the case of **PDB** format, 'HETA' also is assumed to define an atom. If the user wishes to use fractional coordinates, a CELL definition must be supplied before ATOMs are input to the program. If no CELL definition is input, the program assumes that the input coordinates are cartesian coordinates. Since MOLECULEs are independent objects, each molecule may have its own CELL definition and if the user so desires, it own color scheme. In addition to atoms, a MOLECULE contains BONDs and if so desired, may also contain variable TORSION angles. Most of the information needed to define a MOLECULE may be input via MOLECULE READ commands. Many of the attributes of a MOLECULE may be assigned via MOLECULE READ commands as well as interactively by utilizing the various submenus. A file suitable for use with **VIEW** may also be used by **DOCK**, thus the sample file shown in Table 2 may be used in both programs. The TITL instruction found at the beginning of the file SAMPLE.PIC is ignored by the MOLECULE READ in **DOCK**, the remaining commands are digested appropriately.

Table 2 also serves to illustrate that the preparation needed to get started in **DOCK** is relatively easy. In our own laboratory, we have modified almost all programs which supply coordinate files to give coordinates in **CRYSNET** format including the coordinate output routines of the **MULTAN** package(Main, Fiske, Hull, Lessinger, Germain, Declercq, and Woolfson 1980) . It is a trivial matter, therefore, to use **DOCK** to aid in the solution of crystal structures. In fact, if one has access to such a system, one tends to become very spoiled indeed.

2.3 DOCK Menus

When the user has invoked the program **DOCK** by typing DOCK on the terminal keyboard, The following menu will appear on the screen:

    DOCK:
        WORKSPACE
        SCRIPT
        CONTOUR
        SYMMETRY
        MOLECULE
        SURFACE
        REFINE
        TORSION

BOND
ATOM
STOP
LATER

By using the graphics tablet or the numeric keypad on the terminal, the user may select the desired option and, with the exception of STOP and LATER, the user will then be presented with the appropriate submenu. If this is the initial session with a given problem, the MOLECULE option should be invoked, whereupon the user is presented with the MOLECULE menu. Table 4 contains the long menu listings for all the main **DOCK** menus. A brief review of the functions performed within these menus follows:

Workspace The items in this menu are used to control the overall picture. It is in the WORKSPACE where global rotations, translations, centering and other overall functions are controlled. By selecting Clear the user may delete the current **DOCK** contents, or by selecting Checkpoint, the current information in **DOCK** may be saved on a checkpoint file for use at another time.

Script This menu controls scripted motion. Scripts may be created, read or executed while within this submenu.

Contour The items in this menu governs the creation and characteristics of electron density contours to be displayed.

Symmetry Creation and control of symmetry groups is done through this menu and its associated submenus.

Molecule Creation and control of characteristics of a Molecule is done here. Here, Molecules may be created, plotted, identified, reset to original orientation and even deleted.

Surface This control the use of dotted spherical representations of the atoms. If the default atom radii have been used, space filling views are displayed.

TABLE 4

DOCK menus

DOCK Menus				
Workspace	Script	Contour	Symmetry	Molecule
Clear	Read	Contour file	Create	Read
Attach	Execute	Add contour	Attach	Write
Center	Build	Delete contours	Relation	Write frac
RCenter	Exit	Change levels	Reset	Attach
Plot		New Colors	Exit	Reset
Local plot		New Draw planes		Merge frag
Checkpoint		Status		Delete
Restore		Exit		Color
Lsq				Clone
Split on/off				Plot
Reset				Identify
Zoom				Plane
Intensity				Center
AutoCkp On/Off				Rcenter
AutoCkp Time				Exit
Exit				

Surface	Refine	Torsion	Bond	Atom
Sphere all	Unfix all	Create	Create	Add
Unsphere all	Fix mol	Uncreate	Delete	Names
Sphere mol	Unfix mol	Angle?	Length	Distance
Unsphere mol	Fix rigid	Setangle	Bond angle	Bond angle
Sphere rigid	Unfix rigid	Attach	Torsion angle	Torsion angle
Unsphere rigid	Fix atoms	Sense	Clear	Identify
Sphere atoms	Unfix atoms	Reset	Range	Delete
Unsphere atoms	Adjust wts.	Reverse	Radius	Color
Change template	Refine	Pucker	Type	Radius
Exit	Add criteria	Color	Change color	Type
	Exit	Exit	Exit	Center
				Rcenter
				Exit

Refine This option is still in the planning stage and has not yet been implemented.

Torsion Creation, deletion and control of the Torsion angles is handled here. The user may create and control as many torsion units as can be maintained through the otherwise unused dials. If a user were already using Dial sets 1-3 for the control of 2 molecules and the Workspace, the potential still exist for attaching the remaining 8 dial sets (64 devices) to Torsion angles.

Bond Control of bonding including the assignment of radius of bonds for later plotting by **VIEW** is accomplished here. The user may query **DOCK** on the current values of bond lengths, bond angles and torsion angles. Here the user may also make changes in the bonding or connectivity of the ATOMs in the structure.

Atom This controls what is done at the atomic level. New atoms may be added, specified atoms may be identified, distances between atoms calculated, atomic radii changed, and deleted while in this menu.

The above summary is intended to give a general idea of the use of the various **DOCK** submenus. Many of the selections which may be made in the menus can lead to further menus and choices. For example, if a user selects the Delete option in the ATOM menu, the user is requested to point to the atom to be deleted or as an alternative, the user may elect to select the menu item 'FROM KEYBOARD', in which case the user is prompted to type in the name of the atom to be deleted. Thus the user is able to mix selection methods as best suits the users style.

2.4 Example

The following is an example of the use of **DOCK** to display two different molecules and to overlay the second molecule on the first using the dials and the least squares procedure:

DOCK   Initiate the program by typing **DOCK** .

Select **WORKSPACE**

Select **Attach**

Press key '(shift)F1' this sets dial set to 1

Select **All Dials** attaches WORKSPACE rotation, etc., to dial set 1.

Select **Exit** returns to **DOCK** menu.

Select **MOLECULE**

Select **Read**

Type **NAME** M1 name this molecule M1.

Type **FILE** SAMPLE initiates the reading of the file of MOLECULE contents from SAMPLE.PIC.

Type **NEWMOL** signals the end of definition of the first molecule and initiates reading of a second file.

Type **NAME** M2 name second molecule M2.

Type **FILE** SAMPLE2 reads file SAMPLE2.PIC.

Type **CENT** Y directs that the second molecule is to be centered on the screen.

Type **ATCH 2.1,2.2,2.3,2.5,2.6,2.7** instructs that MOLECULE rotation and translation be attached to the specified devices which in this case is dial set 2.

Type <cr> signals that current MOLECULE Read is complete.

Select **EXIT** return to **DOCK** menu.

The user may now rotate the WORKSPACE to obtain the view of the first molecule. By typing (shift)(F1) or (shift)(F2), one can select the dial set for the WORKSPACE or the second MOLECULE. By rotation and translation of molecule 2, on should be able to obtain a reasonable fit of the two molecules.

Select **WORKSPACE**

Select **LSQ**

Type **DIST** M1:C1,M2:C1,0.0   These commands set the

Type **DIST** M1:C2,M2:C2,0.0   target distance between

                    .   .   equivalent atoms in the two

                    .   .   molecules to 0.0 Å.

Type **DIST** M1:C17,M2:C17,0.0   The commands could have been

                                typed in advance on a separate file

                                and read by a FILE command.

Type <cr>

The non-linear least squares minimization will run until satisfied the minimum is reached. All attached dials (other than the WORKSPACE) at the current level are allowed to vary in the minimization. The distances between the atoms specified in the above specifications will be reported on the terminal screen and the second molecule will be moved to its refined position. At this point, the user will be returned to the WORKSPACE menu.

Select **Checkpoint**

Type M1M2   checkpoint current WORKSPACE onto a

                        file named M1M2.CHK.

Select **EXIT**   return to **DOCK** menu

Select **LATER** *or* **STOP**   If LATER is selected, a checkpoint file

                        will be written to username.CHK so that

                        when **DOCK** is invoked in the same

                        VAX directory, the checkpoint file will be

                        automatically restored as it was left. If

                        STOP is selected, no auto-checkpoint file

                        is saved. In both cases, the user will be

                        returned to VMS control.

The above example is intended to show only a small portion of the capabilities of **DOCK**. Within the limitations of this tutorial and the limitations on time and space, it is not possible to discuss every aspect of the use of **DOCK**. The sample session and the preceding discussion will, I hope, provide the prospective user of **DOCK** with at least the ability to get started in the use of the program and the curiosity to determine the programs fullest capabilities.

3 Final Remarks

The program **VIEW**, while written by this author (HLC), makes use of the same graphics package utilized in **DOCK**. The hidden line routines used in the plotting were written by R.K. Stodola. The routines for the use of the TEKTRONIX 4107 as a display, were written by J.P. Gibbons of the programming staff at the Fox Chase Cancer Center. The program **VIEW** represents my first, and I hope, last attempt at writing a graphics program. The cooperation between the crystallographers and the professional programming staff at our institution has made the development of both **DOCK** and **VIEW** possible.

Acknowledgments

This work was supported by USPHS Grants CA-10925, CA-06927, RR-05539 and CA-22780 from the National Institutes of Health, and by an appropriation from the Commonwealth of Pennsylvania. I also wish to thank Drs. Jenny Glusker and Miriam Rossi for their helpful comments and to Frank Manion for his help in preparing this manuscript.

REFERENCES

Berman, H. M., Carrell. H. L., Stodola, R. K., Andrews, L. C., Bernstein, F. C., Bernstein, H. J., Koetzle, T. F., Meyer, E. F., and Morimoto, C. N. (1974). *The* **CRYSNET** *system*. X International Congress of Crystallography, Abstract 23.1-7.

Bernstein, F. C., Koetzle, T. F., Williams, G. J. B., Meyer, E. F. Jr., Brice, M. D., Rodgers, J. R., Kennard, O., Shimanouchi, T., and Tasumi, M.

(1977). *J. Mol. Biol.* , **112**, 535-542.

Carrell, H. L. (1976). **VIEW**: *A computer program for producing molecular diagrams.* Institute for Cancer Research, The Fox Chase Cancer Center, Philadelphia, PA, USA.

Huml, K. (1985). In *Crystallographic Computing 3: Data Collection, Structure Determination, Proteins, and Databases* (ed. G. M. Sheldrick, C. Kruger, and R. Goddard) Clarendon Press, Oxford, U.K.

Johnson, C. K. (1965). **ORTEP**: *A Fortran thermal ellipsoid plot program for crystal structure illustrations.* ORNL Technical report 3794, Oak Ridge National Laboratory, TN, USA.

Main, P., Fiske, S. J., Hull, S. E., Lessinger, L., Germain, G., Declercq, J. P., and Woolfson, M. N. (1980). *MULTAN80. A System of Computer Programs for the Automatic Solution of Crystal Structures from X-ray Diffraction Data.* Universities of York, England and Louvain, Belgium.

Stodola, R. K., Manion, F. J., Berman, H. M., Wood, W. P., and Badler, N. (1987). **DOCK**: *A computer program for interactive molecular graphics.* Institute for Cancer Research, The Fox Chase Cancer Center, Philadelphia, PA, USA.

Appendix: Summary of VIEW commands.

HELP Prints instructions to the terminal screen for the specified command.

FILE Causes the named file to be processed by **VIEW**.

CELL Unit cell parameters, a,b,c,α,β,γ (default is 1.,1.,1.,90.,90.,90.)

ATOM Atomic parameters including name, coordinates, drawing radius and atom type. The atoms may be drawn as open, filled or stippled circles.

RNGE Range for bond calculation, the radius of the bond and the bond type. Stick bonds may be drawn as the tapered outline or solid (dark) bonds.

LINK Form bond between atoms sequentially within the specified distance range. Very useful with alpha carbon representation of a protein.

BNDP Calculate bonding by partitioning the atoms. This is useful for very large structures.

DELE Atom deletion.

SYMM Apply the input symmetry operator to the range of atoms given. All bonds already associated with the included atoms will be carried along.

BILD Reads the named file to obtain a range of atoms to be included, the limits (in fractions of the cell edge), and symmetry operations to be used. This allows the user to easily examine crystal packing.

CLRA Clears the atoms array.

CLRB Clears the bonds array. These two commands allow the user to start over without leaving the program.

AXIS Orient the view with the specified crystal axis perpendicular to the display.

AXI* Similar to AXIS except that the reference is to a*, b* or c*. The default orientation is with c* perpendicular to the display.

BOND Orient the view so that the user is looking along the vector between the two named atoms.

PLAN Orient the view so that the user is looking onto the least squares plane of the specified atoms. Currently up to 20 atoms may be used to define the plane.

ROTR Rotate the current view around the specified direction by the specified number of degrees.

SHOW Display the current view on the graphics screen. Bonds are represented by lines and unbonded atoms are represented as lone dots. The user may now utilize the devices available to the particular graphics system to examine and orient the view.

PERS Apply some perspective to the final plot, namely, give a sense of depth to the final plot by making atoms and bonds appear smaller as a function of distance from the viewer. This is useful with a large number of atoms such as in a macromolecule or in a very busy packing diagram.

SCAL The scale factor for the final plot (cm/Å).

SIZE The size of the plot. If scale is specified, this is not used.

TITL Title for plot. This may be changed whenever the user wishes to provide a new description of the view being plotted.

PLOT Write a file of plot instructions for the current view. The actual plotting is carried out by the program VUEPLOT.

CURR Display the current view contents. The number of atoms, bonds and current plot scale factor.

DISP Write a list of the names of the atoms in the current view to the terminal screen.

SAVE Save the atomic parameters for the current view on the file specified.

LIST Write the current view coordinates to the file VIEWPRT.DAT.

DIST Write the contents of the current bond array to VIEWPRT.DAT.

ANGL Calculate the angles between connected bonds and write to VIEWPRT.DAT.

TRSN Calculate the torsion angles about all the bonds and write the angles to VIEWPRT.DAT.

GEOM Calculate bond lengths, angles and torsion angles using the current bonds array and write to the file VIEWPRT.DAT. This is equivalent to entering DIST, ANGL and TRSN in succession. For very large structures, this can produce a very large file.

HALT,QUIT,STOP These commands offer the user the way out of the program without having the plot instructions automatically submitted to the batch stream.

END Leave the program after submitting plot files for plotting.

Database techniques

17

FILE STRUCTURES AND SEARCH STRATEGIES FOR THE CAMBRIDGE STRUCTURAL DATABASE

Frank H. Allen and John E. Davies

Abstract

A new unified file structure and associated search software for the Cambridge Structural Database is described. Strict design criteria of portability, flexibility and efficiency have led to a serial search system, employing bit–screen techniques. The file structure uses three logical records per entry: (a) screens, (b) text and connectivity information, (c) numerical data. Only the screens need to be read for every entry, records (b) and (c) are skipped unless the screen record indicates that the entry is a potential hit. A basic alphanumeric query language, which extends previous conventions, uses a series of 'tests' of individual information fields. These tests may then be combined via logical operators 'AND', 'OR', 'NOT', to form the complete query. Optional extras will include direct access (instead of sequential) files, and graphical input and output. FORTRAN77 is used throughout.

1 Introduction

The Cambridge Structural Database (CSD) stores information relating to X-ray and neutron studies of 'organo-carbon' compounds, i.e. organics, organometallics and metal complexes. The information encompasses bibliographic and chemical text, chemical connectivity representations, and the primary numerical results of each diffraction experiment. The current database contains details of 58,239 complete diffraction analyses, of which 83.5% have atomic coordinates. CSD is fully retrospective, and is updated on a current basis with some 6,500 new entries per annum.

Software for search, retrieval, analysis and display of information has always been an integral part of the complete CSD system. For a number of years this has involved separate searches of bibliographic text or chemical connectivity representations, followed by retrieval and analysis of the relevant numerical data. In this chapter we describe the first part of a major upgrade of CSD software: the generation of a portable, flexible and integrated system for search and retrieval functions.

2 Current Csd Systems

2.1 Files and file structures

The Cambridge Structural Database (CSD) (Allen *et al*, 1979) is currently released as three discrete files of information:

(a) <u>BIB</u>: a text file containing chemical and bibliographic information, e.g. compound names, molecular formulae, authors' names, literature citation, classification.

(b) <u>CONN</u>: packed integers describing the chemical connectivity of the bonded residues of each crystal structure, in terms of atom, bond and residue properties.

(c) <u>DATA</u>: a mixture of numerical and text fields containing crystal data, symmetry, atomic coordinates, precision indicators, evaluation flags, and comment concerning the experiment, errors located in the publication, and details of any disorder.

Two versions of these files are supplied:

<u>Version 1</u> Formatted 80-byte records. These are converted at each user site into sequential files of variable-length, directory-controlled, unformatted (binary) records. Each record contains the relevant BIB, CONN or DATA information for a given file entry, for use with Version 1 software (described below).

<u>Version 2</u> Three discrete indexed-sequential binary files in VAX backup format for use with Version 2 software (described below).

BIB, CONN and DATA repectively represent *ca.* 12%, 25% and 63% of the whole database (a total of approximately 130Mb).

2.2 Current Software Systems

Version 1 (Allen *et al*, 1979) of the Cambridge software system was developed in the late 1970's as a portable FORTRAN IV system operating on the sequential files BIB/CONN/DATA. The program BIBSER performs character–by–character matching of query text with text in the BIB file, whilst CONNSER matches a substructural query connectivity with 'target' entries in the CONN file via a full atom–by–atom, bond–by–bond procedure. Each program generates printed output and a file of CSD reference codes corresponding to the 'hits'. These codes can be used to RETRIEVE a subfile of DATA for processing by the geometrical analysis program GEOM78, or by the plotting package PLUTO78.

More recently Version 2 software was developed from Version 1 specifically for DEC–VAX series computers. It employs FORTRAN77 and the indexed-sequential files permit very rapid retrieval of entries. Some interactive query input is provided, and geometrical and statistical capabilities are enhanced via program GEOSTAT (Murray-Rust and Raftery, 1985).

A number of other search philosophies have been applied to CSD data. For example the Crystal Structure Search Retrieval (CSSR) system (Machin, 1985) operates on a fully inverted file structure, consisting of a series of linked, ordered indexes of searchable items. Other users have applied home–produced or commercially written systems based upon screened files, in which a compact sequential file of bit–screens is used to decide which entries need to be closely examined.

2.3   Problems with Current Systems

Versions 1 and 2 of the Cambridge software both suffer from some basic defects:

(a) Searches are restrictive. Combined interrogation of BIB and CONN files is not possible, and many important DATA items (e.g. crystal data, symmetry, precision indicators, etc.) are not accessible.

(b) Searches are slow, due to the exhaustive approach to the comparison of query coding with information in the sequential files, i.e. every entry is regarded as a potential hit and every search must laboriously read the whole search file (no part of any entry is ever skipped). This problem increases with increasing file size (currently 12-14% per annum).

(c) Individual tests and the logical operators AND, OR and NOT cannot be combined with parentheses. This has created some confusion in the past: certain questions cannot be phrased at all and care is often necessary to define a question unambiguously.

(d) Present file structures are inflexible. New information fields in CSD (see Section 4) cannot be added easily.

(e) There have been problems with program portability. Many variants of Version 1 now exist, making software support difficult.

Some problems associated with fully inverted systems (Machin, 1985) are noted in Section 3.1 below.

3 The New Cambridge Search System: Version 3.1

3.1 Design Criteria

At the outset — during the initial planning for the new search file and its associated software — the following important design criteria were already established as a result of experience with the old BIB/CONN/DATA system:

1. Portability: The new system must accommodate the needs of many different users and it must run on a wide range of different machines.

2. Flexibility: Ideally, the search software should enable any search query to be constructed easily and unambiguously. The program should be able to test all data items within an entry and it should allow any selection of such tests to be combined with the logical operators 'AND', 'OR' and 'NOT' with at least one level of parentheses. It should be possible for a user with some programming experience to add extra (purpose-built) tests easily, without risking major damage to the whole program.

3. Efficiency: The search program should operate as efficiently as possible, within the constraints set by (1) and (2) above. To complete any search, it should not be necessary to combine and/or intersect the results of several different "sub-searches": one question should suffice.

4. Graphics: There should be graphics facilities for coding search fragments, displaying chemical diagrams for 'hit' entries, etc. These facilities should be optional ('layered') products, because some users lack graphics hardware.

5. Compatibility: As far as possible, instructions for the new program should be compatible with the instructions associated with the old BIB/CONN/DATA software.

6. Program language: The new software should be written entirely in FORTRAN77. The improved character handling facilities of this language (compared with earlier versions of FORTRAN) enable the program to be much more portable than the old BIB/CONN/DATA software.

While some important design decisions — like the choice of program language —were made quite easily, others were more difficult. Any attempt to satisfy simultaneously criteria (1)–(4) above inevitably leads to certain compromises. For example, the ability to create 'direct-access' or 'indexed-sequential' files can significantly improve search efficiency: unfortunately, not all users have the ability to do this and others cannot afford to maintain the necessarily large (60–150Mb) disc files.

Inverted files appear to be an attractive alternative, since they offer almost instantaneous access to CSD via interactive query construction. However (Machin, 1985) there are drawbacks here too:

(a) The interactive system can only search those fields which the designer considers to be 'useful' as search terms.

(b) Operation of the files and interactive access are machine dependent.

(c) There are problems caused by increasing database size. The computer time required to create the inversions can be considerable, as is the disc space needed to store the results. Since the files form a linked hierarchical structure, all files must be present in the system.

After carefully considering the factors outlined above, it was decided to adopt the common alternative strategy of serial searching using bit screens for the new search system. For some searches, this approach is less efficient than fully inverted systems but it does have distinct advantages in terms of portability and flexibility.

3.2   Query Screening via Bit–Maps

A screen may be regarded as a simple switch, which can take the value 1 or 0. '1' indicates the presence of a particular item of information in the entry, '0' indicates that the item is absent. Simple TRUE/FALSE switches of this type are ideally suited for storage as bit-maps: in a machine which can store 32-bit integers, 31 such bits can be stored in each word and used as switches or 'bit screens' (bit 32 stores a sign and cannot be used for a screen). So far (April 1987) 22 words of screen information (682 bits) have been calculated for each database entry. The new search program derives the corresponding screens from the query posed by the user. Each word of the query bit-map may be compared with the relevant word of the entry bit-map by a single FORTRAN77 instruction. For example, bit screens 1–31 in the CSD search software store information concerning element groups present. If an entry contains the elements C,H,Br,O,Rh,N then the integer I, containing all 31 element group screens for that entry, is 1750188288, that is —

$$I = 11010000101000111000001000000000 \quad (binary)$$

$$\uparrow \underline{\qquad} \quad bit\ 1$$

and this number will be stored within the SCREEN record of that entry. In this example, 9 bit screens are set, as follows:

Bit	Element group present
9	GroupY8 (Co,Rh,Ir)
15	GroupA5 (N,P,As,Sb,Bi)
16	GroupA6 (O,S,Se,Te,Po)
17	GroupA7 (F,Cl,Br,I,At)
21	GroupTr (transition metal)
23	GroupR2 (2nd row trans.metal)
28	GroupM4 (metal present)
30	GroupHD (either H or D)
31	GroupHH (H)

Note that the integer screen I is given by:

$$I = 2^8 + 2^{14} + 2^{15} + 2^{16} + 2^{20} + 2^{22} + 2^{27} + 2^{29} + 2^{30}$$

and also note that a single element may set more than one bit ('Rh' for example automatically sets bit screens 9,21,23 and 28). Now consider a search query designed to locate all entries which (a) contain a transition metal and (b) contain a GroupA6 element. Any 'hit' must have bit screens 16 and 21 set. Before testing any entry, the program will code this search by calculating a single integer J, as follows

$$J = 2^{15} + 2^{20} \qquad = 1081344$$

$$= 00000000001000010000000000000000 \quad (binary)$$

To test each entry, the program must decide whether all the bit screens set in J are also set in I. The FORTRAN77 statement —

$$IF(J-IAND(J,I).EQ.0)GOTO\ n$$

will do this very efficiently. This test will succeed for our example (I=1750188288, J=1081344) as both numbers have bit screens 16 and 21 set —

$$I = 11010000101000111000001000000000$$

$$J = 00000000001000010000000000000000$$
$$\uparrow \quad \uparrow$$

If the question is reversed (hit all entries with neither a transition nor a GroupA6 element), then all bit screens set in J must not be set in I and the program may decide this as follows

$$IF(IAND(J,I).EQ.0)GOTO\ n$$

3.3   The Screening System: An Overview

The use of bit screens in bibliographic and text databases is well known (van Rijsbergen, 1975), and the optimization of chemical structure screens has been the subject of considerable research effort (see e.g. Adamson, Lynch and Town, 1971; Dittmar *et al.*, 1983; Ash, Chubb, Ward, Welford and Willett, 1985). For CSD the 22 words so far assigned (April 1987) in the bit-map cover the following information types:

(a) Element groups (1 word) covering 31 groups of elements derived from subdivisions of the periodic table. Some examples of these settings are described above. Each group has a pseudo-element symbol which can be used throughout the query coding for Version 3.

(b) Entry information (4 words) covering such items as presence of non-mandatory fields, precision, evaluation flags, crystal data summary and experimental details.

(c) Text of author and compound name fields (3 words) based on hash-coding (Harrison, 1971) of contiguous letter pairs onto a 29 bit (authors) or 62 bit (compound and synonym names) 'signature'. The query text is analysed identically and the two signatures compared (via IAND).

(d) Connectivity screens (14 words) derived from many previously published schemes, in conjunction with a statistical analysis of CSD connectivity representations. Target structures in CSD are, of course, completely defined in chemical terms, whereas query substructures may be defined at increasing levels of precision. Thus the screening methods must be designed to cope with very general (Markush) fragments, involving specification of, for example, variable element and bond types, up to fragments having very few degrees of freedom. The screens employed are summarized under general group headings, which indicate an approximately hierarchical ascent from generality to precision:

(i)   Bonded elements and counts

(ii) Simple connections, any bond type

(iii) Bond types, cyclicity, acyclicity

(iv) Bonded pairs, bond type specified

(v) Bonded triplets

(vi) Atom-centred fragments

(vii) Aromatic sequences

(viii) Bond-centred units (quadruples and higher)

(ix) Miscellaneous (atomic charges, bond counts)

(x) Rings and ring assemblies

Because of the nature of the file, many connectivity screen definitions refer (decreasingly) to C, O, N, S, P, halogens; fragments involving rarer elements are very effectively screened via (a) above. Ring screens are of special importance as more than 80% of CSD entries contain one or more cyclic substructures. A complete ring analysis, within a limit of 50 rings per residue or query, is accomplished by extension of the algorithm of Wippke and Dyott (1975). This is based on Boolean combination of ring-bond sets, derived from a canonically renumbered connectivity tree (Morgan, 1965).

3.4   The Version 3.1 File Structure

The new file structure associated with Version 3.1 software involves three sequential logical records per entry:

(a) SCREEN is a short fixed-length record containing the bit maps already described, together with some essential integer information relating to commonly searched fields.

(b) TEXTCONN is a variable length record containing information which may be regarded as 'searchable', i.e. the bibliographic and chemical text, and the chemical connectivity tables.

(c) <u>DATA</u> is a second variable length record containing the bulk of the numeric information for each entry, i.e. symmetry operators, atomic coordinates, etc.

The structure of the Version 3.1 file may be summarized as:

Record	Item	
1	File Header	
2	SCREEN	1st entry
3	TEXTCONN	1st entry
4	DATA	1st entry
5	SCREEN	2nd entry
6	TEXTCONN	2nd entry
7	DATA	2nd entry

.

.

.

where the 'file header' identifies both the file and the information it contains. This file structure is a flexible one: the content may be tailored for a particular user or application and new information items may be added easily, as required. Finally the sequential structure described above can be stored on magnetic tape or disc, or (most importantly) the TEXTCONN and DATA records can be converted into direct-access files leading to increased search efficiency as described below.

3.5 Search Strategy

Let us now return to the query posed in Section 3.2: find all entries which contain a transition element and an element of Group A6. The program will search the file according to the following general strategy:

STEP 1.   Read and encode user's search query.

STEP 2.   Read HEADER from CSD file

STEP 3.   Read the SCREEN record for next entry
           Go to STEP 8 if last entry already read.

STEP 4.   If SCREEN eliminates this entry, skip
           TEXTCONN and DATA (if present) and go
           to STEP 3.

STEP 5.   Read TEXTCONN

STEP 6.   Examine entry in detail. If the entry
           is not a hit, skip DATA (if present) and
           go to STEP 3.

STEP 7.   Read DATA, print information for this
           entry and go to STEP 3.

STEP 8.   Wind up, print statistics etc

For the example query STEP 1 calculates the integer $J=1081344$ and is compared with the element-group SCREEN integer I for each entry at STEP 4. The efficiency of this search process depends upon the efficiency of steps 3-7, particularly the efficiency with which the SCREEN record can eliminate entries at STEP 4 (i.e. remove the need to read and examine TEXTCONN records). With a sequential file, this efficiency further depends upon how quickly a machine can skip TEXTCONN records (and DATA records, if present) — for on all machines it is quicker to skip than to read a record; one or two good screens which together eliminate the need to read past the SCREEN record of, say, 95% of all entries will significantly reduce the total cpu time. If the TEXTCONN and DATA records are direct access files, the time taken to skip these records disappears altogether and efficiency is further increased.

Note that with this type of serial searching, the efficiency of the program depends almost entirely on its ability to scan a comparatively small file of information, and the ability to use this information to eliminate a large proportion of entries (typically $> 90\%$) as potential search hits.

The simple bit-screen example above is sufficient to describe the general strategy adopted within the program to improve the efficiency of most searches. Many of the screens, especially those associated with text and connectivity searches, are activated automatically by the program during query analysis. Some must be user-activated, but all may be accessed individually, and some or all of them may be switched off if required.

3.6 Search Query Structure

When running in its basic mode (sequential file, no graphics), a search query is defined by a set of instructions (the 'instruction document'). This document defines one search query and may re-set some of the program's default parameters. Multiple searches (different unrelated questions in the same job) are not allowed: the program should enable any search to be completed with a single question.

For example, the following instruction —

SCREEN 16 & 21

will hit all entries containing at least one transition element and at least one group A6 element. Alternatively, the instructions —

T1 *ELEMENT Tr
T2 *ELEMENT A6
QUESTION T1.and.T2

define the same question using two separate element-group tests (T1,T2) combined with the 'AND' operation. The number of hits may be further reduced by adding extra tests in the same general way:

> eg. COMMENT All entries with group Tr & A6 elements
> COMMENT studied in temperature range 100-200K
> COMMENT by authors 'Cook' and/or 'Bligh'
> SCREEN 16 & 21
> T1 *TEMPERATURE 100-200
> T2 *AUTHOR 'COOK'
> T3 *AUTHOR 'Bligh'
> QUESTION T1.and.(T2.or.T3)

If given, the QUESTION line must occur once only and must be the last in the instruction document. It defines a search query in terms of the simple tests Tn, the logical operators '.OR.'/'.AND.'/'.NOT.' (or ','/'+'/'-') and parentheses ( ). It will be executed *after* any SCREen instructions and *only* if the SCREen tests succeed.

The search program will evaluate logical expressions in the QUESTION line according to the following order of precedence:

.NOT.	(or '−')	First (highest)
.AND.	(or '+')	Second
.OR.	(or ',')	Third

Operators of equal rank will be evaluated from left to right; unnecessary tests will *never* be performed. This means that, for fast searching, the most efficient tests should be executed first. Consider the following query:

> QUEStion (T1.or.T2) .and. (T3)

Here T1 will be tested first. If it succeeds T2 will not be executed and the program will skip directly to T3. If T1 and T2 both fail, T3 will not be executed.

Many different types of test may be defined via 'Tn *SUBKey —' lines: these include —

(a) Tests of 682 bit screens

(b) Tests of 19 text field types

(c) Tests of 38 integer field types

(d) Tests for elements and element/count combinations within sum formulae or residue formulae

(e) Connectivity tests for specified chemical fragments (based on original coding by W.D.S.Motherwell)

(f) Tests based on reduced cell parameters (based on original coding by R.F.McMeeking)

In particular, connectivity tests are specified with a

Tn *CONN

line, followed by a connectivity query coded in a similar manner to that required for the old CONNSER program.

eg. COMMENT Connectivity test example
T1 *CONN
Q FIND ALL NON-FUSED EPOXY RINGS
AT1 O 2 E
AT2 C 2
AT3 C 2
BO 1 2 1 C
BO 2 3 1 C
BO 1 3 1 C
NOCS
END
QUESTION T1

Facilities for fragment definition have been enhanced in the new software. These enhancements include: use of the complete range of 31 element group symbols (either singly, or in combination with element symbols); provision of a NOT operator for element definition; specification of coordination numbers and hybridization states; specification of terminal atoms as cyclic or acyclic; improved specification of the secondary environment of a fragment; searches for two or more discrete fragments within the same bonded residue. Interactive graphics facilities will be available as an alternative to the alphanumeric coding of query fragments.

Any test — including *CONNSER tests as described above — may be combined with other tests using AND, OR, NOT operators in the QUESTION line. For example, combining a formula test and/or an element test with a reduced-cell test can be a powerful way of locating an entry in CSD. Although one test alone (formula, element or cell) can probably locate the desired entry, the constraints or tolerances necessary to ensure a hit may produce an unacceptably large number of spurious hits. Two or more well-chosen tests operating together can often eliminate all spurious hits and — perhaps more importantly — can allow wider (more relaxed) constraints and tolerances for individual tests.

For example, suppose that an unidentified substance contains C, H, O, only, that analysis indicates a formula C(10) H(7–9) O(4) and that preliminary X-ray data indicate a primitive orthorhombic cell with rough dimensions a=11.65, b=13.86, c=11.07Å. A search for this substance may be constructed as follows —

```
T1 *PCELL ORTHorhombic 11.650 13.860 11.070 90 90 90 0.5
T2 *RESIdue C10 + H7–9 + O4
QUEStion T1.and.T2
```

where the final '0.5' in test T1 is a cell-edge search tolerance in Angstroms, and where the '+' symbols in test T2 indicate that no elements may be present in the RESIdue, other than those specified. This search will be one of the fastest that the program can execute (as test T1 operates on the SCREEN record and will eliminate most entries). In fact, it eliminates all but the one true hit (in the file dated April 1987) despite the generous cell and H-count tolerances. T2 alone takes four times longer to execute (Cambridge IBM) and produces nine hits while T1 alone produces 186 hits.

4 Summary

At the time of writing (April 1987), the new CSD search system already meets most of the design criteria listed in Section 3.1. The program is entirely FORTRAN77 and has been tested with IBM, Fujitsu and VAX compilers. Running in its basic mode (no graphics, sequential file) the only machine-dependent code is one very small subroutine which must accomplish an efficient character UP-case operation. The program has been designed to operate in several different modes — sequential or direct access file, optional graphics input and output — but in all modes any search must scan the complete SCREEN file (one SCREEN record per database entry): the time to scan all SCREEN records is therefore a fixed overhead and defines a minimum possible time for any search on a particular machine. This strategy is (hopefully!) a reasonable compromise between flexibility and efficiency. Some search times (seconds) for the New compared with the Old (BIB/CONN/DATA) systems are as follows, assuming a fragment search with 95% screenout in the new system:

	New	Old
IBM 3081 (Cambridge)	5	120
VAX 11/750 (CSD)	120	2100

The program includes several subroutines which enable a user to add purpose built tests easily and to modify the program output. In particular, a user may add tests without altering the tricky part of the program concerned with encoding and executing the AND/OR/NOT logic of the QUESTION line.

An important aspect of the new file structure is that it includes a number of information fields unavailable in the old system. Furthermore, the architecture of the new file allows new fields to be added easily in the future, if necessary. The new fields available in Version 3.1 are:

(a) Two-dimensional coordinates (x,y) of the atomic nodes of the chemical connectivity representation (a chemical diagram). This now permits us to add a graphics output routine (in preparation) which will produce high-quality chemical diagrams in response to a search query.

(b) Estimated standard deviations of individual atomic coordinates for post-1984 structures. These will be of considerable use in any extension of the geometrical analysis facilities.

(c) Results of a 1:1 computerized mapping of the chemical and crystallographic connectivity representations. This has been successful for about 80% of entries (disordered structures, polymers and entries containing coordinate errors present special problems). This too has very considerable implications for geometrical analysis in the future, since the crystallographic coordinates corresponding to the results of a chemical connectivity search can be accessed directly, without further re-location of the fragment in the crystallographic connection table.

The file structure already allows the DATA record for any entry to be 'retrieved' immediately the search program hits an entry. Eventually this factor, plus the 1:1 connectivity matching described above, will allow the program to (a) test DATA information such as bond lengths, torsion angles etc, (b) store the DATA records for all hits and (c) perform geometrical and statistical calculations — *all in the same job*, thus eliminating the need for separate RETRIEVE and GEOMETRY and/or STATISTICS steps.

REFERENCES

Adamson, G.W., Lynch, M.F., and Town, W.G. (1971). *J. Chem. Soc. (C)*, p 3702.

Allen, F.H., Bellard, S.A., Brice, M.D., Cartwright, B.A., Doubleday, A.,

Higgs, H., Hummelink, T., Hummelink-Peters, B.G., Kennard, O., Motherwell, W.D.S., Rodgers, J.R., and Watson, D.G. (1979). *Acta Crystallogr.* **B35**, 2331.

Ash, J.E., Chubb, P.A., Ward, S.E., Welford, S.M., and Willett, P. (1985). *Communication, Storage and Retrieval of Chemical Information.* Horwood, Chichester, UK, and Wiley, New York, USA.

Dittmar, P.G., Farmer, N.A., Fisanick, W., Haines, R.C., and Mockus, J. (1983). *J. Chem. Inf. Comput. Sci.*, **23**, 93.

Harrison, M.C. (1971). *Comm. Assoc. Computing Machinery*, **14**, 777.

Machin, P.A. (1985). *Crystallographic Computing* 3, (ed. G.M. Sheldrick, C. Kruger and R. Goddard), p 106. Oxford University Press, Oxford, UK.

Morgan, H.L. (1965). *J. Chem. Documentation*, **5**, 107.

Murray-Rust, P. and Raftery, J. (1985). *J. Mol. Graphics*, **3**, 50 and 60.

van Rijsbergen, C.J. (1975). *Information Retrieval.* Butterworths, London, UK.

Wippke, W.T. and Dyott, T.M. (1975). *J. Chem. Inf. Comput. Sci.*, **15**, 140.

18

PRINCIPLES OF DATABASES FOR CHEMISTRY

G.Bergerhoff

Seven millions of compounds in chemistry and more than 400,000 papers per year demand for new methods of information. Methods specially designed for the demands of chemists. This means that structure and properties are items which should be retrievable immediately and not only through a literature reference. Databases for facts or numerical databases are the instruments of the future which will free the scientist from the slavish search in the library.

Crystallography is a field where the importance of this instrument can be shown in a special way. There is a bulk of data for every structure and even the specialist cannot evaluate immediately from this data what he wants to know. In the meantime several well known groups collected data for crystallographic databases. We took care of the inorganic compounds. Thus we had the opportunity to demonstrate how we try to reach the five properties which a database of high quality should have. Databases should be 1. complete, 2. up-to-date, 3. correct, 4. versatile, and 5. user-friendly.

To be complete:

Once the user has recognized the advantages of a numerical database he wants to have all data in one place. Thus we try to be complete for inorganic structures in respect to the following principles. The ICSD contains all:—

1. Structures which have no C-C-, no C-H-bonds in any one of their residues. (Cambridge has the opposite definition). Structures which have incorporated at least one of the non-metallic elements. This means: H, He, B, C, N, O, F, Ne, Si, P, S, Cl, Ar, As, Se, Br, Kr, Te, I, Xe, At, Rn. (the metals file (Crystallographic Databases 1987) has in principle the opposite definition)
2. Structures, the atomic coordinates of which have been fully determined. Coordinates of hydrogen and vagabond ions like in zeolites can be missed.
3. Structures to which refer Strukturbericht, Structure Reports, Crystal Data and Landolt-Bornstein.

This aim now has been fully reached. The file contains 26 045 datasets. 22 056 of them contain the full numerical information. In spite of these numbers I am sure we have not all known structures because they are widely spread over many journals in many countries. Please check the database and help us to make it complete.

To be up-to-date:
Databases in principle have the possibility to be up-to-date if the new results reach them without delay. Many references we get through regularly scanning of Chemical Abstracts and of the ten most important journals:

Acta Crystallographica A, B, C
Journal of Solid State Chemistry
Zeitschrift für anorganische und allgemeine Chemie
Zeitschrift für Kristallographie
Zeitschrift für Naturforschung B
Journal of the Less-Common Metals

Materials Research Bulletin
Revue de Chimie Minerale
Kristallografiya
American Mineralogist

But the input could be much more up-to-date and complete if it would be possible to find an arrangement how data could reach directly the database producers without the long way through typescript ⇒published paper ⇒reference journal. It is a good thing that several journals such as *Acta Cryst., J.Chem.Soc. Chem.Comm., Angewandte Chemie* already cooperate with the database producers to provide them with data which normally will not be published. The annual increase of the ICSD is about 1200 structures but the time lag may be one year and more.

To be correct:

Information which is not correct is worthless. This is especially true for databases. The user not only wants to work with correct data but also a successful retrieval process depends on correct data. There are three kinds of errors.

1. Input errors, which are our fault and should be avoided.
2. Printing errors, which are the author's faults and are eliminated by correspondence.
3. Errors by misunderstanding of the authors intention. They make most trouble and could be avoided by strictly using the recommended standard data structure.

In any case it is impossible to eliminate all errors by means of inspection by eye. Therefore it is highly necessary to develop programs which allow the computer to check as many data as possible. For non-numerical data these programs are in development. Numerical data are strictly checked for consistency where the program uses the following scheme.

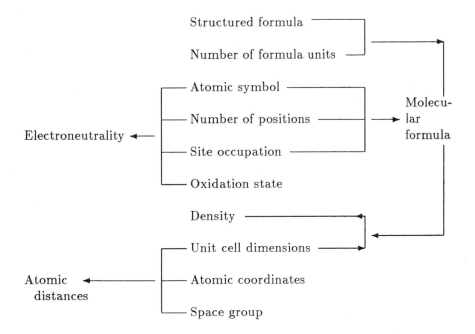

The check results are included into the database. Typical examples are: calculated density anomalous, given molecular formula different from formula calculated from number of positions and site occupation for each atom, no electroneutrality, temperature factors not plausible, atomic distances largely deviate from sum of ionic radii. Once more users of the database are asked to help to correct data when they are able to do so.

To be versatile:

Once all data have been collected and corrected as far as possible the database should answer all imaginable questions of the user. And this should be done in the most convenient manner: in a dialog retrieval at the computer terminal. For this reason Rolf Sievers and Rolf Hundt at our institute developed the retrieval system CRYSTIN. The system not only allows the access to the ICSD but also handles simultaneously the Cambridge and the Metals File in such a manner that the user does feel all crystallographic databases as one unit. Thus the border lines between the crystallographic databases, cited above, no longer exist for the user of CRYSTIN and he must not decide which area of chemistry his compound belongs to.

To formulate queries to the database the user can combine descriptors of various types by logical connections. The following table of descriptors gives the possibilities existing today.

    Chemical data
     Chemical element
     Chemical element and oxidation state
     Chemical element and stoichiometric index of molecular formula
   Element group
     Part of name and part of formula (by string search)
     Number of different elements
     Mineral name
     Origin of minerals (by string search)
     Refcode (Cambridge Crystallographic Data File only)
     Chemical Class (Cambridge Crystallographic Data File only)
   Structural data
     Unit cell volume
     Formula type (e.g. AB, A2B3, A2BX4)
     Pearson symbol
     Defect structure (in general or polytype, twinned and modulated structures, mixed crystals)
     Shortest distance between two atoms in a structure
   Symmetry data
     Space group (by Hermann-Mauguin symbol or number of IT)
     Crystal class
     Laue class
     Crystal system
     Polarity and centrosymmetry

Methodic data

    Reliability index

    Diffraction method

    Temperature and pressure of measurement

    Test result

Bibliographic data

    Author

    Journal (through CODEN or title or country)

    Publication year

This table will be expanded and any proposal is welcome. The attentive reader will miss one descriptor especially important for inorganic chemists: the structure type. We omitted this with full intention. Any insider knows the great confusion existing in this field caused by the lack of standardization of structure descriptions. We intend to incorporate into the database system the method for structure standardization developed by Parthe and Gelato (1984). Doing this we follow the line to avoid descriptors not evaluated from the data by programs. Because descriptors should be as correct as data should be.

To connect descriptors and find out the results the following commands exist:

display	show all data of a desired descriptor and their frequencies in alphabetical order on the display
find	define subset by expressions built from descriptors by logical (Boolean) operators including ranges
string	select datasets with desired character strings in the defined subset
save	store the defined subset
show	show desired categories of the defined subset on the display
dist	calculate atomic distances and bonding angles for specified atoms and show them on the display (or store them on independent device)

plot   draw stereopicture of the structure either automatically or after input of specifying commands

findcodn   find out journal CODEN from a journal title or from parts of it or find out full journal title for a given CODEN

help   give short information for all commands, categories or descriptors in German or English (a detailed printed manual in German and English is available)

Of special importance is the command print. It allows not only to print all wanted information from the display it also can transfer data to a special file in a fixed format. Thus the user can connect his own program to evaluate data as he wants.

Categories which can be calculated and displayed or printed give the next table:

Name of Compound

Mineral name and origin

Chemical formula in structured form

Title of publication

Authors

Citation with full journal title or CODEN, volume, year of publication, first page, last page, (issue number)

Unit cell dimensions, unit cell volume, number of formula units

Measured density

Space group symbol (Hermann-Mauguin)

Atomic parameters

Atomic symbol

Oxidation state

Number of positions and Wyckoff symbol

Atomic coordinates

Site occupation

Isotropic or anisotropic temperature factors

Reliability-Index

Remarks about method of measurement, specialities of the structure, etc.

Atomic distances

Bonding angles

Stereopicture (in many cases)

Powder diagram

Test results

Refcode (Cambridge Crystallographic Data File only)

Chemical Class (Cambridge Crystallographic Data File only)

In our experience the handling of the system can be learned in a few minutes. Some more time is necessary to think how to build the expressions from descriptors in the best way. This topic will be the main purpose of the database course. Some examples may illustrate the way. The number of answers from ICSD (Release 2.8) or CCDF (Release May 84) you find in brackets.

Chemical queries:

What is the structure of μ-Carbonato-di-μ-hydroxo-bis(triamminecobalt(III)) sulfate pentahydrate?

```
find c and o and h and n and co+3 and s and elc=6     (5)
string mue name                                       (4)
```

What is the structure of a alloy with aluminium, iron and samarium and the stoichiometry 2:7.5:9.5?

```
find al and fe and sm and elc=3                       (1)
```

What is the structure of Benzo-2,1-3-selenadiazole?

```
find c and h and n and se and elc=4 and chcl=42       (9)
string diazole name                                   (2)
```

In which binary oxides iron has several oxidation states?

```
find O and ((fe+2 and fe+3) or (fe+2.03 to fe+2.97))
        and elc=2                                        (24)
```

What is the structure of the artificial zeolite A?

```
display minr=zeo        (to detect the correct writing)
find minr=zeolite                                    (222)
string ' A ' m                                        (56)
keep
string artificial m                                   (42)
```

Structural queries:

Which simple cyclo-phosphates have been investigated?

```
find p and o and met and elc=3                       (340)
string cyclo name                                     (11)
```

Are there structures of the type ABX3 with defects in the anionic sublattice?

```
find anx=abx3 and rem=defs                            (58)
show p        (and inspect)
```

Which structures related to CsCl are rhombohedral distorted?

```
find sgr=143 to 167 and anx=ax                        (39)
show p     (and inspect coordinates by eye)
find last and (alk or Fe)        (only relevant)       (7)
```

How often the space group P213 has been realized in case of ternary fluorine compounds?

```
find sgr=p213 and f and elc=3                          (1)
```

Can one find a significant difference in symmetry of the carbonate group between simple carbonates of the main group and the transition elements?

```
find c and o and (ale or alk) and elc=3
string carbonate n                                    (34)
save main
find c and o and trm and elc=3                        (94)
string carbonate m                                     (8)
dist angles from c to o
find usrn=main
dist angles from c to o
```

Analytical queries:

Which cubic crystallizing halogenides have a cell dimension of about a = 5.3Å?

```
find hal and syst=cub and cvol=144                    (13)
s e f n        (inspect)
```

Physical queries:

Which iron sulfides could be ferroelectric?

```
find sypr=pol and fe and s not o                      (10)
```

Methodical queries:

Are there very good investigations about N-H-O-hydrogen- bridging bonds by neutron diffraction at low temperatures?

```
find n and o and h and rem=tem and (rem=nds or rem=ndp)(16)
find last and d=n-h-2.0 and d=o-h-2.0                  (9)
find last and rval=0 to 0.05                           (5)
dist from h to n o dmax=2.5
```

Bibliographical queries:

Which papers N.V.Belov has published in American Mineralogist between 1960 and 1970?

```
findcodn mineralogist   (to find out CODEN)
display aut=bel      (to detect correct writing)
find codn=ammia and aut=belov and year=60 to 70        (1)
```

To find an answer to a query one should not always try to specify the query as strong as possible. Instead of this one should make a compromise between the expenditure on time to formulate the query and the number of answers which could be expected. Having done the find command the system always gives the number of true answers. Then the user can decide if he wants to restrict his query.

Further on it happens that the user will be astonished about the big number of answers which he receives. He didn't realize that his query would cover a much wider field than he intended. On the one hand some show commands will help him to exclude the not wanted answers. On the other hand this way has some effect of browsing in so far as the user gets more information than he had expected.

This may be demonstrated by the following example:
Determined structures of sodium sulfates and their hydrates

```
find na and s and o and h and elc=4 not rem=nprm       (25)
found formulae:   Na H S O4              Na3 H (S O4)2
                  Na2 S O4 (H2 O)10      Na H S O4 (H2 O)
                  Na2 S2 O3 (H2 O)5
                  Na2 S2 O6 (H2 O)2      Na2 S4 O6 (H2 O)2
                  Na2 S (H2 O)5          Na2 S (H2 O)9
 find na and s and o and elc=3 not rem=nprm             (15)
found formulae:   Na2 S O3               Na2 S O4
                  Na2 S2 O4              Na2 S4 O6
```

I think there are unexpected formulae as well as those formulae which could be expected but have not been determined until now.

Finally it will also happen that the user cannot find what he has in mind. Unsuccessful queries only based on checked data (see above) show that the database has no relevant entry. In all other cases the user should check the asked descriptor through the display command because e.g. names can be spelled very different.

To be user-friendly:

Access to the Inorganic Crystal Structure Database can happen in two ways:

1. Direct access at the host INKA. All what you need is an accession number. You will get it from Fachinformationszentrum Energie Physik Mathematik, D–7514 Eggenstein-Leopoldshafen 2. The connection is possible by simple acoustic coupling of your telephon apparatus or much cheaper by coupling your terminal to a data network like DATEX-P (in FRG), PSS (in Britain), EURONET (in Europe), Transpac (in Canada), STN (Scientific and technical information network)(in USA, Japan). For occasional use it is the best way because you always access the newest version of the database without any trouble for implementation etc. The common use of all three crystallographic databases is restricted to certain users*.

2. For multiple use of the database and if you wish to join your own evaluation programs it would be better to implement the system on your own mainframe computer. The implementation has already been checked for computers of IBM, Siemens, VAX. But be aware that it needs 33 Mbyte storage capacity. If you also want to add the Cambridge Crystallographic Data File, it needs 120 Mbyte more.

The considerable work of updating, checking and programming could only be done by the help of numerous scientists and students in Bonn and outside. Thus we got data from Clausthal, Darmstadt, Delft, Erlangen, Gottingen, Hamilton, Marburg, Osaka and Parma. But as mentioned above all crystallographers can help to complete and to improve the database by giving additional structures and corrections. Continuation has been ensured by the Fachinformationszentrum Energie Physik Mathematik if the scientific community will accept databases by intensive use.

This use is not restricted to simple queries which give you an overview about the literature. At long term it is our aim to expand the system

---

*for conditions write to Fachinformationszentrum

in such a way that all questions which chemists ask the structure of a compound can be answered by programs. Sometimes such systems are called 'Expert systems'. In the moment we want to be more modest. Even the automatical drawing of a structure is no simple task with lattice structures. To recognize isotypic structures is another old task in crystal chemistry which could probably be solved by strict standardization of the description of the structure. In other structures it might be more valuable to find strong bond systems like in silicates. A program should be able to evaluate from coordinates only a layer, a framework, or a chain structure. Finally we know that electric, electronic, magnetic, mechanical, optical, and many other properties depend on the structure of the crystal; but just not only on the crystal structure but also on the structure of the crystals. This includes the microstructure, the imperfections, the stacking faults, the grain sizes, the interfaces, etc. With the Crystal structure database we have a real base for all these endeavours but we are still in the beginning and many things have to be done.

REFERENCES

Crystallographic Databases (1987). (ed. F. H. Allen, G. Bergerhoff, and R. Sievers). Data Commission of the International Union of Crystallography, Chester.

Parthe and Gelato (1984). *Acta Cryst.* **A40** 169–183.

19

MULTIVARIATE ANALYSIS OF STRUCTURE DATA

Karel Huml and Wolfgang Hummel

1 INTRODUCTION

A growing number of entries in data banks leads to the situation that a chemical fragment under study may occur in several hundred published crystal structures. Therefore any detailed study of the fragment necessitates the use of multivariate statistical methods. Some of them will be discusssed here briefly in connection with a study of static and dynamic properties of molecules.

2 INPUT STRUCTURE DATA

2.1 Level of measurement

When we talk about measurement, we usually mean assigning of numbers to observations in such a way that the numbers are amenable to analysis by manipulation or operation according to certain rules. In order for us to be able to make certain operations with numbers that have been assigned to observations, the structure of our method of mapping numbers (assigning scores) to observations must be isomorphic to some numerical structure which includes these operations. If two systems are isomorphic, their structures are the same in the relations and operations they allow, i.e., they are isomorphic to their algebra.

The theory of measurement consists of a set of separate

theories, each concerning a distinct level of measurement. The
operations allowable on a given set of scores are dependent on
the level of measurement achieved. Usually four levels of
measurement—nominal, ordinal, interval and ratio are discussed
(Stevens, 1946; Siegel, 1956, p.21; Sachs, 1974, p.107; Ježek,
1976; Lukasová & Šarmová, 1985, p.19). The nominal
(classificatory) scale is the weakest one when numbers or other
symbols are used simply to classify an object, or a characteristic. The only relation allowed is that of equivalence. The
ordinal (ranking) scale incorporates not only the relation of
equivalence but also the relation "greater than". We speak about
the interval scale when a scale has all the characteristics of
an ordinal scale, and in addition the distances between any two
numbers on the scale are of known size. Finally, when a scale
has all the characteristics of an interval scale and in
addition has a true zero point as its origin, it is called
a ratio scale. For details in connection with hypothesis testing
see (Huml, 1986).

2.2 Crystallographic databases

There are now seven major crystallographic databases (Table I)
with areas of interest which can be delineated on crystallographic or chemical criteria (Allen, 1984; 1985):
- Powder Diffraction File (PDF) : JCPDS International Center for
Diffraction Data, Swarthmore, Pa., USA : Powder patterns, cell
parameters etc.
- Crystal Data File (CDF) : National Bureau of Standards,
Washington, D.C., USA : Crystal data from powder or single-
crystal studies.
      These two sources are chemically comprehensive, but have
limited structural information (Mc Carthy, Mighell, Hubbard &
Nichols, 1984); the additional atomic coordinate data are
available in five chemical divisions:
- Cambridge Structural Database (CSD) : CCDC, Univ. of

Cambridge, UK : Organics, organometallics, complexes (Taylor, 1985).

- Inorganic Crystal Structure Database (ICSD) : Univs. of Bonn, FRG/McMaster, Canada: Inorganics (Bergerhoff, 1985; Bergerhoff, Hundt, Sievers & Brown, 1983).

- Metal Data File (MDF) : National Research Council of Canada, Ottawa: metals, inter-metallics (CISTI, 1986).

- Protein Data Bank (PDB) : Brookhaven National Lab., Upton, NY, USA : Macromolecules (Abola, Bernstein & Koetzle, 1985).

- OD Data Bank (ODDB) : Zentralinstitut für physikalische Chemie, Berlin-Adlershof, GDR: order-disorder grupoid etc. (Backhaus, Grell & Schrauber, 1984).

TABLE I

Databases

Name	Number of entries	Increment per year	Year of information
JCPDS PDF	45,000	2,000	1986
NBS CDF	70,000		1984
CSD	56,000	4,000	1987
ICSD	26,000	1,000	1987
NRC CRYSMET	10,000	300	1987
BROOKH.-PDB	300	20	1987
ODDB	160		1987

For our further discussion we shall refer mainly to the CSD which has the most relevant data to our analysis. It is also the fastest growing structure databank, where we estimate that 100,000 structures will be reached before the end of the century (Taylor, 1985; Allen, Kennard & Taylor, 1983). Software system

CAMAL which is part of the CSD utility programs gives also an
ample list of statistical methods available for our purposes
(Taylor, 1986b).

Similar trends can be found in other databases as well
(Altermatt & Brown, 1985; Brown & Altermatt, 1985).

2.3 Coordinates

The existence of structure within multidimensional data sets
depends largely on the variables that span the multidimensional
space and on the metrics defined in this space. Furthermore, for
classification procedures one has do decide on the meaning of
similarity and dissimilarity of categories. Leaving out relevant
variables naturally makes a meaningful analysis impossible.
Adding variables that are not very relevant to the purpose of
the analysis but induce nevertheless a partition of the data is
clearly misleading (Naumann & Schiller, 1984, p.H1). A proper
choice of the system of coordinates is problem dependent and
should reflect the symmetry of the system.

From theoretical mechanics it is known that a set of N
independent points in 3-dim space has 3N-6 degrees of freedom.
Hence, geometry of a set of N atoms should generally be
described by 3N-6 mutually independent coordinates which are
also called parameters or descriptors of the system.

Coordinates can be classified in different ways. We shall
distinguish between external coordinates and internal ones.
External coordinates are defined usually by unit cell vectors or
other set of vectors dependent on them. On the other hand,
internal coordinates are defined by the molecular geometry
itself and are invariant to rotation and translation of the
fragment, e.g., bond distance and bond angles. Other criterion
is linearity. We say that a coordinate system is rectilinear
when the basis vectors have fixed direction and lengths, e.g.,
the Cartesian system, and curvilinear systems, when the
directions and lengths of the basis vectors may change from

point to point, e.g., cylindrical and spherical system (Sands , 1982).

In this study we shall recognize different classes of geometrical parameters according to the number of atoms involved in the calculation of an individual parameter. However, in general also other parameters than geometrical may be included into the analysis. For details and application see the adjacent references.

(a) Parameters of individual atoms

i. Rectilinear coordinates:

- Fractional (secondary, crystallographic) coordinates - related to unit cell vectors. External type of coordinates.

- Cartesian coordinates - as external when derived from fractional coordinates, or internal when related to internal molecular axes (Murray-Rust & Raftery, 1985).

- Orthogonal (orthonormal) homogeneous coordinates - usually derived from fractional coordinates. Used in graphics (Newman & Sproul, 1979; Granát & Sechovský, 1980).

ii. Curvilinear coordinates:

- Spherical coordinates - for angular distribution of bonds, etc. (Sands, 1982, p.186; Taylor, Kennard & Versichel, 1983; Einspahr & Bugg, 1980; 1981).

- Cylindrical coordinates - for helical molecules, etc. (Sands, 1982, p.146; Holmes & Blow, 1966, p.210).

(b) Parameters given by couples of atoms

- Matrix of interatomic distances - contains usually Euclidean distances between couples of atoms (Liebman, 1982; 1985). For the calculation of Cartesian coordinates from the matrix of interatomic distances see (Crippen & Havel, 1978).

- Matrix of intermolecular distances - similar to the previous case but deals with a set of molecules (Murray-Rust & Raftery, 1985; Albertsson & Svensson, 1978).

(c) Parameters given by more atoms

- Valence (Chemical, primary, intrinsic) coordinates: bond

distances, bond angles and torsion angles (Murray-Rust & Raftery, 1985; Norskov-Lauritsen & Bürgi, 1985). Transformations from and to Cartesian coordinates are given by Gavuzzo, Pagliuca, Pavel & Quagliata (1972) and by DiNola & Brosio (1982).

(d) Generalized coordinates

Combinations of other (usually chemical) parameters. Transformation from and to Cartesian coordinates are given by Mackay (1974).

- Principal components - dimension reduction with minimum loss of information of configuration (Murray-Rust & Bland, 1978; Murray-Rust & Raftery, 1985; Morrison, 1976, p.266; Kowalski & Bender, 1972; 1973).

- Symmetry (normal) coordinates - description of molecules slightly distorted from a symmetrical reference configuration (Murray-Rust, Bürgi & Dunitz, 1978a; 1978b; 1979; Murray-Rust & Raftery, 1985).

- Out-of-plane bending and twist - linear combination of torsion angles (Winkler & Dunitz, 1971; Bürgi & Shefter, 1975; Dunitz, 1979, p.329).

2.4 Data matrix

Let us assume that we have a set of N molecules, where the k-th molecule (k=1,2,...,N) is described by its column vector of P coordinates $(x_{1k}, x_{2k}, \ldots, x_{Pk})^T$. Then the complete data set is given by the data matrix, M, of the dimension (P,N)

$$M = (x_{jk}); \quad j = 1,2,\ldots,P; \quad k = 1,2,\ldots,N.$$

Each molecule can be represented as a point in a P-dim parameter (configuration) space (Bürgi, 1986). Similarly, each property given by its row vector can be represented as a point in an N-dim space.

If we assume that a row vector is a statistical sample of a size N, we can estimate characteristics of the corresponding sample distribution, incl. sample mean, sample variance, and sample covariance. To simplify further calculations standardized data matrix, Z, is usually expected, where

$$z_{jk} = \frac{x_{jk} - m_j}{s_j}$$

with the sample mean, $m_j = \sum_{k=1}^{N} x_{jk}/N$ and the sample variance $s_j = \sum_{k=1}^{N} (x_{jk} - m_j)^2/N$.

Elements of the Pearson correlation matrix $R = ZZ^T/N$.

2. 5 Symmetry

Often the structure problem has an inherent symmetry. The structural data sample retrieved from the data bank, however, usually does not show this symmetry precisely. Particularly small samples may not reveal the inherent symmetry at all and false results can be derived from the data. Therefore careful treatment of the symmetry at the very beginning of the analysis is necessary.

Application of symmetry operations can be described in two ways. We can think of a data matrix constructed from the original data and subsequently act on by mathematical programs which include corresponding symmetry operations. But almost none of the available statistical program packages treats symmetry explicitly. This deficiency may be compensated for by introducing the symmetry into the raw data and constructing the data matrix from the transformed data (Norskov-Lauritsen & Bürgi, 1985). To elucidate this idea several examples are given below.

a) Aperiodic distributions

Murray-Rust (1982) discussed the case of a linear triatomic XYX fragment in different crystal and molecular environments. It was supposed that each case is completely described by two parameters, the bond distrances r_1=XY and r_2=YX, which in general will not be equal. If a large number of cases were taken, we would expect r_1 and r_2 to have identical distributions, since there is no way of distinguishing between them. The

labeling problem can be treated by explicitly applying the
operation of permutation group S_2 to the input data, i.e.,
permuting the labels i and j of r_i and r_j. Then we proceed in our
analysis with the transformed data matrix containing twice as
many points as the original one.

Murray-Rust (1982) published a study of distortions of
tetrahedral molecules as an illustration of a distribution with
high symmetry which can be described by permutation group S_4.
Before proceeding to further analysis the point group T_d, which
is isomorphic to S_4, was applied to the input data.

b) Periodic distributions

Molecular transformation studies, e.g., conformational
interconversions, lead to periodic multidimensional distributions
of data points due to the use of torsion angles as the main
parameters of the fragment. The permutation group approach
described above for aperiodic distributions will not work in
these cases. The space group approach, however, does provide
an appropriate description of the symmetry of periodic
distributions (Dunitz, 1979, p.465; Bürgi, Dunitz & Shefter,
1973). This approach considers the fragment as a rigid frame
with one, two, or more rotating groups attached to it. The
rotational periodicity of these groups transforms to
translational periodicity of the n-dim parameter space, and the
symmetry of the parameter space can be described in terms of
an isomorphic n-dim space group. General positions of this space
group correspond to a set of isometric structures, and special
positions correspond to structures with higher symmetry (Dunitz,
1979, p.450). The picture produced by this method may be called
a conformational map.

A two-dim example: Diphenylmethane. This is a molecule
with two torsional degrees of freedom. It is convenient to
regard the three central carbon atoms as a rigid frame (having
C_{2v} symmetry) and the phenyl groups as 2-fold rotors. The
overall symmetry is then given by a semi-direct group product

H = R^F, where F is the frame group and R corresponds to flips (rotations of phenyl rings by 180°). The conformational map has the symmetry of the plane group cmm (Dunitz, 1979, p.450).

A three-dim example: Triphenylmethane. An arbitrary conformation of the molecule is specified by three torsion angles, each referring to a given two-fold rotor. The frame symmetry in this case is C_{3v}, and the conformational map has the symmetry of the three-dim space group R32 (Dunitz, 1979, p.466).

Available statistical analysis packages do not automatically recognize identity given by periodicity in the parameter space. To avoid truncation effects we have to include more than one period in each of the n dimensions. A drawback of this procedure is an increasing demand on computer time. Especially in high dimensional parameter spaces it is useful to construct a torus in the 2n-dim space by bending each parameter axis into a circle with a circumference corresponding to one period. Coordinates are then to be expressed in terms of trigonometric functions. This only doubles the amount of the data to be analyzed. $XYM(PPh_3)_2$ is an example of two groups with three mutually interlocked 2-fold rotors. The conformation of a fragment is represented by a point in the 8-dim parameter space, one dimension for each of the torsion angles. Truncation effects were avoided by transforming the problem into the 16-dim space, where the data points lie on a 8-dim surface of a torus (Norskov-Lauritsen & Bürgi, 1985).

3 STATIC PROPERTIES OF MOLECULES
This topic covers methods of representation, classification, and quantification of differences in the geometry of a set of fragments. Furthermore, special interest is paid to correlations between molecular parameters and other physicochemical properties of the molecules.

A general strategy of data analysis can be recommended:
First of all, the inherent symmetry of the problem under

investigation should be treated as described in the previous
chapter. After that, the first insight can be reached by plotting
one-dim histograms and two-dim scattergrams revealing some
information about possible clusters and mutual interdependence
of a chosen couple of variables. Visual display of the data
should be employed on any possible occasion as the human eye is
the best pattern recognizer in the familiar 2- or 3-dim space
(Kowalski & Bender, 1972; 1973). In cases of high dimensionality
cluster analysis has to be used to find a (small) number of
groups of fragments with widely differing conformations (Taylor,
1986a). If there is any suspicion that some of the clusters are
weakly connected, then factor analysis and scatterplots of
factors can be carried out for selected aggregations of clusters
(Murray-Rust & Raftery, 1985). For pitfalls of cluster analysis
see Schweizer (1985).

Three types of clustering will be considered separately.

3.1 A single cluster

In this case all molecules have nearly the same geometry; i.e.,
they can be represented by small deviations from a mean geometry.
We are usually interested in finding a mathematical model of the
data distribution and in estimating its parameters, mainly the
vector of the mean parameters and variance-covariance matrix.
a) Mean values of fragment parameters characterize the "standard
geometry" or "preferred conformation" of the fragment. It is
invaluable in model building, parametrization of empirical force
fields (Taylor, 1986a), phase problem solution, rigid body
refinement (Sheldrick & Akrigg, 1980) and interpretation of the
final structural data (Taylor & Kennard, 1982; 1983; 1986; Allen,
Kennard & Taylor, 1983; Allen, 1986).
b) Variances characterize the width of the corresponding
potential energy well.
c) Covariances may show interrelations between the parameters
studied. High linear and/or non-linear correlations reflect

possible physicochemical relations which control the link between
the structure and properties of the molecule.

Significant correlation may be exploited in two ways:
i. It can be explicitly expressed in an analytical form using
the methods of regression analysis.
ii. It can be used to reduce the dimensionality of the problem
removing correlation in the data. This may be achieved by linear
mapping techniques such as factor and principal component
analysis (Murray-Rust & Bland, 1978) and/or by nonlinear mapping
techniques (Kowalski & Bender, 1972; 1973).

3.2 Several well-resolved clusters
The molecules in the set have several distinct conformations,
i.e., they can be represented by clusters where the intercluster
variance is much higher than the intracluster variance. Clustering techniques should be applied first and then followed by
methods applicable to the individual clusters separately, as
mentioned above. For a survey of relevant methods see Murray-
Rust & Raftery (1985) and Taylor (1986a).

3.3 Clusters are diffuse and tend to overlap
There may be a large, continuous, variation between widely
different geometries (Murray-Rust & Raftery, 1985). The
situation may be difficult to analyze automatically. A typical
example is the reaction path, which will be discussed in the
next chapter. Bridges between clusters may be expressed in an
analytic form using methods of regression analysis.

4 DYNAMIC APPROACH
The study of structural data gathered in the databanks can also
provide important information about the dynamic aspects of
molecular structure. Particularly about relative motions of
atoms during molecular transformations, i.e., about conformational and configurational changes along the reaction path.

In this chapter we shall describe how to extract this dynamic information from crystallographic structural data by the structure-correlation method introduced by Bürgi & Dunitz (1983) and extensively used by other authors (Bertolasi, Bellucci, Ferretti & Gilli, 1984; Gilli, Bertolasi, Bellucci & Ferretti, 1986).

4.1 Structure correlation principle

The basic assumption behind the structure-correlation method is that a distribution of sample points corresponding to observed structures will tend to be concentrated in low-lying regions of the (Born-Oppenheimer) potential energy surface (Bürgi & Dunitz, 1983). This can be true when the interaction energy between the molecule or molecular fragment of interest and its various crystal or molecular environments is a small perturbation relative to the total molecular potential energy. Distributions where the sample points are concentrated in a narrow deformation path corresponding to a steep-sided energy valleys are in a good agreement with this assumption. On the other hand, significantly scattered sample points witness wide, shallow valleys, where relative strong perturbations due to the environment take part. Then the scattergram maps response path rather than the minimum potential energy path. In such a case the basic assumption is not fulfilled and the structure correlation method cannot be recommended.

These ideas were formulated by Bürgi (1973) and Dunitz (1979) as the Structural Correlation Principle: If a correlation can be found between two or more (geometrically) independent parameters describing the structure of a given structural fragment in various environments, then the correlation function maps a minimum energy path in the corresponding parameter space.

In this article we shall use the term "correlation function" equivalently to the deformation path and the minimum energy path equivalently to the minimum energy reaction path, or simply to the reaction path (Eyring & Polanyi, 1931; Frost & Pearson, 1961).

4.2 Conformational interconversion

The structure correlation method has been applied to map reactions for several types of the isomerization process.

a) First, let us show several examples where the molecular parameters have not a cyclic character. We have to mention here an analysis of the Berry process (Muetterties & Guggenberger, 1974; Bürgi & Dunitz, 1983; Auf der Heyde & Nassimbeni, 1984). Similarly, it was the pericyclic closure which was described by Bürgi, Shefter & Dunitz (1975). Structural studies of crystalline enamines are given in (Brown, Damm, Dunitz, Eschenmoser & Hobi, 1978) and discussed by Müller (1980). Extensive attention was paid to the stereoisomerization of the $R_1R_2C-NR_3R_4$ fragment by Gilli and coworkers (Gilli & Bertolasi, 1981; Gilli, Bertolasi & Belucci, 1986; Gilli, Bertolasi & Veronese, 1983). The rearrangement process in ML_9 complexes was described by Guggenberger & Muetterties (1976).

b) The case of cyclic molecular parameters is an interesting one. An analysis of two 2-fold rotors is represented by study of diphenylmethane (Dunitz, 1979). Two 3-fold rotors are represented by propane, di-tert-butylmethane and bis(9-triptycyl) methane (Bürgi, Hounshell, Nachbar & Mislow, 1983). Triphenylmethane is the case of three 2-fold rotors (Dunitz, 1979; Bye, Sweizer & Dunitz, 1982). Chandrasekhar & Bürgi (1983) published an example of three 3-fold rotors like Wilkinson's catalyst. Tetraphenyl-methane is a typical case of four 2-fold rotors (Dunitz, 1979). Relatively complicated is the square planar $XYM(PPh_3)_2$ molecule which represents two groups of three 2-fold rotors characterized by eight torsional degrees of freedom (Norskov-Lauritsen & Bürgi, 1985).

4.3 Chemical reaction path

Under this heading we shall talk about the structure correlation method in connection with a study of configuration and constitution changes.

a) Systems XYX were studied both theoretically and experimentally. Results concerning O-H...O are given, e.g., by Schuster, Zundel & Sandorfy (1976). Similar distributions of structural fragments are obtained for opposite bond distances in other three-center-four-electron systems X-M...Y where X,Y, are the Lewis bases (nucleophilic centers) and M is a Lewis acid (electrophilic center). The cases I_3^- and S-S...S were given by Bürgi (1975a; 1975b). There are also several examples of associative ligand-exchange reactions which occur during the Walden inversion (S_N2 displacement with inversion of configuration), e.g., the interaction of hydride ions with methane. For a review see Müller (1980). Pathway for the S_N2 reaction at Cd(II) was published by Bürgi (1973), for the S_N2 and S_N3 substitution at Sn(IV), by Britton & Dunitz (1981), and for the S_N1 reaction of tetrahedral molecules, by Murray-Rust, Bürgi & Dunitz (1975). For a discussion of the nucleophilic addition/elimination at the carbonyl carbon see lecture notes by Bürgi (1986) and papers by Bürgi, Dunitz & Shefter (1973; 1974).

b) Special interest was taken in a study of the directional preference of atomic approach to the molecules. Britton & Dunitz (1980) studied the approach of nucleophiles to sulfonium ions, Chakrabarti & Dunitz (1982) of ether O atoms towards alkali and alkaline earth cations, Murray-Rust & Gluster (1984) that of H atoms to sp^2- and sp^3- hybridized O atoms, and Wallis & Dunitz (1984) studied the approach of directions of the nucleophilic attack on N,N triple bonds.

c) Until now we have discussed the application of the structure correlation principle to different systems, where the structure-structure relationships, between geometrical parameters, such as bond lengths and angles, were studied. But there are also systems where exist sufficient kinetic as well as structural data to take the analysis a stage further, and to describe the structure-reactivity relationships. Kirby and coworkers (Briggs, Glenn, Jones, Kirby & Ramaswamy, 1984; Allen & Kirby, 1984; Jones & Kirby, 1984) described the correlation between bond length and reactivity of acetals and glucosides.

4.4 Interactions of molecules and macromolecules

Finally, we should mention a group of rather ambitious efforts to simulate interactions of molecules with macromolecules, usually of biological importance. Most of the problems have been solved by calculating the conformational energy (Šantavý & Kypr, 1984; Karfunkel, 1986). But there have also been attempts to use the information from structural databases to solve problems of docking, intercalation, etc., in combination with other methods, including graphics, molecular mechanics, etc. This approach is of great interest to medicinal chemistry for the prediction of binding sites, design of drugs and an analysis of surface recognition (Rosenfield & Murray-Rust, 1982; Murray-Rust, 1984; Rosenfield, Swanson, Meyer, Carrel & Murray-Rust, 1984; Lesk, 1986).

REFERENCES

ABOLA, E. E., BERNSTEIN, F. C. & KOETZLE, T. F. (1985). In *The Role of Data in Scientific Progress* (ed. P. S. Glusker) pp. 139-144. Elsevier Science Publishers B.V., North-Holland.

ALBERTSSON, J. & SVENSSON, C. (1978). *Acta Cryst.* A34,S-17.

ALLEN, F. H. (1984). *Acta Cryst.*A40, C 441.

ALLEN, F. H. (1985). In Proc. of the 9th ECM in Torino pp. 645-646.

ALLEN, F. H. (1986). *Acta Cryst.* B42,515-522.

ALLEN, F. H. & KIRBY, A. J. (1984). *J.Amer.Chem.Soc.* 106, 6197-6200.

ALLEN, F. H., KENNARD, O. & TAYLOR, R. (1983). *Acc.Chem.Res.* 16, 146-153.

ALTERMATT, D. & BROWN, I. D. (1985). *Acta Cryst.* B41, 240-244.

AUF DER HEYDE, T. P. E. & NASSIMBENI, L. R. (1984). *Acta Cryst.* B40, 582-590.

BACKHAUS, K. O., GRELL, H. & SCHRAUBER, H. (1984). *Acta Cryst.* A40, C-445.

BERGERHOFF, G. (1985). in *Crystallographic Computing 3: Data Collection, Structure Determination, Proteins, and Databases* (ed. G. M. Sheldrick et all.) pp. 85-95. Clarendom Press, Oxford.

BERGERHOFF, G., HUNDT, R., SIEVERS, R. & BROWN, I. D. (1983). *J.Chem.Inf.Comput.Sci.* 23, 66-69.

BERTOLASI, V., BELLUCCI, F., FERRETTI, V. & GILLI, G. (1984) *Acta Cryst.* A40, C107.

BRIGGS, A. J., GLENN, R., JONES, P. G., KIRBY, A. J. & RAMASWAMY, P. (1984). *J.Amer.Chem.Soc.*106, 6200-6206.

BRITTON, D. & DUNITZ, J. D. (1980). *Helv.Chim.Acta* 63, 1068-1073.

BRITTON, D. & DUNITZ, J.D. (1981). *J.Amer.Chem.Soc.* 103, 2971-2979.

BROWN, I. D. & ALTERMATT, D. (1985). *Acta Cryst.* B41, 244-247.

BROWN, K. L., DAMM, L., DUNITZ, J. D., ESCHENMOSER, A. & HOBI, R. (1978). *Helv.Chim.Acta* 61, 3108-3135.

BÜRGI, H. B. (1973). *Inorg.Chem.*12, 2321-2325.

BÜRGI, H. B. (1975a). *Angew.Chem.* 87, 461-475.

BÜRGI, H. B. (1975b). *Angew.Chem.*Int.Ed.14, 460-473.

BÜRGI, H. B. (1986). In *Crystallographic Computing* (ed. P. Paufler, V. Geist and D. Klimm) pp. 134-147. Karl-Marx-Univ. Press, Leipzig.

BÜRGI, H. B. & DUNITZ, J. D. (1983). *Acc.Chem.Res.*16, 153-161.

BÜRGI, H. B., DUNITZ, J. D. & SHEFTER, E. (1973). *J.Amer.Chem. Soc.* 95, 5065-5067.

BÜRGI, H. B., DUNITZ, J. D. & SHEFTER, E. (1974). *Acta Coyst.* B30, 1517-1527.

BÜRGI, H. B., HOUNSHELL, W. D., NACHBAR, R. B.,Jr., & MISLOW, K. (1983). *J.Amer.Chem.Soc.* 105, 1427-1438.

BÜRGI, H. B. & SHEFTER, E. (1975). *Tetrahedron* 31, 2976-2981.

BÜRGI, H. B., SHEFTER, E. & DUNITZ, J.D. (1975). *Tetrahedron* 31, 3089-3092.

BYE, E., SCHWEIZER, W. B. & DUNITZ, J. D. (1982). *J.Amer.Chem. Soc.* 104, 5893-5898.

CHAKRABARTI, P. & DUNITZ, J. D. (1982). *Helv.Chim.Acta* 65, 1482-1488.

CHANDRASEKHAR, K. & BÜRGI, H. B. (1983). *J.Amer.Chem.Soc.* 105, 7081-7093.

CISTI (1986), CAN/SND - *Scientific Numeric Database*, National Research of Canada.

CRIPPEN, G. M. & HAVEL, T. F. (1978). *Acta Cryst.* A34, 282-284.

Di NOLA, A. & BROSIO, E. (1982). *J.Appl.Cryst.* 15, 129-132.

DUNITZ, J. D. (1979). *X-Ray Analysis and the Structure of Organic Molecules.* Cornell.Univ.Press, Ithaca, NY.

EINSPAHR, H. & BUGG, C. E. (1980). *Acta Cryst.* B36, 264-271.

EINSPAHR, A. & BUGG, C. E. (1981). *Acta Cryst.* B37, 1044-1052.

EYRING, H. & POLANYI, M. (1931). *Z.Physikal.Chem.B.* 12, 279-311.

FROST, A. A. & PEARSON, R. (1961). *Kinetics and Mechanism,* pp.77-102. J.Wiley & Sons, New York, London.

GAVUZZO, E., PAGLIUCA, S., PAVEL, V. & QUAGLIATA, C. (1972). *Acta Cryst.* B28, 1968-1969.

GILLI, G. & BERTOLASI, V. (1981). *Acta Cryst.* A37, C 85.

GILLI, G., BERTOLASI, V. , BELLUCCI, F. & FERRETTI, V. (1986). *J.Amer.Chem.Soc.*108, 2420-2424.

GILLI, G., BERTOLASI, V. & VERONESE, A. C. (1983). *Acta Cryst.* B39, 450-456.

GRANÁT, L. & SECHOVSKÝ, H. (1980). *Počítačová grafika.* SNTL, Praha.

GUGGENBERGER, L. J. & MUETTERTIES, E. L. (1976). *J.Amer.Chem. Soc.* 98, 7221-7225.

HOLMES, K. C. & BLOW, D. M. (1966). *The use of X-ray diffraction in the study of protein and nucleic acid structure.* John Wiley & Sons, New York, London, Sydney.

HUML, K. (1986). In *Crystallographic Computing* (ed. P. Paufler, V. Geist and D. Klimm) pp. 157-170. Karl-Marl-Univ.Press, Leipzig.

JEŽEK, J. (1976). *Univerzální algebra a teorie modelů.* SNTL, Praha.

JONES, P. G. & KIRBY, A. J. (1984). *J.Amer.Chem.Soc.* 106, 6207-6212.

KARFUNKEL, H. R. (1986). *J.Comput.Chem.* 7, 113-128.

KOWALSKI, B. R. & BENDER, C. F. (1972). *J.Amer.Chem.Soc.* 94, 5632-5639.

KOWALSKI, B. R. & BENDER, C. F. (1973). *J.Amer.Chem.Soc.* 95, 686-693.

LESK, A. M. (1986). *Acta Cryst.* A42, 83-85.

LIEBMAN, M. N. (1982). In *Molecular Structure: Biological Activity* (ed. J. F. Griffin and W. L. Duax, Elsevier Sci. Publish., New York.

LIEBMAN, M. N. (1985). Pre-Meeting on Molecular Systematics ECM-9 in Torino.

LUKASOVÁ, A. & ŠARMANOVÁ, J. (1985). *Metody shlukové analýzy.* SNTL, Praha.

MACKAY, A. L. (1974). *Acta Cryst.* A30, 440-447.

Mc CARTHY, G. J., MIGHELL, A. D., HUBBARD, C. R. & NICHOLS, M. C. (1984). *Acta Cryst.* A40, C 442.

MORRISON, D. F. (1976). *Multivariate Statistical Methods.* 2nd ed., Mc Graw-Hill Kogakuska, Ltd., Tokyo, etc.

MUETTERTIES, E. L. & GUGGENBERGER, L. J. (1974). *J.Amer.Chem.Soc.* 96, 1748-1756.

MÜLLER, K. (1980). *Angew.Chem.* 92, 1-14.

MURRAY-RUST, P. (1982). *Acta Cryst.* B38, 2765-2771.

MURRAY-RUST, P. (1984). *Acta Cryst.* C 56.

MURRAY-RUST, P. & BLAND, R. (1978). *Acta Cryst.* B34, 2527-2533.

MURRAY-RUST, P., BÜRGI, H. B. & DUNITZ, J. D. (1975), *J.Amer. Chem.Soc.* 97, 921-922.

MURRAY-RUST, P., BÜRGI, H. B. & DUNITZ, J. D. (1978a). *Acta Cryst.* B34, 1787-1793.

MURRAY-RUST, P., BÜRGI, H. B. & DUNITZ, J. D. (1978b). *Acta Cryst.* B34, 1793-1803.

MURRAY-RUST, P., BÜRGI, H. B. & DUNITZ, J. D. (1979). *Acta Cryst.* A35, 703-713.

MURRAY-RUST, P. & GLUSKER, J. (1984). *J.Amer.Chem.Soc.* 106, 1018-1025.

MURRAY-RUST, P. & MOTHERWELL, D. S. (1979). *J.Amer.Chem.Soc.* 101, 4374-4376.

MURRAY-RUST, P. & RAFTERY, J. (1985). *J.Mol.Graph.* 3, 50-59.

NEUMANN, Th. & SCHILLER, H. (1984). In *Formulae and Methods in Experimental Data Evaluation* (ed. R. K. Bock et al. Published by European Phys.Soc., Cern.

NEWMAN, W. M. & SPROUL, R. F. (1979). *Principles of Interactive Computer Graphics.* 2nd ed., McGraw-Hill, New York.

NORSKOV-LAURITSEN, L. & BÜRGI, H. B. (1985). *J.Comput.Chem.* 6, 216-228.

ROSENFIELD, R. E. & MURRAY-RUST, P. (1982). *J.Amer.Chem.Soc.* 104, 5427-5430.

ROSENFIELD, R. E.,Jr., SWANSON, S. M., MEYER, E. F.,Jr., CARREL, H. L. & MURRAX-RUST, P. (1984). *J.Mol.Graphics* 2, 43-46.

SACHS, L. (1974). *Angewandte Statistik.* Springer-Verlag, Berlin, Heidelberg, New York.

SANDS, D. E. (1982). *Vectors and Tenzors in Crystallography.* Addison-Wesley Publishing Co., London.

SCHUSTER, P., ZUNDEL, G. & SANDORFY, C. (1976). Eds.*"The Hydrogen Bond"*, Vol.2,Ch.8. North Holland, Amsterdam.

SCHWEIZER, W. B. (1985). In *Crystallographic Computing 3: Data Collection, Structure Determination, Proteins, and Databases* (ed. G. M. Sheldrick et al.) pp.119-127. Clarendom Press, Oxford.

SHELDRICK, B. & AKRIGG, D. (1980). *Acta Cryst.* B36, 1615-1621.

SIEGEL, S. (1956). *Nonparametric Statistics for the Behavioral Sciences*. McGraw-Hill Book Co., Inc., New York, Toronto, London.

STEVENS, S. S. (1946). *Science* 103, 677-680.

ŠANTAVÝ, M. & KYPR, J. (1984). *J.Mol.Graph.* 2, 47-49.

TAYLOR, R. (1985). In *Crystallographic Computing 3: Data Collection, Structure Determination, Proteins, and Databases* (ed. G. M. Sheldrick et al.) pp.96-105. Clarendom Press, Oxford.

TAYLOR, R. (1986a). *J.Mol.Graph.* 3, 123-131.

TAYLOR, R. (1986b). *J.Appl.Cryst.* 19, 90-91.

TAYLOR, R. & KENNARD, O. (1982). *J.Amer.Chem.Soc.* 104, 3209-3212.

TAYLOR, R. & KENNARD; O. (1983). *Acta Cryst.* B39, 517-525.

TAYLOR, R. & KENNARD, O. (1986). *J.Chem.Inf.Comput.Sci.*26, 28-32.

TAYLOR, R., KENNARD, O. & VERSICHEL, W. (1983). *J.Amer.Chem.Soc.* 105, 5761-5766.

WALLIS, J. D. & DUNITZ, J. D. (1984). *Acta Cryst.* A40, C 106.

WINKLER, F. K. & DUNITZ, J. D. (1971). *J.Mol.Biol.* 59, 169-182.

Program systems

20

THE XTAL SYSTEM:
AN APPLICATION PRIMER

Brian Skelton, Sydney Hall, and James Stewart

Summary

The XTAL System is a library of crystallographic programs for doing crystal structure calculations (Hall, Stewart, and Munn 1980) . This talk will provide a brief introduction to the general design concepts of XTAL and how the programs are used for crystal structure studies.

1 Introduction

The XTAL System has been developed as a cooperative programming effort involving a number of different laboratories and authors (Stewart and Hall 1986). XTAL is modular and will run on any machine with the following minimum facilities: 32-bit integer and real numbers; 512 Kb of RAM available to the user, or VMS; 20 Mb of disc storage; a Fortran77 compiler and a 1600 bpi tape drive. The distributed XTAL source is in the language Ratmac. The supplied Ratmac compiler RFPP converts the distributed source into Fortran77 code which is appropriate for the target machine. Portability and specificity are achieved with powerful macro-editor which is built-in to the preprocessor. XTAL macros are currently available for the following machines:

Apollo Computers	International Business Machines
Concurrent Computers	Prime Computers
Control Data Corporation	National Advanced Systems
Cray Research	Siemens
Digital Equipment Corporation	Sperry Corporation
Honeywell	Unix-based operating systems

Details of the implementation, operation, and the use of the pre-processor have been reported elsewhere (Hall, 1984), and are detailed in the RFPP User's Manual (Hall, 1986). Future preprocessors will be used to generate other languages such as C and Fortran88.

This presentation will focus on the application of the XTAL package for crystallographic calculations. The basic functions of the system are controlled by set of routines which are common to all calculations. These routines are referred to as the XTAL nucleus or kernal. The user instructs the nucleus on the calculation(s) to be done via the input line file. A line file is a typical character file that may be manipulated with the local line editor. XTAL is a 'file-driven' system as opposed to an interactive 'menu-driven' system. The file-driven approach is preferred because batch mode is the most efficient method of executing most crystallographic calculations, even when very fast hardware is available. Nevertheless, so as to take advantage of the helpful aspects of the menu approach, a special prompting-editor PREPX is supplied to prepare an XTAL input line file.

2   The Nature of Xtal Files

Because XTAL is a file-driven system it is essential to understand how the various files are used. The two primary files are the line input and line output files. The line input file contains the character data that controls which calculations XTAL will perform, and the line output file contains the character data describing the results of the calculations. The line input file is referred to by the mnemonic unitcd: (which is short for cardreader unit or device). According local machine requirements this logical file is assigned to device number (e.g. 5) or a physical file (e.g. FTN005). The same holds for the line output file unitlp: (which is short for the lineprinter unit or device). These two files contain character or coded data that may be easily manipulated or examined with the local editor or printer. Other character files used in XTAL are the punch file unitpch: and the general purpose ascii file unitasc:.

The other type of file used in XTAL is the binary file. Binary files are used to store information in a form that does not require encoding and decoding during input or output. They may be accessed as sequential or random fixed-length records. In the XTAL system these referred to as binary data files, or simply bdf's. Understanding the use of the bdf is central to the elegant and efficient use of the XTAL system. For most calculations the bdf manipulation is under the control of the nucleus, however, for non-standard chaining operations it is often desirable that the user to control the assignment of bdf's. There are, conceptually, two categories of bdf's.

The first is used to accumulate information about the crystal under study and it is updated as the solution and refinement of the structure progresses. For this reason it is generally referred to as an 'archival' bdf. Data such as atomic coordinates and reflection intensities are stored in this bdf.

The the second category of bdf has the same internal format, but is only used to transfer specific data from one calculation to another. This bdf is never updated and is generally referred to as an 'auxilliary' or 'temporary' bdf. Electron density maps are are typical of the type of data stored in an auxilliary bdf. They exist only during a single sequence of calculations, but may need to be retained if the sequence is broken or is to be repeated.

Digression about file assignations. *XTAL supports eight binary files, labelled fileA, fileB, ..., fileH, and four character files unitcd:, unitlp:, unitpch; and unitasc:. Before the start of a calculation sequence each of these file units must be 'at-*

tached' to a physical file or device where the input or output data will reside. In the Fortran tradition each of these physical units is addressed within XTAL by an integer. For example at some installations unitcd: and unitlp: are referred to by the device integers 5 and 6, while others use the integers 10 and 1 1. Local operating systems handle these device numbers in a variety of ways. Some will require that the number is explicitly assigned to the appropriate devices (e.g. CARD: and LPT:) or filenames (e.g. INPUT.DAT and PRINT.OUT), while others will do the assignations automatically (e.g. FTN005 and FTN006). The assignment device integers to the bdf files is more complicated because they are often interchanged during a calculation sequence. For example logical units fileA and fileB might be assigned to device numbers 8 and 9 at the start of execution, but are interchanged if a calculation updates fileB. With the exception of the program STARTX, fileA must always contains the input bdf and the output bdf must always be written to fileB.

The XTAL bdf's are controlled either by the nucleus using standard procedures, or by the user entering a FILES control line. As with any computer program, XTAL files must be attached to the executable task using the local command language. After execution is finished all files must be disposed of in an appropriate way. How this is done depends very much on the local operating system. When the XTAL system is installed, 'procedures' can be set up to greatly simplify this task for the user.

3   Xtal Control Lines

As discussed above, the line input file is used to control an XTAL task. Once execution has been started the XTAL nucleus is directed from the line input stream. Three levels of command lines are recognised in the input file. They are: 1) System control, 2) Program initiation and 3) Program data input.

3.1 System Control Lines.

There are eight system control lines:

TITLE	to specify the ouput page header
REMARK	to insert comments into the printed output
FILES	to control the use of line and binary files
MEMSET	to set memory allocation
SETID	to specify line identifiers of undefined line data
FIELD	to permit fixed format input lines
ORDER	to reorder input line data
FINISH	to signal the end of the input line file

The most important of these lines are the FILES and FINISH lines. The others are either for convenience or for special line input manipulations. Often the local editor can be used in place of the SETID, FIELD and ORDER lines. The TITLE and REMARK lines are optional, but research supervisors are well advised to insist upon their routine use.

Digression about the use of FILES lines. *The FILES line allows the user to connect the 'logical' file units used by the XTAL programs and the 'physical' file units recognised by the local computer installation. Here are some typical applications of the FILES line. If the most up-to-date bdf is, say, FTN009, then at the start of a new calculation sequence the user can enter the line 'FILES 9 8' so that fileA is assigned to FTN009 and fileB to FTN008 (this assumes that the default assignations are normally the reverse). This could also be done externally to XTAL using OS commands, but the FILES line is quicker and neater! FILES lines are also useful for saving auxilliary bdf's during the calculation sequence. For example the* *program GENSIN outputs the structure invariant relationships to fileD. If a certain sequence of calculations needs to use this file more than once, another integer can be assigned to fileD by inserting a FILES line at the appropriate place in the input line file (probably after the END line that follows GENTAN or SIMPEL lines).*

3.2 Program Initiation Lines.

XTAL calculations are selected by a program initiation line. Each line contains at the beginning a mnemonic which identifies the desired calculation. For example, to calculate structure factors a line starting with FC is entered. Or, to calculate normalised structure factors, a GENEV line is entered. The program initiation line may also contain signals about the options to be used during the calculation. Here are the mnenonics and brief descriptions of the programs available in the XTAL2.2 version of the system. An asterisk (*) indicates a macromolecular routine.

ABSCAL		scale diffractometer data for psi-scan absorption
ABSORB		apply Gaussian and analytical absorption corrections
ADDATM		load atom parameters to binary file
ADDREF		place and reduce reflection data onto binary file
ATABLE		prepare table of atomic parameters for publication
BAYEST		estimate Bayesian statistics
BDFIN		load, expand, or collapse binary file records
BFOURR	*	prepare special Fourier coefficients for FOURR
BONDAT		generate atom coordinates from molecular geometry
BONDLA		calculate bond lengths and angles
CHRIN	*	create a binary file from a character file
CHROUT	*	create a character file from a binary file
CONTRS		contour Fourier maps
CRITIQ	*	determine the most variable reflections
CRYLSQ		refine atomic parameters by least-squares

DIFDAT		process CAD4 and Nicolet diffractometer tape
DUMCOP		print contents of binary file
FC		calculate structure factors
FINDKB	*	get K and B for protein data
FODIFF.	*	map difference between two fourier maps
FOGEN	*	generate a volume from an asymmetric set
FOSTAT		provide fourier map density statistics
FOURR		map Beevers-Lipson and Fast Fourier Transform
GENEV		normalize structure factors
GENSIN		generate structure invariant relationships
GENTAN		estimate phases using generalized tangent procedures
LISTFC		list structure factor data for publication
LOADAT		load and edit atomic parameters on binary file
LSQPL		calculate least squares planes
MERGDS	*	select data for addition or replacement
MERGOB	*	merge data from two binary files
MIR	*	apply multiple isomorphous refinement techniques
MODEL		search for molecular sites in Fourier map
MODHKL		modify reflection data on binary file
MULIST	*	list mult.-obs. reflection data
NEWMAN		calculate Newman projection data for selected atoms
NIKNAK		process profile data on Nicolet diffractometer tape
ORTEP		calculate thermal ellipsoid data for plotting
PEKPIK		search for density max/min in Fourier map
PHONYD		generate idealized intensity data
PLOT,Y,Z		plot outputs of ABSORB, CONTRS and ORTEP
PRECED	*	load the protein refinement constraint information
PROATM	*	load atomic data from the protein data base
REFORM		convert between XRAY, XTAL and SCFS file formats
REMSET	*	remove data set from mult. data set binary file
RERWT		estimation of least-squares weights
REVIEW		review structure invariant phase relationship statistics

RIGBOD		transform atom parameters to a different unit cell
RSCAN		scan structure factor agreement by R-factors
SCALE1	*	places independently measured data sets on same scale
SFLSX		refinement using structure factor least squares
SIMPEL		estimate phases using symbolic addition procedure
SLANT		interpolate slanted map from FOURR map
SORTRF		sort and merge reflection data on F, F**2 and I
STARTX		create an initial binary file

The program initiation line is always matched with a following END line. Additional data and signals controlling the program are placed between the program initiation line and the END line. When the XTAL nucleus encounters a program initiation line, it branches to the routines in memory appropriate for this calculation. Within these routines the program initiation line is scanned for control signals and data. If none are entered default values are assumed. Then program data lines are entered until an END line is encountered. Concurrent with or subsequent to the input of data from these lines or the attached bdf, the specified calculations are performed.

3.3   Program Data Lines

Each program data line has a mnemonic which uniquely identifies its function and this is used to direct the data to its proper place in memory and bdf. Often the order of entry of these data lines is arbitrary. However, some programs require that certain data must be entered first. For example, overall temperature factors must precede individual atom thermal parameters. The entry rules for each line is given in the XTAL User's Manual (2). During both the data entry and calculation process, all new and updated information is archived to the attached output bdf. This means that as the crystal study progresses fewer and fewer data lines are needed to run certain calculations.

4 Application Example

The example input shown in Appendix I is a typical sequence of input lines for structure analysis run. Appendix II is an abbreviated sunmmary of the output from this run. This example shows the main steps in the solution and refinement of a small molecule. In practice these calculations would be carried out in stages. The calculation sequence can be viewed as being analogous to the use of a complex editor. The archival bdf is the edited file and after each editing session it is stored with all the 'corrections' applied. The typical 'editing session' might consist of a series of least-squares structure-factor refinement cycles in which the input file contains rough atomic parameters and the output bdf contains refined parameters, calculated structure factors and estimated errors of the atomic parameters.

Let us now examine some of the calculations in the example. STARTX is the program that creates a new bdf containing information about the symmetry and the unit cell. It is this bdf which will, after successive editing and copying, contain all the data about the solved crystal structure. ADDREF is one of several programs in XTAL which load reflection intensity data. Other programs, such as DIFDAT and NIKNAK, load reflection information by converting diffractometer data from a local format to an XTAL format. The program ABSORB applies Gaussian or analytical absorption corrections to the intensity data. These programs are not shown here. In addition to the reflection loading, ADDREF reduces net intensities to measured structure factors. The ADDREF calculation adds reflection data to the bdf created by STARTX. The SORTRF program sorts reflection data into a specified order of hkl and merges multiply-observed data into an asymmetric set. The XTAL system always operates on a unique asymmetric set of reflections and atoms.

> **Digression about the status of the bdf's.** *Let us now review what has happened to the bdf's. STARTX wrote a new bdf on fileA, ADDREF read fileA and wrote a new bdf to fileB. The bdf's attached to fileA and fileB were then interchanged automatically by the nucleus so that the most up-to-date bdf will*

reside on fileA. Thus if at the start of this sequence the user attached FTN008 as fileA and FTN009 as fileB, this has now been reversed with FTN009 attached to FileA and FTN008 attached to FileB. SORTRF read fileA and overwrote fileB, and interchanged the bdf's again. The file FTN008 now contains the sorted hkl data needed for subsequent calculations. XTAL users must become familiar with bdf assignments. The main rule is that calculations writing or updating fileB automatically interchange the bdf's attached to fileA and fileB on completion. The user must be certain which physical file (e.g. FTN008 or FTN009) was last updated successfully. When starting a new run this file that must be assigned to fileA. The fileA and fileB assignments are done either via the local command language, or through the use of FILES *lines.*

The next calculations to be considered involve the solution of the phase problem. To use conventional Patterson methods, one would compute an $|F^2|$ map using the FOURR program. In this example direct methods programs are employed. The program GENEV calculates normalised structure factors, as well as an $|F|$ scale factor and an overall isotropic temperature factor. These are output to fileB. This bdf is read (after being interchanged to fileA) by GENSIN and used to generate triplet and quartet relationships among the highest E-value reflections. GENSIN outputs the structure invariant relationships to the auxilliary bdf on fileD.

Although the structure invariant relationships are essential to subsequent calculations, such as GENTAN or SIMPEL, they represent a large amount of data which are usually only used once per analysis. It is therefore neither convenient nor efficient to archive these to fileB. For difficult structures the user may save the bdf assigned to fileD for separate attempts to calculate structure factor phases. The program GENTAN has a wide range of options for applying the multi-solution tangent phasing methods. The program SIMPEL (not applied here) provides a symbolic addition approach to determining phases.

Digression about the application of direct methods .
*It is useful at this point to give some hints on the optimal
use of GENEV, GENSIN or GENTAN. Note that the input
lines used to invoke these routines in the example are very
sparse. Only the program initiation and END lines are used.
This is because the default controls are appropriate for routine
analyses. For difficult structures additional control signals may
be required. Here is a summary of the direct methods 'recipe'
used in a laboratory that solves more than 150 structures a year
(molecular weights from 250 to 1600 Daltons) using XTAL.*

** As a first attempt use the default values for a structure so-
lution sequence
(i.e.STARTX⇒ADDREF⇒SORTRF⇒ GENEV⇒GENSIN⇒
GENTAN⇒ FOURR⇒PEKPIK⇒MODEL). This solved 90%
of the structures automatically.*

** SIMPEL can be used in place of GENTAN and is about four
times faster. It is most effective for centrosymmetric structures.*

*For structures that do not solve with default options, try the
following alternatives. Note that judicious saving of the bdf's
permits you to start calculations at the program in which the
option is changed. Don't always go back to STARTX, or even
GENEV!*

** Increase the number of starting phases in GENTAN using
the* SELECT *line (particularly for low symmetry structures).*

** Decrease Emin in GENSIN to 1.5 - 1.3 (particularly for tri-
clinic).*

** If there are sufficient triplets (i.e.>10 times the number of
E's), increase Emin to about 1.6 in GENTAN (or GENSIN).*

** Increase the number of triplets and quartets by reducing Amin
and Bmin to about 0.8 and 0.4 on the GENSIN lines* TRIP *and*

QUAR. *Even lower values can considered for large structures.*

* *If a fragment is found in the space group* $P\bar{1}$ *with a possible origin translation, redo STARTX as P1 and solve using the fragment in FC, FOURR, etc. Then return to* $P\bar{1}$ *for the refinement.*

* *Exclude quartets from the GENTAN calculation using the* INVAR *line.*

* *For noncentrosymmetric structures, change the* MAGIC *option to* PERMUTE *on the* SELECT *line of GENTAN. Alternately, use the* RANDOM *option instead of* PERMUTE.

* *Apply the* BLOCK *and/or* W2 *options on the* REFINE *line of GENTAN.*

There are other combinations of options available in GENSIN and GENTAN if the above fail. However, you should now critically assess your reflection data, both in terms of its precision and high angle limit. In some cases structures are solved automatically by recollecting the data with better precision and/or higher angle data. In at least one case it was necessary to collect low-temperature data (there were very high B-values!). It cannot be emphasised enough that these methods are very dependent on good E-values. GENEV does a good job with good data, but if the Wilson plot shows any anomalies at high angles it is better to exclude this data from the structure solution (by setting Smax in GENEV). ALWAYS check the Wilson plots carefully if the default run fails.

GENTAN outputs several sets of phases to fileB. The FOURR program uses these phases to produce an E-map. FOURR is a general reciprocal-space to direct-space Fourier transform program which uses either Beevers-Lipson or FFT summation algorithms. It outputs the E-map density points to the auxilliary bdf on fileC. The program PEKPIK reads fileC and searches for a unique assymetric set of highest peaks in the density map. These are output to the bdf on fileE. This file is read by the program MODEL which establishes connections between the peaks which satisfy specified distance and angle criteria. From this MODEL produces produces a table of bond lengths and angles, and a printer plot of the connected peaks. It also ouputs atom parameter lines to the punch file for use in subsequent calculations such as BONDLA, ADDATM or LOADAT.

REFERENCES

Hall, S. R. (1984). In *Methods and Applications in Crystallographic Computing* (ed. S. R. Hall and T. Ashida) pp. 343–352. Clarendon Press, Oxford.

Hall, S.R. (1986) *RFPP Ratmac to Fortran77 Preprocessor: User's Manual*. Crystallography Centre Report, University of Western Australia.

Hall, S. R., Stewart, J. M., and Munn, R. J. (1980). *Acta Cryst.*, **A36**,979–989.

Stewart, J. M. and Hall, S. R. (1986). *XTAL User's Manual*, Technical Report TR1364.2, Computer Science Center, University of Maryland.

Appendix I

```
TITLE    STICK6   C23H2806     P212121 Z=4
STARTX STICK6
CELL    17.076   16.604  7.425  90. 90. 90.
CELLSD  .004   .002 .007
SGNAME  P 2AC 2AB   : space group P212121
CELCON  C  92
CELCON  O  24
CELCON  H 112
CELCON  N  0
END
ADDREF STICK6 LIST NOFFAC RSABS NOBAY SMAX NOFR SIGF 5
REDUCE ITOF RLP4 XRAY 12.2
HKLIN   HKL   COD        IREL   SIGI
HKL  0  0  2   1        21023   644
HKL  0  0  4   1           40    5.97
HKL  0  0  6   1          252   11.4
HKL  0  0  8   1         33.4    4.9
HKL  0  1  1   1        19600   604
HKL  0  1  2   1         5910   186
.........reflection data omitted for brevity
HKL 19 10  0   2            0    8
HKL 19 10  1   2          8.8    3.54
HKL 19 10  2   2            0    8
HKL 19 11  0   2         .214   3.33
END
SORTRF STICK6   HKL   NOPRINT
END
GENEV   STICK6
END
GENSIN  STICK6
END
GENTAN  STICK6
END
FOURR STICK6 EMAP
END
PEKPIK STICK6
END
MODEL STICK6
END
ADDATM STICK6
SCALE       1.02033          1
ATOM C1  .48236 -.11502  .93088   3.3299  1.0000  .00023  .00024  .00053
BETA C1  .00244  .00391  .01162  -.00008  .00009 -.00021
ATOM O1  .47364 -.18767  .83988   3.8520  1.0000  .00017  .00015  .00039
BETA O1  .00378  .00323  .01594  -.00027 -.00053  .00071
.........atom data omitted for brevity
ATOM H6  .75500 -.10800  .93600   5.7000  1.0000  .00000  .00000  .00000
ATOM H7  .64700 -.14500 1.20600   6.8000  1.0000  .00000  .00000  .00000
```

```
ATOM H81 .81360 -.11540 1.23320    8.4000    1.0000 .00000    .00000    .00000
ATOM H82 .79070 -.20340 1.18740    8.4000    1.0000 .00000    .00000    .00000
ATOM H83 .78040 -.17080 1.37730    8.4000    1.0000 .00000    .00000    .00000
END
SFLSX STICK6   CY 2   AD   FM WS   FU 0.7
NOREF  H
END
BONDLA STICK6
ATRAD  C  1.5   1.0
ATRAD  N  1.5   1.0
ATRAD  H  1.    .5
ATRAD  O  1.5   1.0
END
RSCAN  STICK6
END
FOURR STICK6 FDIF R2 0
END
PEKPIK STICK6
END
LSQPL  STICK6
PLANE
DEFINE  C22 C23 C24 C25 C26 C27
NONDEF  O2  C21
PLANE
DEFINE  C32 C33 C34 C35 C36 C37
NONDEF  O3  C31
PLANE
DEFINE  C6 C7 O6
NONDEF  C5 C8
END
ORTEP STICK6
EXEC AUTO CELL EXCL      : a complete cell
PLOTP 12 12 0. 1.2  AB
ELLIPS 5 1.0
SEQ  REPLACE  4
INST   401  1 555 1 -29 565 1  1 555 2 -29 665 2  1 456 3 -29 556 3
INST   401  1 655 4 -29 655 4
END
PLOT STICK6
COLOR NOFIL   NOCOL
STARTP VDU
END
ORTEP STICK6
EXEC AUTO MOLE SPHE      : a single molecule
PLOTP 12 12 0. 1.2
ELLIPS 5 1.0
END
PLOT STICK6
COLOR NOFIL
STARTP VDU
END
```

```
BONDAT STICK6    PC
CALCAT  TETCHN  .95  05 C1  C2  H1  H12
CALCAT  TETERM  0.95  C1  01  C11 H111  H112 H113
CALCAT  TETCHN  0.95  C1  C2  C3  H21   H22
CALCAT  TRIGON  0.95 C32 C33 C34  H33
CALCAT  TRIGON  0.95 C33 C34 C35  H34
CALCAT  TRIGON  0.95 C34  C35  C36  H35
CALCAT  TRIGON  0.95 C35  C36  C37  H36
CALCAT  TRIGON  0.95 C36  C37  C32  H37
CALCAT  TETCHN  0.95  C3  C4  C5  H41  H42
CALCAT  TETCHN  0.95  C4  C5  05  H51  H52
CALCAT  TETERM  0.95  C6  C7  C8  H81  H82  H83
END
TITLE  STRUCTURE FACTOR AMPLITUDES (10*F0,10*FC,10*SIGF0)  C23 H28 06
LISTFC  STICK6  S M F NOFL  120 64
END
FINISH
```

Appendix II

****** Output Listing ******

```
*** BEGIN STARTX ***

CELL CONSTANTS      A         B         C       ALPHA      BETA     GAMMA
                 17.0760   16.6040   7.4250   90.0000   90.0000   90.0000
                                     COSINES   -.0000    -.0000    -.0000


STANDARD DEVIATIONS ----
CELL CONSTANTS      A         B         C       ALPHA      BETA     GAMMA
                  .0040     .0020     .0070  -.4000+21 -.4000+21 -.4000+21
                                     COSINES   .0000     .0000     .0000


CELL CONSTANTS      A*        B*        C*      ALPHA*     BETA*    GAMMA*
                  .0586     .0602     .1347   89.9999   89.9999   89.9999
                                     COSINES   .0000     .0000     .0000


TRANSFORMATION MATRIX
               17.0760    -.0000    -.0000
                 .0000   16.6040    -.0000
                 .0000     .0000    7.4250


INVERSE TRANSFORMATION MATRIX
                 .0586     .0000     .0000
                 .0000     .0602     .0000
                 .0000     .0000     .1347


       OBSERVED DENSITY          CELL VOLUME
           NOT GIVEN              2105.211

METRIC TENSOR FOR CONTRAVARIANT VECTORS DIRECT SPACE
          .2916+03    -.1814-03    -.8111-04
         -.1814-03     .2757+03    -.7887-04
         -.8111-04    -.7887-04     .5513+02

METRIC TENSOR FOR COVARIANT VECTORS RECIPROCAL SPACE
          .3429-02     .2256-08     .5046-08
          .2256-08     .3627-02     .5189-08
          .5046-08     .5189-08     .1814-01

SYMTRY X,Y,Z
SYMTRY 1/2-X,-Y,1/2+Z
SYMTRY 1/2+X,1/2-Y,-Z
SYMTRY -X,1/2+Y,1/2-Z


LATTICE TYPE  ACENTRIC P
```

```
MATRIX     COLUMN 1        COLUMN 2        COLUMN 3       TRANSLATION
SYMOP    R11 R21 R31     R12 R22 R32     R13 R23 R33     T1  T2  T3
  1        1   0   0       0   1   0       0   0   1       0   0   0
  2       -1   0   0       0  -1   0       0   0   1       6   0   6
  3        1   0   0       0  -1   0       0   0  -1       6   6   0
  4       -1   0   0       0   1   0       0   0  -1       0   6   6
```

ATOMIC CONTENT OF CELL

```
 NO.  TYPE  AT.WT. AT.NO. ELECTRONS  BOND R. CNTCT R.   DFR    DFI
  92   C    12.011    6       6        .770   2.020    .002   .002
  24   O    16.000    8       8        .740   1.990    .008   .006
 112   H     1.008    1       1        .370   1.620    .000   .000
   0   N    14.007    7       7        .740   1.990    .004   .003
```

CALCULATED DENSITY 1.264 GRAMS PER CUBIC CENTIMETRE

... EXIT STARTX ...

*** BEGIN ADDREF ***

INPUT FILE FROM UNIT 01. FILE HISTORY..
STICK6 STARTX 27/01/87 12:56:28

ADDREF CONTROL PARAMETERS

```
LIST DATA FOR ALL REFLECTIONS      YES
APPLY BAYESIAN CORR TO F**2        NO
CALCULATE EXTINCTION FACTOR        NO
REMOVE SYS ABS REFLECTIONS         YES
INCLUDE FORM FACTORS IN LRREFL     NO
REQUESTED DATA CONVERSION          ITOF
LORENTZ-POLARIZATION METHOD        RLP4
RADIATION TYPE USED                XRAY
INCLUDE 1/LP IN LRREFL             NO
TREAT FRIEDEL PAIRS                NO
SIGMA FOR VOID INTENSITIES         5.0
WAVELENGTH IN ANGSTROMS            .71069
TWO-THETA ANGLE OF MONOCHROMETER   12.163
MONOCHROMETER CRYSTAL PERFECTION    .500
```

```
FROM HKLIN LINES --      ITEM IDNUM
                         HKL     1
                         RCOD  1308
                         IREL  1300
                         SIGI  1301
```

INPUT FILE FROM UNIT 01. FILE HISTORY..
STICK6 STARTX 27/01/87 12:56:28

```
 H  K  L  STOL MUL EPS  RLP RCOD    IREL    SIGI    FREL   SIGF
 0  0  2 .1347  1   2 .1940   1 .210+05 .644+03   63.83    .98
```

```
 0  0  4 .2694   1   2 .4038   1 .400+02 .597+01    4.02    .30
 0  0  6 .4040   1   2 .6462   1 .252+03 .114+02   12.76    .29
 0  0  8 .5387   1   2 .9381   1 .334+02 .490+01    5.60    .41
 0  1  1 .0738   2   1 .1053   1 .196+05 .604+03   45.42    .70
.........output data omitted for brevity
19  9  1 .6225   4   1 1.149   2 .355+01 .341+01    2.02    .97
19  9  2 .6333   4   1 1.178   2 .333+01 .360+01    1.98   1.07
19 10  0 .6326   2   1 1.176   2 .000+00 .800+01     .00   5.00
19 10  1 .6362   4   1 1.186   2 .880+01 .354+01    3.23    .65
19 10  2 .6468   4   1 1.214   2 .000+00 .800+01     .00   5.00
19 11  0 .6475   2   1 1.216   2 .214+00 .333+01     .51   3.97
```

```
INTERCHANGE FILES    A  B  C  D  E  F  G  H
                     02 01 03 04 05 06 07 08


OUTPUT FILE CONTAINS -- ITEM IDNUM      MAX        MIN
                        H       1   19.0000       .0000
                        K       1   21.0000       .0000
                        L       1    9.0000       .0000
                        STOL    2     .6487       .0420
                        EPSL    3    2.0000      1.0000
                        MULT    3    4.0000      1.0000
                        RCOD 1308    2.0000      1.0000
                        IREL 1300  197000.1       .0000
                        SIGI 1301   6080.001      .0893
                        FREL 1304   171.6989      .0000
                        SIGF 1305     5.0080      .1502


OUTPUT FILE CONTAINS   2670 REFLECTIONS
                       1763 WITH RCODE 1
                        915 WITH RCODE 2
REJECTED FROM INPUT       0 REFLECTIONS


TIME OF DAY HH/MM/SS    CPU SECS ADDREF    CUMULATIVE CPU SECS
            12:58:25          112.0001             117.0001
WORDS OF MEMORY USED BY ADDREF   MAXIMUM MEMORY USED THIS RUN
            9781                          9781


... EXIT  ADDREF ...


*** BEGIN SORTRF ***


INPUT FILE FROM UNIT 02. FILE HISTORY..
STICK6 STARTX 27/01/87 12:56:28  ADDREF 27/01/87 12:57:34


SORTRF CONTROL PARAMETERS
-------------------------

SORT IN FOLLOWING ORDER     H SLOW  K MED   L FAST
HKL SORT BASED ON /HSLO/ SIGN /HMED/ SIGN /HFAS/
MERGE FRIEDEL EQUIV HKL          YES
FRIEDEL HKL IN SEP BDF PAK       NO
```

```
MERGE INTENSITY DATA              NO  SCHEME  NONE
NUMBER OF REFLECTIONS SORTED    2678
PASSES REQUIRED FOR SORTING       1


INTERCHANGE FILES    A  B  C  D  E  F  G  H
                    01 02 03 04 05 06 07 08


TIME OF DAY HH/MM/SS     CPU SECS SORTRF      CUMULATIVE CPU SECS
            12:58:38           13.00001              130.0001
WORDS OF MEMORY USED BY SORTRF   MAXIMUM MEMORY USED THIS RUN
            35408                          35408


... EXIT  SORTRF ...



*** BEGIN GENEV  ***


INPUT FILE FROM UNIT 01. FILE HISTORY..
STICK6 STARTX 27/01/87 12:56:28  ADDREF 27/01/87 12:57:34
        SORTRF 27/01/87 12:58:25


CONTROL PARAMETERS
------------------

PROCESS BDF DATA SET NO. 1
MAX SINT/L PROCESSED          .650
SCATTERING FACTOR TYPE   XRAY
USE FF**2 SUMS ON BDF    NO
APPLY BAYESIAN STATS.    NO
FILL IN MISSING FRELS    NO
SCALE PRESET FOR /E2/    NO
B(OV) PRESET FOR /E2/    NO
SCALING MODE FOR /E2/    LINEAR
HKL RESCALES FOR /E2/    YES
LIST HKL F /E/S ESDS     NO
LS INFLEXN. PTS. S2 =    .15    .26
BDF FREL SCALE APPLIED    1    1.0000


UNIT CELL CONTENTS
------------------

ATOM TYPE    C      O      H      N
ATOM/CELL   92     24    112      0


F(0,0,0)     856.   E(0,0,0)    12.15
```

LINEAR LEAST SQUARES FIT OF LN(I(OBS)/I(EXP)) .VS. S**2

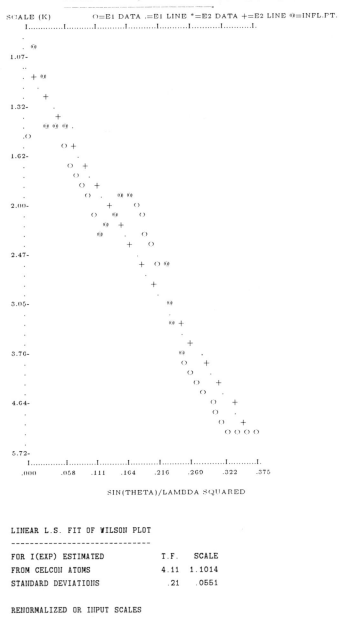

SCALE (K)   O=E1 DATA .=E1 LINE *=E2 DATA +=E2 LINE @=INFL.PT.

SIN(THETA)/LAMBDA SQUARED

LINEAR L.S. FIT OF WILSON PLOT

FOR I(EXP) ESTIMATED	T.F.	SCALE
FROM CELCON ATOMS	4.11	1.1014
STANDARD DEVIATIONS	.21	.0551

RENORMALIZED OR INPUT SCALES

OVERALL SCALES	FOR /E1/	SUMRM	FOR /E2/	SUMRM
	1.1223	9387		

```
INDEX GROUP SCALES
GROUP  (H+K+L)MOD=N    (H+K+L)MOD=N    (H+K+L)MOD=N   FOR /E2/  SUMRM
  1    1    2  1      1    2  0      1    2  0     1.0619   1176
  2    1    2  0      1    2  1      1    2  0     1.1456   1154
  3    1    2  1      1    2  1      1    2  0     1.1866   1178
  4    1    2  0      1    2  0      1    2  1     1.0546   1168
  5    1    2  1      1    2  0      1    2  1     1.0645   1190
  6    1    2  0      1    2  1      1    2  1     1.1997   1174
  7    1    2  1      1    2  1      1    2  1     1.1566   1180
  8    1    2  0      1    2  0      1    2  0     1.1413   1167
```

```
INPUT FILE FROM UNIT 01. FILE HISTORY..
STICK6 STARTX 27/01/87 12:56:28  ADDREF 27/01/87 12:57:34
       SORTRF 27/01/87 12:58:25
```

```
FREL SCALE STORED ON BDF   1.1223    1
```

```
UPDATED LR20 ON UNIT 02 CONTAINS ITEMS...
    1      2      3   1308   1300   1301   1304   1305   1600   1601
 1602   1603
```

E-STATISTICS

```
                  MEAN /E/     /E**2/   /E**2-1//E**2-1/**2  /E**2-1/**3
THEORY RANDOM P1B    .798     1.000      .968      2.000      8.000
THEORY RANDOM P1     .886     1.000      .736      1.000      2.000
FOR /E1/ ESTIMATE    .849     1.000      .808      1.453      5.981
FOR /E2/ ESTIMATE    .851     1.000      .804      1.390      5.110
```

E-DISTRIBUTION

```
 PERCENTAGES /E/.G    .00  1.00  1.20  1.40  1.60 1.80 2.00 2.50 3.0
THEORY RANDOM P1B 100.00 31.73 23.01 16.15 10.96 7.19 4.55 1.24  .2
THEORY RANDOM P1  100.00 36.79 23.69 14.09  7.73 3.92 1.83  .19  .0
FOR /E1/ ESTIMATE 100.00 34.45 21.82 13.76  8.43 5.21 2.91  .70  .1
FOR /E2/ ESTIMATE 100.00 34.87 22.67 13.94  8.75 5.04 2.80  .77  .1
```

POPULATION

```
FOR /E1/ ESTIMATE   2678    909    586    381    237    146     87     21
FOR /E2/ ESTIMATE   2678    920    609    385    244    142     81     24
```

STANDARD DEVIATIONS

```
MEAN ESD    /E/.GT .00  1.00  1.20  1.40  1.60  1.80  2.00  2.50  3.0
FOR /E1/ ESTIMATE  .14   .17   .18   .19   .20   .22   .23   .27   .2
FOR /E2/ ESTIMATE  .15   .17   .19   .20   .21   .23   .25   .28   .3
```

```
MEAN ESD    S2/L2  .014  .048  .094  .141  .187  .235  .282  .328  .37
FOR /E1/ ESTIMATE  .06   .05   .06   .07   .10   .12   .13   .18   .2
```

```
FOR /E2/ ESTIMATE    .07    .06    .06    .07    .11    .12    .14    .19    .2
```

*** BEGIN GENSIN ***

INPUT FILE FROM UNIT 02. FILE HISTORY..
STICK6 STARTX 27/01/87 12:56:28 ADDREF 27/01/87 12:57:34
        SORTRF 27/01/87 12:58:25 GENEV 27/01/87 12:58:45

GENSIN CONTROL PARAMETERS

```
PROCESS BDF DATA SET NUMBER        1
NUMBER OF UNIQUE REFLECTIONS     2678
OVERALL SIN THETA/LAMBDA LIMIT   .649
GENERATOR /E/-TYPE  /E/-MIN       E1     1.56
/E/S GT 3.3 ARE ATTENUATED       YES
NUMBER OF GENERATORS ACCEPTED    273
NUMBER OF PSI(0) GENERATORS      100
GENERATOR SIN THETA/LAM RANGE   .042 TO  .649
PROB. FACTOR SIG3/(SIG2)**1.5   .0924
PRINT GENERATORS WITH NGEN  LE   273
PRINT INVARIANTS FOR N1 RANGE      0 TO    0
PRINT INVARIANTS WITH NGEN  LE   273
```

ORIGIN DEFINITION PARAMETERS

```
MINIMUM NUMBER OF ODR REQUIRED     3
STRUCTURE SEMINVARIANT VECTORS    +H  +K  +L
STRUCTURE SEMINVARIANT MODULI      2   2   2
ENANTIOMORPH SPECIFICATION REQUIRED
```

TRIPLET STRUCTURE INVARIANTS

```
NUMBER OF UNIQUE INVARIANTS      3479
RANGE OF PROBABILITY FACTORS     1.00 TO  5.26
MAXIMUM INVARIANTS/GENERATOR      121
NUMBER OF PSI(0) TRIPLETS         140
```

QUARTET STRUCTURE INVARIANTS

```
NUMBER OF UNIQUE INVARIANTS      1932
RANGE OF PROBABILITY FACTORS      .67 TO  3.54
MAXIMUM INVARIANTS/GENERATOR      640
QUARTETS WITH XVSUMS IN RANGE     .25 TO  2.50  155
QUARTETS WITH XVSUMS IN RANGE    4.00 TO  8.63 1777
```

```
NGEN   H    K    L   /E/  SIGE   U  V  W  POS ODR  RESTR.  NT3  SUM A  NQ4
   1  14    0    1  3.40   .30   0  0  1  YES      90/270   57  140.5  346
   2   7    0    2  3.39   .22   1  0  0  YES       0/180   67  160.0  370
   3  15    6    0  3.39   .33   1  0  0  YES      90/270   45  109.4  199
   4  14   14    1  3.30   .39   0  0  1  YES        NONE   54  132.3  175
   5   2    2    1  3.29   .17   0  0  1  YES        NONE  114  259.1  418
```

```
 6   1   6   2  3.00  .18   1 0 0   YES   NONE    116  260.1  640
 7  15   0   1  2.97  .26   1 0 1   YES   90/270   43   96.5  216
 8   2  14   7  2.95  .39   0 0 1   YES   NONE     46   93.4   60
 9   0   5   4  2.93  .20   0 1 0   YES   90/270   46   95.4  233
10   1  14   0  2.90  .24   1 0 0   YES   90/270   48  113.2  250
11   3   8   5  2.87  .24   1 0 1   YES   NONE     77  165.9  189
12   6   8   5  2.81  .25   0 0 1   YES   NONE     80  170.2  172
13   1  16   0  2.79  .26   1 0 0   YES   90/270   48  100.2  106
14   8   0   1  2.75  .17   0 0 1   YES   90/270   50  100.7  109
15   5   8   5  2.72  .23   1 0 1   YES   NONE     78  157.8  120
16   5   3   9  2.68  .35   1 1 1   YES   NONE     40   77.9   18
17  10   5   5  2.63  .24   0 1 1   YES   NONE     66  123.4   73
18   3   3   8  2.56  .28   1 1 0   YES   NONE     46   81.6   13
19   8   6   4  2.53  .20   0 0 0   NO    NONE     82  157.8   95
20   3   3   9  2.52  .32   1 1 1   YES   NONE     46   83.0   15
21  13  16   0  2.51  .32   1 0 0   YES   90/270   25   44.9    9
22   4  15   0  2.47  .22   0 1 0   YES   0/180    14   25.2   12
23  14   6   1  2.47  .22   0 0 1   YES   NONE     73  131.6   48
24   8   0   6  2.46  .22   0 0 0   NO    0/180    26   48.7   26
25   0  13   1  2.44  .19   0 1 1   YES   90/270   49   90.0   56
26   0  14   0  2.41  .20   0 0 0   NO    0/180    43   74.8   49
```

*** BEGIN GENTAN ***

INPUT FILE FROM UNIT 02. FILE HISTORY..
STICK6 STARTX 27/01/87 12:56:28 ADDREF 27/01/87 12:57:34
       SORTRF 27/01/87 12:58:25 GENEV 27/01/87 12:58:45

GENTAN CONTROL PARAMETERS

```
START PHASING PROCEDURE        MAGIC
TANGENT PHASING PROCEDURE      CASCADE
TANGENT WEIGHTING SCHEME       W2
PROCESS BDF DATA SET NUMBER     1
GENERATOR /E/-TYPE /E/-MIN      E1   1.56
NUMBER OF GENERATORS PHASED    273
REDUCE A AND B VALUES BY       1.00 SIGMA
EXPECTATION VALUE FORCED >      .50
FRAGMENT PSI TO BE APPLIED      NO
MAX ALLOWED TRIPLETS/GENER     100
NUMBER OF TRIPLETS ENTERED     3428     AMIN= 1.0
MAX ALLOWED QUARTETS/GENER     100
NUMBER OF QUARTETS ENTERED     451      BMIN=  .7
POSITIVE QUARTETS (PQ4)        366      XVSM> 4.0
NEGATIVE QUARTETS (NQ4)         85      XVSM< 2.5
NQ4 USED ONLY IN FOM, PSI=     180
```


STARTING PHASE INFORMATION

INVARIANTS USED IN SELECTION   T3 + PQ4

```
SELECTION OF MAXEXT BASED ON    MAX GENS

NGEN    H    K    L    /E/  PHI RESTR  TYPE  WGT  FIX ASSIGNED  PERMUTE
 14     8    0    1   2.75   90   90   ODR   1.0   30 MAXCON
 13     1   16    0   2.79   90   90   ODR   1.0   30 MAXCON
 25     0   13    1   2.44   90   90   ODR   1.0   30 MAXCON
  7    15    0    1   2.97   90   90   EDR   1.0    5 MAXCON
 19     8    6    4   2.53   45   NR   PER   1.0    5 MAXCON   18.0 *  4
 15     5    8    5   2.72   45   NR   PER   1.0    5 MAXCON   18.0 *  5
142     8    8    3   1.82   45   NR   PER   1.0    5 MAXEXT   18.0 *  3


START SET CONNECTS  93 OF TOP 100 SORTED GENERATORS IN FIRST PASS
MAX PHASE SETS TO BE TESTED    20


PHASE REFINEMENT PARAMETERS
---------------------------

FIX PHASES WHEN BLOCK REFINED YES
MAXIMUM REFINEMENT CYCLES       30
ACCEPT PHASE IF AL**2 MAX >   2.49
REDUCE AL**2 MAX AT RATE OF    .70
TERMINATE IF AL**2 SHIFT <      2 %
PHASE ACCEPTED IF WEIGHT >     .15


FIGURE OF MERIT PARAMETERS
--------------------------

NUMBER OF PSI(0) TRIPLETS      140
OPTIMAL RFOM AND AMOS WEIGHT  1.00  1.00
OPTIMAL RFAC AND AMOS WEIGHT   .25  1.00
OPTIMAL PSIO AND AMOS WEIGHT   .75  2.00
OPTIMAL NEGQ AND AMOS WEIGHT  60.0  1.70
EXPECTED AVERAGE PHI (APHI)    166


SUMMARY OF PHASE SET RESULTS
----------------------------

NSET NPHI APHI DPHI RMSD NEGQ NINV NNQ4 NCYC  RFOM RFAC PSIO
  1   273  157    1   63   95 14854  420   26  1.36  .22  .87
  2   273  167    0   57   88 14691  405   23  1.20  .16  .97
  3   273  161    3   73  118 12745  373   17  1.66  .29  .94
  4   273  159    0   60   95 15112  417   18  1.23  .19 1.06
  5   273  161    3   67   96 12501  311   20  1.50  .26  .65
  6   273  167    1   56   88 14619  402   20  1.20  .16  .96
  7   273  175    1   65   94 14135  380   20  1.35  .22  .82
  8   273  172    1   69  113 13860  368   19  1.51  .26  .65
  9   273  174    2   64  104 14393  401   18  1.31  .22  .92
 10   273  169    2   73  120 12149  338   19  1.72  .31  .94
 11   273  175    2   65   95 14266  386   25  1.35  .22  .83
 12   273  158    0   55   79 15649  434   21  1.21  .15  .92
 13   273  170    2   67  110 12048  290   22  1.46  .26  .72
..........data omitted for brevity


PHASE SET ASSESSMENT (IN ORDER O/P TO BDF)
```

```
-------------------------------------------
PSET   NSET   RFOM   RFAC   PSIO   NEGQ   *CFOM*   AMOS(%)
  1     12    1.21    .15    .92     79     .78       76
  2      6    1.20    .16    .96     88     .66       69
  3      2    1.20    .16    .97     88     .65       68
  4      5    1.50    .26    .65     96     .65       68


INPUT FILE FROM UNIT 02. FILE HISTORY..
STICK6 STARTX 27/01/87 12:56:28  ADDREF 27/01/87 12:57:34
       SORTRF 27/01/87 12:58:25  GENEV  27/01/87 12:58:45


*** BEGIN FOURR ***


INPUT FILE FROM UNIT 01. FILE HISTORY..
STICK6 STARTX 27/01/87 12:56:28  ADDREF 27/01/87 12:57:34
       SORTRF 27/01/87 12:58:25  GENEV  27/01/87 12:58:45
       GENTAN 27/01/87 13:18:08


FOURR CONTROL PARAMETERS
------------------------
FOURIER TRANSFORM TYPE    E-MAP
FOURIER TRANSFORM PHASE   PHASE SET 1
ORIENTATION OF  X  AXIS   DOWN   WITH GRID  70
ORIENTATION OF  Y  AXIS   LAYER  WITH GRID  68
ORIENTATION OF  Z  AXIS   ACROSS WITH GRID  32
MAP BOUNDARY IN X  GRID      0 TO  70    INCR.=  1
MAP BOUNDARY IN Y  GRID      0 TO  17    INCR.=  1
MAP BOUNDARY IN Z  GRID      0 TO  32    INCR.=  1
MAP BOUNDARY IN X COORDS  .000 TO 1.00   A/PNT=  .24
MAP BOUNDARY IN Y COORDS  .000 TO .250   A/PNT=  .24
MAP BOUNDARY IN Z COORDS  .000 TO 1.00   A/PNT=  .23
MAP FRACTION OF CELL VOL  .2500


SUMMATION ALGORITHM       BEEVERS-LIPSON
SUMMATION ORDER OF HKL    2ND 1ST 3RD
DATASET NUMBER APPLIED    1
F(0 0 0) TERM  APPLIED       12.2
SYMMETRY EQUIV. MAXHKL    19    21     9
SIN(THETA/LAM) MIN MAX       .0420    .6487
INPUT GROUP SCALE(S)       1         1.1223
LAYERS SUMMED PER BDF PASS    18
REFLECTIONS APPLIED -     260
REFLECTIONS REJECTED      2418 NO COEFF.S


*** BEGIN PEKPIK ***


INPUT FILE FROM UNIT 03. FILE HISTORY..
STICK6 STARTX 27/01/87 12:56:28  ADDREF 27/01/87 12:57:34
       SORTRF 27/01/87 12:58:25  GENEV  27/01/87 12:58:45
       GENTAN 27/01/87 13:18:08  FOURR  27/01/87 13:18:13
```

```
PEKPIK CONTROL PARAMETERS
-------------------------
FOURIER MAP TYPE PROCESSED    EMAP
MAXIMUM UNIQUE PEAKS OUTPUT    58
PEAK OVERSTORE BUFFER HOLDS   250
MINIMUM SEPARATION OF PEAKS   .75   ANGSTR
NUMBER OF SYMOPS APPLIED        4
PRESET THRESHOLD DENSITY      2.00
```

ATOM OR PEAK SITES FOUND IN MAP

ATOM-PEAK	X	Y	Z	PEAK	GRIDX	GRIDY	GRIDZ
1	.4694	.0466	.3390	32.385	32.86	3.17	10.85
2	.5268	.1876	.1638	29.516	36.88	12.75	5.24
3	.5182	.1171	.0654	29.440	36.27	7.97	2.09
4	.4417	.1079	.0000	26.647	30.92	7.33	.00
5	.5311	.0449	.2031	26.402	37.18	3.05	6.50
6	.6044	.0510	.2844	26.182	42.31	3.47	9.10
7	.3839	.1118	.1163	25.050	26.87	7.60	3.72
8	.0212	.0186	.9531	22.766	1.48	1.27	30.50
9	.3922	.0444	.2533	20.990	27.46	3.02	8.10
10	.7898	.0495	.7072	17.802	55.29	3.37	22.63
11	.3318	.0526	.3961	17.757	23.22	3.58	12.68
12	.3029	.1070	.0000	16.824	21.20	7.27	.00
13	.2959	.0477	.8960	16.694	20.71	3.24	28.67

.........peaks omitted for brevity

55	.6238	.1072	.3236	7.312	43.66	7.29	10.36
56	.1243	.2106	.1660	7.297	8.70	14.32	5.31
57	.0381	.0824	.0216	7.141	1.97	5.60	.69
58	.7740	.1078	.7869	7.117	54.18	7.33	25.18

*** BEGIN MODEL ***

MODEL CALCULATION PARAMETERS

```
MAXIMUM SITES PROCESSED          58
BOND LENGTH LIMITS MIN/MAX     1.10    1.95 ANGSTR.
BOND ANGLE  LIMITS MIN/MAX     85.    145.DEGREES
MAXIMUM BONDS PER ATOM            4
HEAVY ATOM SITES EXPECTED         0
MAX RATIO LIGHT ATOM PEAKS     2.00
SUBCLUSTER SEARCH WEIGHTED       NO
```

ATOM AND PEAK SITES INPUT FROM FILEE

SITE	X	Y	Z	ATOM-PEAK
1	.46939	.04661	.33899	32.39
2	.52679	.18757	.16381	29.52
3	.51816	.11715	.06539	29.44

.........peaks omitted for brevity

```
56   .12427   .21063   .16596        7.30
57   .02809   .08242   .02159        7.14
58   .77404   .10781   .78691        7.12
```

SITES CONNECTED WITHIN LENGTH LIMITS ARE GROUPED IN CLUSTERS.
SITES CONNECTED WITHIN ANGLE LIMITS ARE GROUPED IN SUBCLUSTERS.

INTERPRETATION 1 OF CLUSTER 1

FOM BASED ON CONNECTS 1.19
FOM BASED ON FRAGMENT   .00
NUMBER OF SUBCLUSTERS   1

```
SITE   SUBCL   CONNECTED TO....                ASSIGNED   GROUP
  1      1      5   9   8  31
  2      1      3  14
  3      1      5   4   2  38
  4      1      3   7  37
  5      1      1   3   6
  6      1      5  16  36
  7      1      9   4  12  27
.........peak connections omitted for brevity
 56      1     28
 57      0
 58      0
```

L-S PLANE PROJ SITE CLUSTER 1 SCALE= 2.26CM/A   SITES: *RING +BOND .LONE XCELL

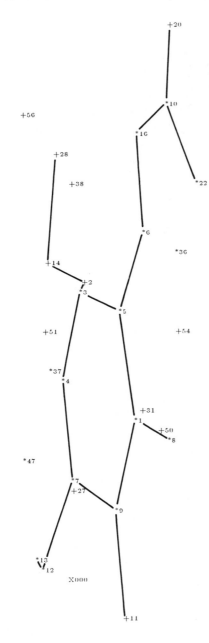

21

CRYSTALS, FOR TEACHING AND RESEARCH

David Watkin

The program CRYSTALS is a product of over 20 years of development. Its origin was a machine code program subroutine library (on paper tape — the subroutines were spliced together to form the program!) designed around a data structure definition which is substantially the same as that we use today. At that time, most crystallographers were also amateur programmers, so that copies of the program were always being modified to treat special problems or try out new ideas.

As time passed, the system was re-written, firstly in ALGOL and later in FORTRAN (3 times). At each re-write the structure and appearance of the program changed in response to the demands of the Chemical Crystallography Laboratory. Experience showed the major technical features required, and these became incorporated as standard, so that less tinkering with the code was done by students in general. Since these procedures were standard, it also made sense to be able to call them from the data, rather than having to write short master programs to call subroutines from a library. Also during this period, the excitement and novelty of computers began to wear off, and students and casual users became less tolerant and forgiving of difficult data input strategies — they wanted things to be 'user friendly'.

This trend, of expecting more from the program, continues, and our response to it can be seen in the current and projected facilities available in CRYSTALS. We can crudely divide users into several classes:

1. Novices who want to learn some crystallography.

2. Novices who don't, and only want to get results as quickly as possible.

3. Professionals, who want to get routine results as quickly as possible.

4. Professionals with difficult problems.

5. People with ideas about new algorithms.

The needs of these different classes are met with varying degrees of success by CRYSTALS, and we will start by considering type 2.

1 Getting results as quickly as possible.

This is the weakest link in CRYSTALS, because the strategies needed to achieve the results are not yet well systematised. During the 70's, the excitement and obvious importance of solving stuctures by Direct Methods led to a lot of money and expert man-power being applied to the problem, with the result that we now have programs like MULTAN and SHELXS. (I remark in passing that CRYSTALS contains NO Direct Methods facilities — it would be pointless to compete with either of the above programs, both of which can in any case be called by the user quite simply from CRYSTALS.) These programs have an incredible success rate even with structures at one time thought difficult.

Much less attention has been paid to the refinement of structures. The success of Direct Methods has lead to chemists cheerfully presenting bigger and bigger structures for analysis, leading to problems with data management and structure refinement. Two of the most obvious problems remaining unsolved are

1. Correlation and naming of atoms in the structure from a chemists representation of the structure. Molecular fragments can be identified on the basis of geometry, and used to produce drawings useful to the chemist. There is no serious difficulty in going through a 50

atom structure by hand, and assigning identifiers to atoms found in 'E' or difference maps, but the problem becomes tedious when there are 100 atoms, and very tedious there after. We need good algorithms to permit input of a schematic, perhaps via a graphics tablet, and associate this with the (possibly incomplete or incorrect) model fragment.

2. Automatic refinement of structures showing disorder, shortage of data, space group ambiguity, etc. The individual procedures for handling these problems are in many cases reasonably well understood, and can be applied by hand using current program systems. Few (if any) programs will automatically diagnose these problems and enter procedures to control them. The task remains of determining complex figures of merit which can be applied to all the information available at each point in the analysis — for example correlations between temperature factors, disorder, absorption and extinction corrections. Experienced crystallographers have 'feelings' about these things, and these need quantifying.

These and other similar problems remain to be solved if crystallographic refinement programs are to become as 'expert' as Direct Methods programs. I see them being solved in similar ways, using only traditional computer languages (FORTRAN), and fairly closed logical structures. I do not see crystallographers needing to get into 'Expert Systems' as such, though their systems must become expert.

2   Professional in a hurry.

CRYSTALS has a lot for the professional in a hurry. Of the 100,000 lines of FORTRAN in CRYSTALS, almost half are concerned with the user interface, data input and data management. The effort put into designing this code reflects our belief that a modern program should have a 'character', and that regular users should be able to develop a comfortable relationship with the program, and be able to work quickly and efficiently with it. It is expected that unless the user is constrained to working in batch mode, he will work in the mode most appropriate for the particular stage of the analysis, doing some work on-line, some interactively and some in batch. CRYSTALS aims to provide an operating environment in which complicated crystallographic manipulations and processes can be performed without too much reference to the manuals and in a manner not fatally sensitive to errors.

In 'command driven' mode, data and instructions are presented to CRYSTALS in packets, rather like spoken language paragraphs. These follow definitions given in a pro-forma data file, against which the program checks the user data for defaultable parameters. Because these data packets are structured, the program can give good diagnostics in the event of error. In addition, the user can ask the program to display part or all of a data or instruction packet from the pro-forma file, and so get immediate help about parmeters and defaults. Data need be given to CRYSTALS only once, when they are stored in a direct access file. This is modified or extended as calcultions proceed, building up a proper database for the structure.

In CRYSTALS, the old idea of program subroutine libraries has been replaced by 'data subroutines', in which the user prepares files of commands which can be called from the current control stream like subroutines.

A possible scenario for using CRYSTALS on an in-house computer is to prepare files containing groups of commands which are often used together, (such as those for data reduction), and store these in a 'data subroutine' library. At the beginning of an analysis, the basic crystallographic data would also be stored in a file (and also archived for security). These files can then be processed from CRYSTALS with the following instructions:

#USE basic
#USE reduction

where 'basic' and 'reduction' are the names of the pre-prepared files. Note that either of these files may themselves contain USE instructions (for example to load scattering factors from a library), so that the instruction USE is rather like a FORTRAN CALL statement.

The user then proceeds to enter command sequences, watching the results in the summary displayed on the terminal. If a problem emerges which cannot be solved from the information in the summary, the user can then print out the full listing to perform a post-mortem, otherwise these listing files can be deleted unseen. Once refinement starts, the user can input the controlling information (parameter selection etc) in interactive mode, and then use trivial control files to perform the refinement in batch e.g.

\#SFLS

REFINE

REFINE

CALCULATE

END

\#FOURIER

MAP TYPE=DIFFERENCE

\#FINISH

This is a controlling file for 2 rounds of least squares, a structure factor calculation and a difference map. All the necessary numeric data is fetched from the direct access file. For the professional in a hurry, CRYSTALS provides a consistent input syntax with data being identified either by position or by keyword, an on-line aide-memoire for commands, the ability to pre-prepare data and commands, and a minimal output to examine when things are going well. Every thing which can be done in batch can also be done on-line, so that the user can decide for himself (depending on the type and size of the task, the speed of the computer and its loading) which is the best way of doing a task.

3   Professional with a problem

An outline of the facilities available in CRYSTALS is given at the end of this paper, and they will be seen, with the exception of Direct Methods and computer graphics, to cover most aspects of non-protein crystallography. The philosophy behind the program has always been to make available the best techniques which we can find, and make minimal compromise in the crystallography to hardware limitations, because these disappear so fast. Equally, effort has been taken to make efficient use of the hardware available.

Our aim has been to make CRYSTALS a flexible set of crystallographic tools, which fit together easily and comfortably, but which introduce the minimum restraints on what users can do. Subroutine libraries have great flexibility, but require a lot of user understanding to work well. Separate programs linked through data files are easy to assemble, but often leave the user without a good mechanism for going quickly from one task to another. Single monolithic programs suffer from neither of these difficulties, but may impose restrictions on what the user can do. We have chosen this last alternative as the basis of CRYSTALS, but with a deliberate attempt to keep as many user and programmer options open as possible. This does not lead to the fastest programs possible, but that was not our primary design criteria. The most important thing is to enable users to do good crystallography, without wasting too much of their time wondering how to do it. It turns out that about 80% of cpu time used by crystallographers is spent in least squares, so that is the only real area where great efficiency must be achieved in the code, and since computers are continuing to become both faster and cheaper, there is no reason to compromise facilities for speed even here.

Flexibility of action is achieved by making the program command driven, so that the user can say what he wants to do next, rather than having to follow a menu or dialogue. Flexibility of computation is achieved by only making constant those things which are constant (e.g. π !), and having all other data external to the program.

About half the code in CRYSTALS is concerned with I/O and data management. This code is quite complicated, but because it is only executed occasionally (when compared with inner loops in sfls or Fourier summations) speed has been sacrificed in order to get code that is readable, and which gives users good diagnostics when things go wrong. Some of the code here is included to improve the overall efficiency of the program. For example there are disk caches for handling the traffic between the computations and the data base, which are managed by learning algorithms to help minimise this traffic. Almost all data used during computations is stored in 2 large arrays. By redimensioning these two arrays, CRYSTALS could be made to operate on almost any size of problem, and small proteins have been refined with it. However, because the refinement is performed assuming that much of the normal matrix will be computed, sparse matrix techniques are not used in its storage and the equations are solved via a Choleski inverter, so that CRYSTALS is not ideally suited for problems with more than 1,000 independent atoms.

The code has been shown to be efficient on paged and non-paged machines, and recent experiments on the CYBER 205 have shown that the code is effective on super-computers. The relatively slow disk access times on these machines have lead us to design an internal RAM disk subsystem, which we are currently implementing.

Our particular interest, because of our long association with John Rollett, is still in refinement processses. Increasingly frequently, structures are turning up which are disordered or have other difficulties associated with them. While the chemists who produced them would often be satisfied with the results from the first 'E' map plus a little cosmetic treatment for publication, we are generally concerned to produce refinements which would bear close scrutiny. In addition to a wide range of matrix building techniques for the least-squares, the program also includes many features found in the well known protein processing programs PROLSQ and CORRELS. The program contains many model building aids, including mapping partial or distorted structures onto ideal structures, forcing planarity and conformity to a rigid body thermal motion. The user has a lot of control over the distance - angles routine, and the molecule assembler.

4   The enthusiastic beginner.

They still exist.

When I come across a new system, I like to sit down and browse through the book that goes with it. It seems that this is old fashioned, and modern beginners like to start typing on the terminal from the word 'Go'. This is clearly not very effective with a command driven program, and to cater for this approach we have devised the 'Interrogative Interface'.

This is an in-line processor which is able to interrogate the user, data files, and by sending commands to the command driven interface of CRYSTALS, the users database. The questions asked by this processor are not built into it, but are themselves contained in data files (called SCRIPTS), so that no pre-defined crystallographic strategy is built into the processor. The processor is in fact an interpretor, and the SCRIPTS are programmes written in a PASCAL-like language. The SCRIPT language is block structured, with if-then-else constructs, logical, character and numerical operations, file handling, condition and error handling, and the ability to call other SCRIPTS. The decision making possibilities during SCRIPT processing make the system look a little intelligent, but as yet it is far from an expert system because it neither records how it has arrived at a particular point, nor is able to modify its own decision making procedures. A embryonic scheme exists in which SCRIPTS can write other SCRIPTS, but we have not yet achieved any major results with this.

The importance of SCRIPT processor as a teaching aid comes from its ability both to ask for data, and to display information. The text may either be built into the SCRIPT itself, or be taken from files. These SCRIPTS and files can easily be modified by the tutor or the students. Though the SCRIPT processor is an integral part of CRYSTALS, it communicates with the normal interface by writing normal CRYSTALS commands to the input buffer. In other words, the SCRIPT is a program which can make subroutine calls to CRYSTALS, where the calls take the form of normal commands. This is inevitably much less efficient than programming in FORTRAN and calling the actual CRYSTALS subroutines directly, but it is also much safer since the normal command processor can check the generated commands for errors, and the actual commands are well documented in the USER MANUAL (over 230 pages) and so provide a well defined high level interface.

So for example, a simple SCRIPT may just prompt the user for numerical values and then pass these to CRYSTALS, or it may give a more or less detailed description of what it is about to do, or it may do several different things and ask the user to select the most appropriate outcome. It is very flexible and powerful, but the worst thing it can do to CRYSTALS is pass in erroneous data. Because the SCRIPT itself IS only data, it does not have the potential to corrupt the crystallographic programming.

Because SCRIPTS can call other SCRIPTS, hierarchical decision trees can be devised, and these can lead the student through a particular path in his analysis. However, the SCRIPT processor can itself be made the object of a normal CRYSTALS command, so that experienced user can just pick particular branches or leaves from the tree and execute them without reference to the main tree. SCRIPT mode is in fact subservient to command driven mode.

5   Innovators.

Occasionally we get users who want to do more than just use CRYSTALS, but who want to modify and extend it. If their idea is of a mainly crystallographic nature, such as adding new parameters to the LS refinement, or modifying the weighting schemes, then it is likely that they will have no great difficulty. Simple subroutine calls will get them data from the data base, and the code is very modular and well commented (about 1 card in 3). It is only in the command interpretor that they may have problems. CRYSTALS provides a good environment for trying out ideas.

6   Facilities available

Simple syntax for refering to individual atoms, and to groups of atoms, with or without the application of symmetry.

Data reduction of photographic and diffractometer data, including empirical absorption corrections. Application of decay curves, rejection of systematic absences, merging of equivalent reflections with special treatment of outliers, processing of both X-ray and neutron diffraction data.

Structural parameter editor for modifying individual parameters, atoms or groups of atoms. Modifications available include: addition, subtraction, multiplication, division of parameters; application of matrices to coodinates; renaming, deletion, reordering generation of atoms; application of symmetry operators; selection by type or parameter value; interconversion between U[aniso] and U[equiv]

Structural analysis and model building. Best line and plane through groups of atoms, projection of other atoms onto these planes, forcing planarity onto parts of structure. TLS analysis of groups of atoms, forcing U[aniso] for groups to conform to rigid body motion. Comparison between a group of atoms and another group, a geometric figure or structure from the literature. Model a group of atoms onto another group, figure or structure. Geometric hydrogen and other atom placing.

Reflections to be skipped or included are selected individually or by tests. Testable parameters include F_o, F_c, sigma, I/sigma(I), phase, sin theta, index, zone, etc. The test conditions can be changed at any time without loss of data.

Restrained refinement. Distances and angles restrained to numeric values or their mean values; components of U[aniso] along bonds; planarity of groups; sums and averages of parameters; shift limiting restraints; and user defined FORTRAN-like expressions given as data.

Highly flexible least-squares process, including robust-resistant methods. User can define up to 256 blocks per cycle. Blocks can be as small as one parameter per block, one atom per block, or full matrix, or any intermediate combination of parameters. Any atom parameter shifts (X's,

U's, Occ) can be equivalenced, linked, made to ride. Parameters can be redefined as sums and differences. Atoms can be refined as rigid groups. Shift multipliers can be applied, the current matrix can be re-used, structure factors need only be calculated for partial structures. Data for up to 8 twin components can be refined. Refinable parameters include overall scale, overall Uiso, Dummy Uiso, Larson extinction, Rogers polarity, Flack enantiopole, batch scales, layer scales, twin components, x, y, z, Occ, Uiso, Uaniso.

14 weighting schemes are available, including statistical, unit, Hughes Cruickshank and Dunitz schemes. Weighting optimisation via a Chebychev polynomial can be used to get 'correct' parameter esds. Analysis of residuals by F_o, sin theta, index, class, parity.

Fourier maps are computed with any number of points at any resolution along x, y, z. Map types are F_o, F_c, $F_o - F_c$, $2F_o - F_c$, F_o^2, F_c^2. Maps can be weighted, including Sim weighting. Peak search gives indication of peak quality. Automatic Fourier refinement, peak rejection and molecule assembly. Lineprinter and VDU plots of atoms and peaks.

Comprehensive distance and angle calculation, with selection by atom type, interatomic distance, connectivity, inter molecular, intramolecular. Symmetry included. Esds from the full variance-covariance matrix. Torsion angles, planes, lines, TLS (with bond correction), principal axes of Uaniso.

Publication listings of atomic parameters, distances, angles, torsion angles, reflections.

CRYSTALS is an evolving system, and occasionally metamorphoses. We are particularly indebted to Bob Carruthers (now with Control Data) for implementing the data base management, and to Paul Betteridge (now with Chemical Design Ltd) for building the SCRIPT processor.

'EDITION 1'   (Data base definition)

                (Cruickshank, Freeman, Rollett, Truter, Sime,

                Smith and Wells, 1964)

'NOVTAPE'   (AUTOCODE)

                (Hodder, Rollett, Prout and Stonebridge, Oxford,

                1964)

'FAXWF'   (ALGOL)

                (Ford and Rollett, Oxford, 1967)

'CRYSTALS'   (FORTRAN)

                (Carruthers and Spagna, Rome, 1970)

'CRYSTALS'   (FORTRAN)

                (Carruthers, Prout, Rollet and Spagna, Oxford,

                1975)

'CRYSTALS'   Issue 2 (FORTRAN)

                (Carruthers, Prout, Rollet and Watkin, Oxford,

                1979)

'CRYSTALS'   Issue 7 (FORTRAN)

                (Betteridge, Prout and Watkin, Oxford, 1984)

22

SHELX

Ward Robinson and George M. Sheldrick

The 5000 lines of FORTRAN code which became known as 'SHELX-76' had their origin in about 1970 when the Cambridge University Titan computer was replaced by an IBM-370. Unfortunately the IBM was not familiar with 'Titan Autocode' and the only practical proposition was to write the program in FORTRAN. A simple subset of FORTRAN, which bore a curious resemblance to Titan Autocode, proved adequate for the purpose and saved having to read the FORTRAN manuals. This subset was carefully chosen to be easy to port to other computers, e.g. the ICL machines then flourishing in the U.K., to avoid the trauma of having to write the program ever again. It was never intended to distribute the program outside Cambridge, though eventually after six years of bug exorcism there was sufficient interest to make it worth releasing a one and only official version (SHELX-76). It is believed that this program is still used from time to time in some parts of the world.

   With hindsight, it was fortunate that Titan Autocode was very close to machine language (and hence efficient), and that IBM and ICL FORTRANs were almost completely immiscible. The least-squares routines were based extensively on the general space-group independent algorithms described by Cruickshank (1970). However the SHELX code is convincing evidence that its author had never looked at a FORTRAN program or attended a computing course. The overriding programming consideration was to make sure that every bit of data space and every line of code was working overtime so that the program was as compact as possible. An interesting side effect, which may well account for the survival of the program in its original form, is that it was extremely difficult (even for the author) to understand and hence 'improve' the code. The technique of extremely compact coding proved invaluable in the preparation of the mini-computer version (SHELXTL, incorporated in the Nicolet diffractometer package) which required only 20K 16-bit words of memory for the data and program (which made extensive use of overlays). The dramatic fall in memory costs has now made such considerations of minor importance, even when programming for micro-computers.

1   Current Developments — SHELXS-86 and SHELXL-90

By the 1980's it was clear that improvements in techniques of solving crystal structures and the development of computer architecture would require reconsideration of some of the assumptions made in writing SHELX. A much wider choice of programming languages was available, and good cases could be made for writing in for example C or PASCAL. FORTRAN is didactically and aesthetically so abysmal that it was by no means an automatic choice. On the other hand FORTRAN has been around for long enough for reliable optimising compilers to be developed for virtually all computers, whereas C is to some extent dependent upon having UNIX as an operating system, and many implementations of C and PASCAL handle floating-point operations less efficiently than FORTRAN (e.g. by assuming that all floating point numbers are stored in 64 rather than 32 bits). However the overriding argument for FORTRAN is that the 'FORTRAN-77' standard has now been very widely implemented, and is the only de facto machine independent language. Indeed programs which adhere strictly to the FORTRAN-77 standard are so transportable that there is little advantage in distributing them in preprocessor code such as RATMAC.

Since structure solution and refinement are not normally performed in the same job it was decided to release the new SHELX in two parts, both about the same size as the original SHELX-76. The structure solution part, SHELXS-86, was been 'beta-tested' extensively on a wide variety of computers ranging from Crays to PC's, and a resulting 'final version' has been distributed to several hundred institutions. It is planned to name the structure refinement part 'SHELXL-90'. In the following discussion, all remarks about direct and Patterson methods refer to SHELXS-86, and not to the rather dated methods employed in SHELX-76.

1.1   The User Interface

From the crystallographer's point of view, the ease of communicating with a program is probably the most important factor in determining whether he will use it. When SHELX was first written, communication with FOR-TRAN programs usually took the form of fixed format input on punched cards, and the idea of free format input with extensive use of 'default values' (for parameters not specified by the user) was fairly novel. The input was decoded in standard FORTRAN, avoiding machine-specific routines. The concept of 'default values' is taken a step further in SHELXS-86: the program makes an intelligent guess at each unspecified parameter, on the basis of all other available information. Since these values are based on extensive experience with many types of crystal structure, the program could be termed an 'expert system'. This is particularly true of direct methods, where in most cases the only decision which the user needs to make is how hard to hit the structure, i.e. how many random starting phase permutations are necessary; this is a question of balancing computer resources against the probability of cracking the structure. Of course an experienced user, who knows something about the problem which the program doesn't, can and should intervene by specifying more parameters himself.

A simple criterion is that an experienced user should not normally need to look at the manual (however some SHELXS users take this to extremes, and use BITNET instead !). It is essential to keep the input

SIMPLE, and to reduce the number of files required or generated by the program to an *ABSOLUTE MINIMUM*. For SHELXS-86 these consist of an (ASCII) intensity data file, which is read once at the start of each job and should not normally be altered, and a free format instruction file which contains crystal data, instructions and (possibly) atoms. If it is more convenient, these two files may be combined. Some operating systems may also require a macro or command file for submitting batch jobs.

1.2 Programming for the General Case

Next to a good user interface, the second most important requirement for a crystallographic program (at least for small molecule work) is that all operations should be valid for all space groups, in conventional settings or otherwise, and that there should be no significant restrictions on the number of atoms, atom types, reflection data etc. SHELX-76 is restricted as to the maximum number of atoms and least-squares parameters, but is otherwise very general; an expanded version is also available. SHELXS-86 and SHELXL-90 avoid these restrictions by the technique of dynamically packing all atom parameters etc. into one large real array at the end of blank common; a certain amount of deviousness is required to store atom names and other character strings as floating point numbers without deviating from strict (subset) FORTRAN-77. In many implementations of FORTRAN, the maximum dimension of this array need only be specified in the main program, making it particularly easy to create differently sized versions of the program for different job classes etc. Although this technique was pioneered in SHELX-76, it was not fully exploited there. This 'dynamic memory allocation' also reduces page faulting in virtual memory systems. SHELXS-86 is written in an exact subset of 'subset FORTRAN-77' but in such a way that a change of one (or less !) characters per subroutine converts it to 'FORTRAN-66'. All that is necessary is to comment out the line which defines a few variables as type CHARACTER*1 so that they become (by default) integer variables in which not more than one character is stored per element.

1.3   Efficiency

Despite advances in computer hardware, it is still as important as ever to write efficient code: expectations have increased, and on average crystal structures are getting larger. The art of programming efficiently consists of following certain general rules in the bulk of the code and then identifying and hand optimising the rate determining 'inner loops'. The choice of algorithms is also fundamental. SHELX *NEVER* uses arrays with more than one dimension, and buffers all input-output to the binary scratch files so that only the array name appears in the WRITE or READ statement. 'Inline code' is almost always used in preference to a subroutine or function call; this frequently enables redundant operations to be eliminated. Trigonometric functions etc. are evaluated by table lookup in critical stages. A minimum number of simple variables (as opposed to array elements) are used for temporary storage so that the compiler can use registers rather than memory locations wherever possible. Some discretion is required in applying these rules, since they may make the code less compact and also more difficult to follow; a further disadvantage is that using a higher level of compiler optimisation on SHELX rarely brings much improvement!

1.4   Data Reduction Procedures

SHELX-76 was written primarily for Weissenberg photographic data and two circle diffractometer data, and so for other geometries the user had to include the 'direction cosines' of the incident and diffracted beams in the intensity data file if he wished to perform absorption corrections. The number and locations of the Gaussian grid points were optimised in three dimensions to achieve the desired precision. The linear least-squares method of Rae and Blake (1966) was used to determine 'inter-batch' scale factors. The sort-merge procedure is valid for any number of reflections and is almost linear with the number of data; it makes more disk transfers if less main memory is available. Systematically absent reflections are identified by means of structure factor algebra using the input symmetry operators. Special attention has been given to the intensity statistics printed by SHELXS-86 so that they provide a reliable guide as to whether the full structure and each of the three principal projections are centrosymmetric or not. A further very useful statistic printed by SHELXS-86 is the number of measured, 'observed', and theoretically accessible unique reflections in each resolution range. If the structure is non-centrosymmetric and less than half the theoretical number of reflections in the 1.1 to 1.2Å range are 'observed', then experience indicates that direct methods will not succeed in solving the structure; the assumption of 'atomic resolution' is apparently violated.

1.5 Parallel Processing in Direct Methods

The rate-determining computations in crystal structure determination are generally easy to vectorise, but in the case of direct methods some changes of strategy are required. In the multisolution approach, made popular by the widely used program MULTAN, the phases of the reflections with the largest E-magnitudes are refined (usually by the tangent formula) using a set of phase relations. This is repeated for a large number of starting sets, and the 'best' solution(s) selected according to one or more figures of merit. In SHELXS-86 the phase refinement is based on both the triple phase relations and the negative quartets (Schenk, 1974; Hauptman, 1974) in order to overcome problems in space groups that lack translation symmetry.

Application of the tangent formula, especially when preliminary - screening and negative quartets are incorporated as in SHELXS-86, is not obviously vectorisable. However, since memory is no longer a very scarce resource, it is possible to process a convenient number - let us say 64 - of phase permutations in parallel. The overhead of decoding each (packed) phase relation is then incurred only once per 64 permutations. Separate loops are employed (e.g. for handling reflections that belong to centrosymmetric projections) rather than including a test and jump inside the loop, and the loops are carefully hand-optimised. This technique, which might be called 'software array processing' at least halves the time taken by phase refinement; if parallel processing hardware can be used, there will be a further gain. For example, the phase refinement in SHELXS-86 is automatically vectorised by the CFT compiler on the CRAY-1, improving the performance from 35 VAX to 136 VAX (1 VAX equals a VAX 11/780 with floating point accelerator). The FORTRAN compiler for the FPS array processor also automatically vectorises the critical loops, which are identified by comments for hand programming of other parallel processing hardware.

Partly because of the advances in computer performance, direct methods have gradually changed from a matter of skill and luck to simple computational brute force. We appear to have more success in solving large structures when random starting phases are used for all reflections. It is then simply a matter of running (parallel) phase permutations until the structure cracks. We have recently solved several structures which only gave one correct solution every 10000 (or more) attempts! In retrospect, the technique of selecting origin-determining and a few other starting-set reflections via a 'convergence map' (Germain, Main and Woolfson, 1971) probably increased the rate of success of solving the majority of structures at the cost of solving a few structures never.

1.6 Heavy-Atom Patterson Interpretation

About 30% of all crystal structures are solved by the 'heavy atom method', but SHELXTL and SHELXS-86 appear to be the first widely distributed programs that attempt a fully automatic interpretation of the Patterson in a way that (of course) is valid for all space groups and requires no prior structural information. It is however assumed that for at least one heavy atom, all self (Harker) vectors are present in the peaklist. Many of the difficulties of automation are avoided by not storing the calculated Patterson map! The peaklist contains a great deal of useful information, provided that we remember that peaks may overlap, and that weaker vectors may be missing. As a final check, the Patterson may be recomputed at selected points. Since, subject to the above assumption, the method is exhaustive, two figures of merit are necessary to select the solution which is to be processed further. These are a point atom R-factor based on E-values (RE), which is also employed to assign atomic numbers (from the list of given atom types) to the potential atoms, and R(Patt), which measures the agreement of the observed and predicted Pattersons. Easy structures may then be expanded in the same job to locate the light atoms by E-Fourier recycling; more difficult problems, e.g. involving pseudo-symmetry, require hand interpretation of a 'crossword table' which conveniently summarises the Patterson multiple superposition minimum function and the minimum interatomic distances. Only in this final interpretation is use made of chemical information, so unexpected structures cause no problems. The generation of the crossword table is the slowest stage in the procedure, but vectorises automatically (from 33 to 150 VAX) on the CRAY-1.

There are still two weak links in this procedure. Firstly the assumption that all Harker vectors are present in the peaklist may fail for not-so-heavy atoms (e.g. P, S, Cl) in large high-symmetry cells, and secondly the figures of merit are not (yet) as omnipotent as those employed in direct methods when it comes to identifying the correct solution. This latter problem may be remedied by putting one (or more) heavy atoms from a rejected solution in by hand, but the former can best be resolved by direct methods! (sometimes an expansion to P1 followed by hand interpretation of the crossword table is also successful).

1.7 Protein Applications

Both the direct methods and the Patterson interpretation routines in SHELXS-86 have been used successfully to locate heavy atoms from protein isomorphous delta-F data. Good scaling of the data is important, and it is necessary to experiment with omission of the low angle data (mostly contributed by the solvent) and high angle data (lack of isomorphism) or both. The unit-cell contents should be specified as the correct number of heavy atoms plus the square root of the number of light (native) atoms (which may conveniently be given as nitrogen). For Patterson interpretation it is advisable to ignore vectors less than (say) 6Å from the origin, and in cases of partial occupancy an expansion to P1 may be worth trying. For direct methods the number of phases for refinement must be given (the default would be too large), and the negative quartets may be ignored (though they sometimes help to eliminate the 'uranium atom' solution).

1.8   Parallel Processing in Least-Squares Refinement

The time taken by any least-squares refinement algorithm rises steeply with the number of parameters. One approach to speeding up structure refinement (employed at Goettingen) is to add an array processor so that the matrix buildup (the slowest stage) on the array processor is performed at the same time as the calculation of the derivatives etc. for the next reflection on the host Microvax II. The solution of the least-squares equations may also be performed on the array processor. The Analogic AP-508 array processor always finishes the matrix addition before the Microvax has completed the next reflection, and the solution of the equations only requires about 40 seconds for 800 parameters, so this solution is very effective for large full-matrix problems (the total time is less than twice that for the corresponding number of structure factor calculations).

1.9   Least-Squares Constraints and Restraints

The number of least-squares parameters and hence the time required may also be reduced by means of chemically sensible constraints. For example, it is usually possible to predict hydrogen atom positions more precisely than they can be refined using X-ray data alone. A particularly convenient constraint for CH and CH_2 groups is the riding model (introduced in SHELX-76) defined by the equation:

$$\vec{x}_{(H)} = \vec{x}_{(C)} + constant\ vector$$

which makes the calculation of derivatives for least-squares refinement particularly simple. Although only the C-H distances and H-C-H angles remain rigorously constant, in practice the procedure of alternating coordinate idealization with riding refinement converges very rapidly. If (groups of) hydrogen atom temperature factors are constrained to be equal (by setting them equal to 'free variables') then virtually no extra parameters are required to include the hydrogen atoms in the refinement. Methyl groups may be defined as 'rigid groups' (6 geometrical parameters) or (in SHELXL-90) as 'tied rigid groups' (4 parameters, which may be considered to be x, y and z of the C atom plus a torsion). SHELXL-90 uses the quaternion method (Mackay, 1984) to fit arbitrary fragments to e.g. Fourier peaks in preparation for rigid group refinement. Alternatively a library of standard geometries may be used for the automatic generation of distance and other restraints, which take the form of additional least-squares observational equations.

The available reflection data (and possibly computer resources) may be inadequate to allow full anisotropic refinement of large organic or organometallic structures. On the other hand some allowance for anisotropic motion (or possibly small local disorder induced by partial solvent occupancy) may still be desirable, especially for peripheral atoms. The 'rigid bond' restraint (Hirshfeld, 1976) is easy to apply and chemically the most acceptable, but an additional restraint or constraint to force a bond direction to be one of the principal axes of the thermal ellipsoid of a terminal atom saves two further (effective) parameters per atom, and so is being investigated for SHELXL-90.

2 SHELX on Personal Computers

Since good optimising FORTRAN-77 compilers are now available for MS-DOS/PCDOS based personal computers, SHELX can readily be adapted to micro-computers. In the case of SHELXS-86, a specially optimised pre-compiled version has been distributed; on the faster micros (the Olivetti M24 SP with 10MHz 8086 and 8087 chips is the fastest so far tested) the program runs a little faster than on the VAX 11/730 with floating point hardware. This optimised version does include some assembler routines, but most of the performance improvement resulted from reorganising the memory allocation (the large array in blank common has been split into three) so that *ALL* integer operations are 16-bit. Limited real-time facilities have been included in this version, so for example hitting the escape key causes the program to complete the current batch of phase permutations, then proceed to E-Fourier recycling of the best solution rather than further phase generation (e.g. because it is obvious from the output on the screen that the correct solution has already been found). An alternative cost-effective strategy is to run SHELXS on a satellite 32032 or 68020 processor board which plugs into the PC. Since the PC may also serve as a terminal for a mainframe, and SHELXS requires only two input files, both of which are ASCII, it is particularly convenient to have the same program running on the PC and the mainframe. A single 'floppy' serves as a convenient archive for one crystal structure; the data may be recovered much faster than by using magnetic tape on a mainframe.

3 SHELX Distribution Policy

The public domain programs SHELX-76 and SHELXS-86 have been distributed to a large number of crystallographers without regard to political or religious affiliations or the ability to contribute to the costs involved. SHELXS-86 is supplied on the same tape or diskettes as the Patterson search program PATSEE written by Ernst Egert, which is highly integrated with SHELXS. The programs and documentation are always supplied free to third world countries and others in which it is difficult to raise foreign exchange, and are also supplied free via network transfer (documentation is sent free by post in such cases). Academic crystallographers who insist on receiving tapes or diskettes and are in a position to contribute are asked to donate $99 to the costs, and profit-making institutions are expected to contribute $999; these industrial users thus subsidise the rest, and in particular cover documentation costs. In addition we try to repond to all questions and suggestions from users; BITNET has proved to be by far the most convenient mechanism for such communication (GMS can be reached at UAC10I DGOGWD01), and has the advantage that computer output and data files can also be transmitted if necessary to resolve a problem.

In addition it should be mentioned that an integrated X-ray diffractometer package incorporating all the features of SHELX is available from Nicolet XRD, 5225 Verona Road, Madison WI 53711, USA [(608)273-5048 or toll free (1-800)356-8088]. The system, known as SHELXTL for Data General computers or SHELXTL PLUS for the Digital Microvax series, includes additional proprietary programs for interactive space group determination and instruction file preparation, empirical absorption corrections, preparation of tables for publication, and extensive interactive molecular graphics.

We are particularly grateful to all SHELX users in many countries for their patience, encouragement and suggestions, many of which have been incorporated in the new versions SHELXS-86 and SHELXL-90.

REFERENCES

Cruickshank, D. W. J. (1970). In *Crystallographic Computing* (ed. F. R. Ahmed, S. R. Hall and C. P. Huber) pp. 187-197. Munksgaard, Copenhagen.

Germain, G., Main, P. and, Woolfson, M. M. (1971). *Acta Crystallographica*, **A27**, 368-376.

Hauptman, H. (1974). *Acta Crystallographica*, **A30**, 472-476.

Mackay, A. L. (1984). *Acta Crystallographica*, **A40**, 165-166.

Rae, A. D. and Blake, A. B. (1966). *Acta Crystallographica*, **20**, 586.

Schenk, H. (1974). *Acta Crystallographica*, **A30**, 477-481.

Computing methods

23

PROGRAMMING FOR INTERACTIVE STRUCTURE ANALYSIS

Eric Gabe

1   Introduction

The field of crystal structure analysis, for small molecules at least, is well suited to interactive computing. The computations involved are not very time-consuming for a modern computer and it can be highly productive to solve and refine structures interactively at a terminal. The flexibility of a well-written interactive program combined with the crystallographic skill of a user are a powerful combination in the early stages of the analysis. Interactive graphics are a great advantage at all stages.

The emphasis here, is of course on the term 'well-written interactive program' — and I shall attempt to describe features that I think are important for such programs and outline some of the advantages of interactive computing for structural crystallographers.

In the past, computing schools have tended to emphasise the features and use of specific crystallographic programs, systems and techniques, while not putting a great deal of emphasis on the 'computing' aspect of the topic. It seems to me that a computing school, crystallographic or otherwise, should be concerned with computing itself, as well as new techniques and systems in its particular area of specialization. With this in mind I intend to talk about one aspect of modern computing, namely interactivity.

Many of the ideas I will discuss have been incorporated into the

NRCVAX structure package (Gabe, Lee, and Le Page 1985). However, it is not my intent to talk about the package, but rather the interactive concepts embodied in it.

2 What is Interactivity?

Roughly, interactive computing is taken to mean a real-time dialogue between the output of a running program and the input of an operator — usually by means of a terminal. The dialogue can be a series of questions and answers, or selections from text or pictorial menus, usually interspersed with some form of results.

The manner in which man/machine interaction is acheived can be extremely hardware dependent, particularly for graphics, and while it may be tempting to use specific hardware features the resulting programs are not portable. The interactive concepts I will discuss can be applied on any standard ASCII terminal. The examples are taken from crystallographic routines, though the ideas could equally be applied to any interactive routine.

The ease of use of such programs is very dependent on the amount of programmer effort spent in setting up the interactive sections. The amount of code required to achieve satisfactory interaction can be anywhere between 10 and 50% of the program, depending on the application. Unfortunately, people tend to have different opinions of what they consider satisfactory interactivity. Some people — usually those who have not seriously tried — claim that they do not like interactive computing and others expect such systems to be able to cope sucessfully with the most glaring of operator deficiencies. Most people fall somewhere between these two extremes. They will approach such a system with a basic knowledge of the operations involved and some idea of the order in which these tasks should be performed.

In the case of structure analysis this means that most users will have some understanding of the underlying theory and some appreciation, for example, that a Fourier cannot be computed until some phases are known or that least squares cannot be performed with no atoms. Doubtless systems will eventually be written which will deal with even these situations.

3 Achieving Useful Interactivity

3.1 Reducing the Possibilities for Error

The wide spectrum of operator expertise does cause problems for the writer of interactive systems. If the system is written for the novice, it will be frustrating to the experienced user and if it written for the expert it will be unusable by the beginner. One way out of this dilemma is to take most decisions away from the user. This leads to inflexibility. A better solution is to provide some form of 'help', which is available only if wanted, i.e. by the novice, whereas the expert is allowed to interact quickly and efficiently. Unfortunately, this means more programming effort and adopting different strategies with programs which are run with different frequencies. A frequently run structure factor/least squares routine should be runnable with a minimum interaction, whereas a less frequently run routine, e.g. a powder pattern indexer, can afford to have a more comprehensive set of instructions and questions which will guide the novice and refresh the expert's memory.

From the computer's point of view it is unfortunate that human operators make typing mistakes. It is necessary that the opportunities for these to occur be minimized and that appropriate corrective action be taken whenever they do occur. There are several ways that this may be done.

1. The number of characters required to be typed should be kept to a minimum and the user should never have to input the same quantity more than once. This obviously reduces the chance of error.

2. There should be a liberal and consistent use of clearly indicated defaults. These reduce the typing and give an indication of the expected answer.

3. All numeric typed input should be handled by a screening routine (Hall, 1984) which either ignores or questions unexpected characters in input strings, e.g. an alphabetic character in the middle of the numeric string. This routine should treat all incoming characters as alphameric strings and generate numeric fields itself, rather than relying on an in-built formatter. It is convenient and easy to arrange that real and integer values are generated in cases where there is an ambiguity. An incidental advantage to such a screening routine is that the code becomes much more portable.

4. Most questions can be phrased so that yes/no answers (usually y or n) are all that is required. Again, it is convenient to have a routine which will always provide a standard answer and reject any unexpected input. If the expected default answer is indicated to the routine then the returned answer can always be Y or N. This simplifies the logic which inevitably follows such a question.

5. Questions should be phrased as clearly and concisely as possible. These two requirements are often in conflict. Short questions can be unclear and long ones become annoying if repeated often. If possible the user should never be unsure of the expected form of response, i.e. whether the answer is 1, 2 or 3 or A, B or C. If there is unavoidable ambiguity the coding after the question should take care of it.

6. Checks should always be made that the actual response is a valid choice. With multiple choice questions, if the answer is not a permissible one, the question should be repeated. In cases of numeric input, which cannot be checked, the answer might be echoed for verification, though this can become tedious.

The following examples illustrate some of these points.

```
What is the crystal diameter in microns ?
```

The question is short and clear, no default can be assigned, but it might be worthwhile to check that the answer is between some reasonable limits.

Type the cell parameters

This is not a good prompt. It is fairly clear, but leaves some doubt. Are the values to be given on one line or 6 ? Is it necessary to type redundant parameters and fixed angles ? A better approach would be to ask for only the essential values based on the crystal system or Laue group.

Type the space-group symbol

This prompt also leaves something to be desired, but the solution is not so simple. One needs to know which form of symbol, how complete, and the syntax in which it should be typed. It is probably safest to refer to a writeup, although a longer explanation might be given on the terminal as it is likely to be a rarely asked question.

Another example of such a situation is taken from a routine to calculate hydrogen atom positions. Hydrogen positions can be calculated for several geometries and the required type must be indicated. The following prompt, while lengthy, is clear and should not cause even a novice user much problem.

```
There are 3 types of H-atom generation
2 H-atoms  AB---X--H2     Methylene   (M)
1 H-atom   AB---X--H      Planar      (P)
1 H-atom   ABC---X--H     Tetrahedral (T)
Enter the type of H position (M,P,T) wanted (P)
```

The last line illustrates another point. The default value (P) is spatially separated from the list of valid answers (M,P,T) because the the phrasing

```
......wanted (M,P,T)(P)
```

would be confusing. Also the single letter answers are mnemonic rather than a,b,c or 1,2,3 because such choices lead to fewer errors (Goldstein, 1984).

```
Do you wish to continue (n) ?
```

It should be clear from the context exactly what is to be continued. The default answer is 'n' or 'N' and only the characters n,N,y,Y or <cr> are valid answers.

3.2   Using Defaults

In an interactive program system defaults can have two meanings. A default value can be a value assigned to a program variable, which will be used unless otherwise indicated, e.g. wavelength, atomic radii, scattering factors, map resolution etc. This is the normal form of default. They can also be default answers to the interactive prompts. In this sense they greatly help to reduce the amount of typing necessary and hence the possibility of error. It is important that defaults be assigned wherever they can be and that they be indicated in a consistent and obvious way throughout the system.

The default concept can be expanded in several ways. In a series of similar questions, if the answer to the first question is N for example, then N becomes the default from then on until there is an answer other than the default. Another possibility is to make the program assign the default which is considered to be the most likely answer. In practice, these concepts, while having some merits, are found to be somewhat unsettling to the user. People become very used to system prompts and expect to have to answer them in the same way, unless they deliberately choose a different answer. Changing the rules in midplay is not appreciated.

3.3   Designing Good Dialogue

Questions and prompts eliciting answers and responses are the essence of man/machine interactions. Whereas poorly designed prompts can break a system, clear and well designed dialogue can produce efficient, and equally important, satisfying interactivity. Prompts which are overly compressed or incomplete are quite annoying. The extremely cryptic

I,M,O,S,T ?

which many of you will be familiar with, requires recourse to a writeup for explanation and has no default. It is also incomplete. The most frequently used answer is Q. Questions which disconcert the user are even worse. A question such as

What is the last block number ?

can cause great anguish to even an experienced user. The new user may have no idea what the last block number is and no knowledge of how to find out. The more knowledgeable user may know how to find out, but it may require leaving the program in order to do so. Such questions should be avoided or at least posed in such a way that it is possible to find the answer.

Not the least damaging aspect of poor dialogue is that it creates a bad impression of interactive computing and makes people feel that there are no benefits to be gained from persevering. Clear, well-thought-out dialogue is satisfying to the user and creates efficient interaction. I do not wish to suggest that the dialogue can be sufficient in all cases. An interactive system still needs a writeup, although it can afford to be less comprehensive than for a batch-processing system. One problem with designing interactive dialogue is the same as that for making a good writeup. The programmer knows what the answers to all the questions should be and therefore cannot see the inadequacies. Ideally the programmer should interact with users while designing the program dialogue.

As I have said before, lengthy dialogue becomes annoying in a frequently run program, but is helpful in an infrequently run program. Accordingly routines like least squares should have the option of allowing the knowledgeable user to run the routine with a minimum of essential terminal input, probably only the number of cycles of refinement to perform and accepting all other terminal input defaults, such as file-names, weighting scheme, reflection limitations on theta and Fo values etc. Allowing the user the option of selecting these things produces a flexible mode of operation, but it is safe to accept sensible defaults in most cases. Time-consuming programs like least-squares may in fact be run in batch-mode. This can easily be done with an interactive routine, but the reverse is certainly not true.

With less frequently run routines like Fourier calculations, one could adopt an intermediate strategy of presenting the user with a more complete dialogue of defaults, but still retain the possibility of accepting them with a single keystroke.

Other infrequently run routines should have full dialogue at all times. Examples of these are data reduction, structure solution routines, distance/angle calculations and table preparation routines.

Routines which are by their very nature interactive, such as file editing routines, are in a different category. In any structure system there must be a file which contains all the atomic information and possibly scattering and symmetry data as well. This file will be manipulated by many routines and the user will need to be able to edit the file as the structure solution progresses. The routine that does this is a specialised editor and, like most editors, it should be highly interactive. Users will probably interact with this routine more extensively than any other in the system and so it should be well designed. The routine will have many options and it is necessary that the user should be adequately prompted. However it would be extremely tedious if the routine showed all options after each operation. Some form of help is required. A simple way to achieve this is to make the routine print a synopsis of available commands if the user gives a null or incorrect answer to an option prompt. Thus if the routine prints something like

```
General Atoms Edit Option (Menu) ?
```

and the user responds with <CR>, a menu of the atoms edit options is printed and then the prompt repeated. This same idea can be used to great advantage for interactive graphics.

Another generally applicable concept is to group logically connected items into blocks on the screen. A blank line after each block serves to separate logical entities and gives the user a sense that one phase is completed and another started.

4 Simple Interactive Graphics

Graphics routines are necessarily hardware dependent and, because of this, it is more difficult to lay down rules for interactivity. Writing a graphics routine with specific hardware in mind is tempting because of all the powerful features the hardware may offer, however the result may be completely non-portable. The very sophisticated Evans and Sutherland or Silicon Graphics units are good examples of this, though obviously for large molecule work the benefits outweigh the disadvantages. For small molecule work less complex devices are adequate and a good rule might be 'keep it simple'. The only real rquirements for acceptable small molecule graphics are the capability of clearing the screen and drawing a line between two points. A great deal of work has been done (Goldstein, 1984; Maguire, 1982) on the efficiency of different forms of prompting and response mechanisms. Icons can be used to represent operations and there is some evidence to suggest that this leads to fewer input errors. The use of a mouse can also be more efficient than straight typing, but in both cases (Maclean, Barnard and Wilson, 1985) users tend to revert to typed input if it is available to them.

With this in mind it is reasonable to design an interactive graphics routine to make minimum special hardware demands, use as much of the screen as possible for drawing and leave a minimal alphameric prompt as the normal mode of operation. Just as with the file editor described above, help can be made available with a single keystroke or as a specific command for more comprehensive information. If commands are made short and mnemonic they are easy to learn, e.g. RO – rotate, MA – magnify, ST – stereo etc. users will quickly remember a working subset and find the routine comfortable to use.

5 Benefits of Interactive Computing

I have tried to outline some of the ideas behind the design of interactive routines and I hope it now clearer that writing good programs for interactive use is not just a matter of adding some questions to a batch routine. In view of the fact that interactivity requires so much programming effort, over and above that needed to carry out the essential calculations, one may well ask what are the benefits to be gained and why bother?

There are a few trivial reasons. The disruption caused by having to refer to a writeup is removed. The amount of typing required to perform a task is usually less than that needed to prepare the corresponding batch file. In the event of a major blunder it is easy to stop the task before wasting much user or computer time.

There are other substantial benefits however. Essentially these derive from the use of total computing resources, human and machine. The batch-processing environment we are all used to is really a system designed for the convenience of machines. The traditional justification for batch-processing may be summarised as — machines are expensive, interactive programs imply multi-user operating systems which do not use machine time efficiently, therefore we will use the more efficient batch-processing mode. This argument places little or no value on human time, even less on convenience and none at all on satisfaction.

As time passes this reasoning becomes less and less true even in economic terms. Machines are getting cheaper and human costs are continually increasing. It therefore makes sense to provide people with tools which will increase their productivity while allowing them, if possible, to gain satisfaction from what they are doing. As I said at the beginning of the talk, crystallographic calculations are mostly not very time-consuming and it does not make sense to carry them out in the stop-go manner of batch processing. It is more natural to progress from one operation to another in a smooth continuous flow. There is then no disruption of thought and considerable satisfaction to be gained from controlling the flow of operations in an intelligent way. If the computer response-time is long, interactive computing can become very frustrating. However, this situation is steadily improving and as more powerful micro-computers become available it will be increasingly possible to have a machine, more powerful than a VAX, entirely to oneself.

The idea of deriving satisfaction from performing a task has to do with the psychological concept of 'closure' or the feeling of having completed a task. The more difficult the task the stronger the closure (Maguire, 1982; Goldstein, 1984). An interactive computing system gives the user many opportunities to experience closure in a very immediate sense. Closure and response-time are closely linked. If one performs a trivial computing task e.g. rotating a molecular drawing, one expects more-or-less instant response and it is very dissatisfying if response is slow. For a task where the perceived calculation is much larger, a Fourier calculation for example, one is prepared to wait much longer, but still have a sense of satisfaction from having initiated the operation. This is not some form of confidence trick that the interactive system plays on the user. It is how human beings react to situations and I think this is the major reason why a good interactive system is so satisfying to use.

It is probably in the graphics area that interactivity is most beneficial. The ability to display and manipulate molecular drawings easily and rapidly is extremely useful at all stages of structure analysis. The solution phase is often enhanced with a little judicious pruning of a molecule based on chemical information, especially for 'difficult' structures, During the analysis, geometry can be quickly checked for 'reasonableness', possible hydrogen atom positions can be quickly accepted or rejected and packing diagrams can be examined thoroughly. Preparation of final drawings is much less of a chore if the graphics routine is interactive. The picture can be composed and manipulated on the screen and only sent to the plotter when it is in final form.

I hope I have been able to make you think about the pros and cons of interactive computing. With the proliferation of powerful micro-computers and the advent of voice-controlled systems, more and more computing will be done interactively. Crystallographers have always been among the first to take advantage of advances in the computing field. I hope this will be no exception.

REFERENCES

Gabe, E. J., Lee, F. L. and Le Page, Y. (1985). In *Crystallographic Computing 3*. (ed. G. M. Sheldrick, C. Kruger, and R. Goddard) p. 167. Clarendon Press, Oxford, U.K.

Goldstein, J., (1984). *Report No. UIUCDCS-R-84-1173*, Dept. of Computer Science, University of Illinois, Urbana, Illinois, U.S.A.

Hall, S. R., (1984). In *Methods and Applications in Crystallographic Computing.* (ed. S. R. Hall and T. Ashida) p. 343. Clarendon Press, Oxford, U.K.

Maclean, A., Barnard, P. J., and Wilson, M. D. (1985). In *People and Computers: Designing the Interface,* (ed. P. Johnson and S. Cook) p. 172. Cambridge University Press, Cambridge, U.K.

Maguire, M., (1982). *Int. J. Man—Machine Studies,* **16**, 237.

24

SYMBOLIC PROGRAMMING IN CRYSTALLOGRAPHY

Uri Shmueli

Summary

Interactive computer systems for algebraic simplification have been available for about two decades. We propose to describe some outstanding features of such a system, called REDUCE, and outline in some more detail its participation in a sequence of symbolic programming operations the need of which arose in a crystallographic application. It is concluded that among the most important features of symbolic programming are (i) the facility to expand automatically complicated algebraic expressions and output them in a FORTRAN compatible form, and (ii)programmed programming. Although specially designed symbolic languages are easiest to use in producing such codes, numerically oriented languages, such as FORTRAN, are often capable of producing highly efficient simple symbolic programs.

1 Introduction

Most of crystallographic programming is numerically oriented, and serves computations related to various stages of structure determination, refinement and interpretation. However, simple symbolic manipulations have not infrequently been attempted by programmers. Among well known examples are the symbolic addition procedure (Karle and Karle, 1966), in which symbols are assigned to a few unknown signs or phases and a simple but non-trivial symbol algebra must be incorporated in the computer program, and automated derivation of space groups starting from well defined space-group symbols (Fokkema, 1975; Hall, 1981; Burzlaff and Hountas, 1982; Shmueli, 1984). There are still simpler symbolic manipulations, which are related to preparations of output listings in a desired form and format. All these can be conveniently carried out making use of character-handling facilities which are provided by most numerically oriented high-level computer languages.

Much more extensive symbolic manipulations are needed when even routine algebraic operations are to be performed. Numerically oriented languages, like FORTRAN or PASCAL, may still be employed for this purpose but one usually faces the need to write a more or less intricate program, which is dedicated to a restricted range of algebraic operations. Nevertheless, such dedicated programs often turn out to be most efficient (Shmueli and Kaldor, 1981).

A solution to the problem of automation of algebraic computations seems to be offered by the computer language LISP, which allows one to perform a wide range of symbolic operations. LISP has been in existence since the early 60's (McCarthy, Abrahams, Edwards, Hart and Levin, 1962). However, soon after the invention of LISP it became clear that the construction of a language whose syntax is similar to that of commonly employed numeric computer languages, but which translates to fundamental LISP commands, would make symbolic programming accessible to a wider range of users. An answer to this need was provided by Hearn (1967), who proposed a language with an ALGOL-like syntax, which enables one to program algebraic calculations. This language is called REDUCE, and several versions of it have been prepared over the years. There are also other symbolic languages in existence (*Computers in the New Laboratory - A Nature Survey*, 1981), and others are being developed.

The purpose of this lecture is to illustrate symbolic calculations by examining some simple REDUCE programs and their outputs and conclude with a real application of such symbolic programming to a crystallographic problem.

2   Some Examples of Algebraic Programs

We shall present in this section some examples of actual algebraic programs, written in REDUCE, without entering into the details of the syntax. It appears that participants who have some acquaintance with PASCAL or ALGOL should find the syntax of REDUCE almost familiar.

One of the interesting features of REDUCE is the possibility to define global relations, which will hold throughout the program run; in other words, the user can define his own algebra. For example, suppose that we wish to evaluate the function $cos(\alpha + \beta + \gamma + \delta)$ in symbolic form. There is no built-in algorithm for this, so that we have to teach REDUCE some trigonometry. Since REDUCE is a recursive language, it will be sufficient to define $cos(\alpha + \beta)$, $sin(\alpha + \beta)$ and the symmetry of cos and sin (to be abbreviated by c and s respectively). A self explanatory code follows.

```
OPERATOR C,S;
FOR ALL X,Y LET C(X+Y)=C(X)*C(Y)-S(X)*S(Y);
FOR ALL X,Y LET S(X+Y)=S(X)*C(Y)+C(X)*S(Y);
FOR ALL X LET C(-X)=C(X), S(-X)=-S(X);
    .

    .

F:=C(AA+BB+CC+DD);
    .
```

$$(2.1)$$

REDUCE will now expand the cosine of the sum of the four angles AA, BB, CC and DD in a sum of products of single sines and cosines, as shown below.

```
F   :=   S(DD)*S(CC)*S(BB)*S(AA) - S(DD)*S(CC)*C(BB)*C(AA) -
         S(DD)*S(BB)*C(CC)*C(AA) - S(DD)*S(AA)*C(CC)*C(BB) -
         S(CC)*S(BB)*C(DD)*C(AA) - S(CC)*S(AA)*C(DD)*C(BB) -
         S(BB)*S(AA)*C(DD)*C(CC) + C(DD)*C(CC)*C(BB)*C(AA) .
```

The characters E and I are reserved for the basis of natural logarithms and for $i = \sqrt{-1}$ respectively. There are several operators which are built in; e.g., the kth derivative of y with respect to x:

$$\frac{\partial^k y}{\partial x^k} \quad \text{is written in REDUCE as} \qquad \text{DF(Y,X,K)} . \tag{2.2}$$

Suppose that we want to convince REDUCE to calculate the Hermite polynomials $He_k(x)$ for k = 4, 6, 8 and 10, and would like to transfer these polynomials to a FORTRAN program, without ever retyping them. It appears simplest to use the Rodriguez formula for these polynomials, although the use of recurrence relations might be somewhat faster. We have (Abramowitz and Stegun, 1972)

$$He_k(x) = (-1)^k \, exp(x^2/2) \, \tfrac{\partial^k}{\partial x^k}[exp(-x^2/2)] \tag{2.3}$$

The REDUCE program for this calculation is

```
ARRAY HE(10);
FOR K:=4 STEP 2 UNTIL 10 DO WRITE
HE(K):=E**(X**/2)*DF(E**(-X**2/2),X,K);          (2.4)
```

and its output is

$$\text{HE}(4) \quad := \quad X^4 - 6*X^2 + 3$$
$$\text{HE}(6) \quad := \quad X^6 - 15*X^4 + 45*X^2 - 15$$
$$\text{HE}(8) \quad := \quad X^8 - 28*X^6 + 210*X^4 - 420*X^2 + 105$$
$$\text{HE}(10) \quad := \quad X^{10} - 45*X^8 + 630*X^6 - 3150*X^4 + 4725*X^2 - 945.$$

This certainly suffices for an inspection of the polynomials. We can now make use of the flag FORT, which causes the output to be written in FORTRAN compatible format, and open an output file with the aid of the OUT statement (see REDUCE User's Manual)

```
ON FORT;
OUT FIL;
FOR K:=4 STEP 2 UNTIL 10 DO WRITE HE(K):=HE(K);
SHUT FIL;                                        (2.5)
```

The Hermite polynomials will now be written to file FIL, one polynomial per line or FORTRAN record, starting in column 7 and not exceeding column 72. If longer expressions are involved, REDUCE will take care of a proper arrangement of continuation lines.

This software interface between REDUCE and FORTRAN is a very convenient feature, since we can generate with the aid of REDUCE correct FORTRAN codes for lengthy algebraic expressions, and thus eliminate from the process of programming many tedious and error-prone steps. This is an application of algebraic programming to an automated construction of numerically oriented code.

Symbolic languages (*Computers in the New Laboratory - A Nature Survey*, 1981) usually offer facilities for manipulating algebraic texts, such as factoring out desired terms, controlling the degree of simplification of algebraic expressions and collecting coefficients of successive powers of a given variable. An illustration of the latter facility is in order, and we shall apply it to finding the conjugate, ZC, of a complex algebraic expression, Z, which will be regarded as a polynomial in i. Suppose

$$Z:=(U + I*V)**5; \qquad\qquad (2.6)$$

is the expression. The operator COEFF(F,X,VV) regards F as a polynomial in X and places the coefficients of the successive powers of X in the vector VV0, VV1, ... , where VV0 is the constant term, VV1 is the coefficient of X etc. We thus write

```
COEFF(Z,I,ZZ);
ZC:=ZZ0 - I*ZZ1;                                    (2.7)
```

The need of reducing a complicated algebraic expression to manageable form is common experience. Such problems arose in a study of generalized intensity and structure-factor statistics (Shmueli and Wilson, 1983), and were solved with symbolic-programming techniques. In the calculation of higher moments of a sum of random variables we come across multiple summations of the form

$$\sum_{i\neq}\sum_{j\neq}\sum_{k\neq}\sum_{l} x_i y_j z_k u_l \qquad (2.8)$$

where each index has the same range of values, but no two indices can assume the same value in any of the terms of (2.8). In order to analyze such expressions before their implementation in a numeric program it is usually necessary to decompose the multiple summation in (2.8) into a combination of single summations. The problem is, of course, a correct count of the single summations and their products. Consider the identity involving the corresponding double summation

$$\sum_{i\neq}\sum_{j} x_i y_j = \sum_i x_i \sum_j y_j - \sum_i x_i y_i \qquad (2.9)$$

and define the symbolic operators

$$S(X) \equiv \sum_i x_i \,, \quad S2(X,Y) \equiv \sum_{i\neq}\sum_{j} x_i y_j \,, \ etc.,$$

with a convention that a comma separating two arguments means that their indices are different [as in (X,Y)] while a multiplication sign is equivalent to equal indices and implies a contraction on these indices. The required declarations in the REDUCE program which is supposed to perform a decomposition of (2.8) into single summations are thus:

```
OPERATOR S,S2,S3,S4;
FOR ALL X,Y LET S2(X,Y) = S(X)*S(Y) - S(X*Y);
FOR ALL X,Y,Z LET S3(X,Y,Z) = S2(X,Y)*S(Z) - S2(X*Z,Y)
                - S2(X,Y*Z);
FOR ALL X,Y,Z,W LET S4(X,Y,Z,W) = S3(X,Y,Z)*S(W) -
          S3(X*W,Y,Z) - S3(X,Y*W,Z) - S3(X,Y,Z*W);  (2.10)
```

and so on, if more summations have to be dealt with. For example, the fourfold summation SUM := S4(A,B,C,D); is expanded by REDUCE to

```
SUM  := - 6*S(D*C*B*A) + 2*S(D*C*B)*S(A) + 2*S(D*C*A)*S(B) +
      S(D*C)*S(B*A)
    - S(D*C)*S(B)*S(A) + 2*S(D*B*A)*S(C) + S(D*B)*S(C*A) -
      S(D*B)*S(C)*S(A)
    + S(D*A)*S(C*B) - S(D*A)*S(C)*S(B) + 2*S(C*B*A)*S(D) -
      S(C*B)*S(D)*S(A)
    - S(C*A)*S(D)*S(B) - S(B*A)*S(D)*S(C) + S(D)*S(C)*S(B)*S(A),
```

by substitution of the global declarative statements (2.10).

3   Symbolic programming and space-group symmetry

We conclude with a brief description of an application of symbolic programming to automatic generation of space-group dependent expressions. The actual problem involved the automated preparation of the expressions for the trigonometric structure factors for all the space groups, and arose during the preparation of Volume B of *International Tables for Crystallography* (Shmueli, 1987).

The structure factor is given by

$$F(\mathbf{h}) = \sum_j f_j [A_j(\mathbf{h}) + iB_j(\mathbf{h})] \tag{3.1}$$

where

$$A_j(\mathbf{h}) = \sum_{s=1}^{g} \cos\left[2\pi\mathbf{h}^T(\mathbf{P}_s\mathbf{r}_j + \mathbf{t}_s)\right] \tag{3.2}$$

and

$$B_j(\mathbf{h}) = \sum_{s=1}^{g} \sin\left[2\pi\mathbf{h}^T(\mathbf{P}_s\mathbf{r}_j + \mathbf{t}_s)\right]. \tag{3.3}$$

In Equations (3.2) and (3.3) g is the number of symmetry-equivalent atoms to atom j within the unit cell, $(\mathbf{P}_s, \mathbf{t}_s)$ is the s th space group operation, $\mathbf{h}$ is the diffraction vector and $\mathbf{r}_j$ is the position vector of atom j, referred to the space-group origin. The problem is to expand the trigonometric functions within the summations (3.2) and (3.3), perform the summations and reduce the results to simplified and concise form for all the space groups and all the parities of hkl for which A and B have different functional forms.

The desire to automate this calculation was motivated, in part, by the tedium that it might involve if carried out by hand. Not less important, however, was the fact that hand calculations of this sort are rarely free of human errors and this danger would be avoided if the calculation were automated. Finally, it was found desirable to present the results in a concise form, by representing them in terms of the basic building blocks of the expressions derived. These building blocks are sufficiently well defined to entrust their recognition and separation to a simple computer program. The stages of the calculation of A and B for one space group are outlined below.

(i) Automatic generation of the coordinates of the general positions starting from a computer-adapted space-group symbol (5). The space-group symbol is the only user-supplied input to the calculation.

(ii) Formation of the scalar products appearing in (3.2) and (3.3), and their separation into components depending on the rotation and translation parts of the space group operations:

$$\mathbf{h}^T(\mathbf{P}_s, \mathbf{t}_s)\mathbf{r} = \mathbf{h}^T\mathbf{P}_s\mathbf{r} + \mathbf{h}^T\mathbf{t}_s \equiv S_{rot} + S_{trans}. \tag{3.4}$$

The above separation is performed by a FORTRAN interface which accepts its input from stage (i), and prepares the components in REDUCE format as arguments of trigonometric functions to be susequently expanded.

(iii) Analysis of the translation-dependent parts of the scalar products, and automatic determination of all the parities of *hkl* for which *A* and *B* must be calculated. The corresponding REDUCE assignments are generated by the FORTRAN interface.

A complete REDUCE program for the calculation of *A* and *B* in symbolic form, for a given space group, is then assembled and output by the FORTRAN interface.

(iv) Symbolic expansion of equations (3.2) and (3.3) with the aid of the REDUCE program prepared in (ii) and (iii), and their reduction to trigonometric expressions comparable to those given in Volume I of *International Tables for X-ray Crystallography* (1952).

(v) Representation of the results in terms of a small number of building blocks, of which the expressions for *A* and *B* were found to be composed (12). This interpretive task was did not require very sophisticated pattern matching and was thus carried out by a FORTRAN routine, which accepts as input the trigonometric expressions for *A* and *B* generated by REDUCE and outputs concise symbolic forms of *A* and *B* for this space group.

These abbreviated representations of A and B are the entries in the structure-factor Tables in Volume B of *International Tables for Crystallography* (1952).

The above manner of proceeding from space-group symbol to Table entry was followed for all the space groups except some trigonal and hexagonal ones.

There are two points that ought to be made, in connection with the above examples. Firstly, algebraic computations are very easily programmed in appropriate symbolic computer languages, but the execution of such extensive computations may sometimes be rather slow. If the expressions are simple enough to be handled by high-level languages, and their evaluation is highly repetitive, the use of (say) FORTRAN or PASCAL may give rise to enhanced efficiency. This was strongly suggested by stage (v) of the calculation of A and B. Another point of interest is programmed programming. As shown above, one of the important applications of algebraic computing is an automatic generation of computer code, the manual input of which might be a more or less rich source of typing mistakes and could of course be tedious and time-consuming. Programmed programming appears to be a feasible way of generating computer codes which are linked to certain well-defined mathematical expressions and, as such, it seems to be of interest for various phases of crystallographic computing.

REFERENCES

Abramowitz, M. and Stegun, I. A. (1972). *Handbook of Mathematical Functions*, New York: Dover.

Burzlaff, H. and Hountas, A. (1982). *J. Appl. Cryst.*, **15**, 464–467.

Computers in the New Laboratory — A Nature Survey (1981). *Nature*, **290**, 193–200.

Fokkema, D. S. (1975). *Acta Cryst.*, **A31**, S3.

Hall, S. R. (1981). *Acta Cryst.*, **A37**, 517–525.

Hearn, A. C. (1967). In *Proceedings of the ACM Symposium on Interactive Systems for Experimental Applied Mathematics*, held in Washington, D.C., August 1967.

International Tables for X-ray Crystallography (1952). *Volume I.* (ed. N. F. M. Henry and K. Lonsdale). Birmingham: Kynoch Press.

Karle, J. and Karle, I. L. (1966). *Acta Cryst.*, **21**, 849–868.

McCarthy, J., Abrahams, P. W., Edwards, D. J., Hart, T. P. and Levin, M. I. (1962). *LISP 1.5 Programmer's Manual.* Cambridge (Mass): M.I.T. Press.

Shmueli, U. (1984). *Acta Cryst.*, **A40**, 559–567.

Shmueli, U. (1987). In *International Tables for Crystallography, Volume B.* (ed. U. Shmueli). In preparation.

Shmueli, U. and Kaldor U. (1981). *Acta Cryst.*, **A37**, 76–80.

Shmueli, U. and Wilson, A. J. C. (1983). *Acta Cryst.*, **A39**, 225–233.

25

PROGRAMMING SUPERCOMPUTERS FOR CRYSTALLOGRAPHIC APPLICATIONS

R.P. Millane

1. INTRODUCTION

Crystallographic analyses of large molecules such as proteins and viruses often require high performance computer systems providing high speed computation and large memory capacities, particularly when calculations are performed on large data sets or are performed repeatedly many times. Examples of calculations on large data sets include processing large numbers of optical density data from film packs to obtain observed amplitudes, Patterson searches, structure factor and electron density calculations involving large numbers of reflections, atoms or real space sampling points, and non-bonded contact and packing searches. Repeated calculations are required in, for example, refinement of alternative model structures, density modification or molecular replacement, and refinement using x-ray data and stereochemical restraints. The Cyber 205 at Purdue for example, has been used to refine alternative models of complex low symmetry nucleic acid structures (Millane *et al.*, 1984; Millane *et al.*, 1985; Chandrasekaran *et al.*, 1985) and the determination of a number of virus structures (see Arnold *et al.*, 1987 for example).

The term *supercomputer* is a dynamic concept that refers to the fastest and largest computer operationally available at any particular time. At present, it refers to vector pipeline

machines that include the Control Data Corp. Cyber 205, the Cray
Research Inc. Cray-1, Cray X-MP and Cray-2, as well as some
machines produced in Japan (notably by Fujitsu and NEC). There
are differences between the these machines, but they all achieve
their high performance by the use of *vector pipelines* (or
pipes). The Cyber 205 and Cray-1 have one processor and 1,2 or 4
pipes, the Cray X-MP is available with 1,2 or 4 processors, and
the Cray-2 has 4 processors. The main memory available is
typically 1-16 Mwords (at 64 bits) except the Cray-2 which has 64-
256 Mwords. The *peak performance* is the rate at which the vector
pipes produce results and is measured in millions of floating
point operations per second (MFLOPS). The peak performance of the
Cyber 205 is 50 MFLOPS per pipe, the Cray-1 80 MFLOPS per pipe,
the Cray X-MP 100 MFLOPS (per processor), and the Cray-2 about 1
GFLOP (1000 MFLOPS). The successor to the Cyber 205 will be the
ETA 10 developed by ETA Systems (founded by CDC), will have 8
processors, each with two pipes and a peak performance of about 3
GFLOPS. The Cray-3 is expected to have a peak performance of at
least 10 GFLOPS.

The performance of vector computers can only be realized in
practice if programs and algorithms are taylored for optimum
utilization of their capabilities. Algorithms that are efficient
on scalar machines may need to be completely rearranged to perform
efficiently on a vector machine. It is worth emphasizing that if
scalar codes are implemented, unchanged, on a vector machine the
performance obtained will probably be orders of magnitude less
than that from a properly reconfigured algorithm. New programs
written with the machine architecture in mind lead to the most
efficient code. In many cases however, it is more cost effective
to *vectorize* existing scalar programs than to completely rewrite
them.

The principles of vector programming are described in §2 and
illustrated with reference to the Cyber 205 and Cray-1. In §3
some of the special vector processing facilities typically

available are described with specific reference to the Cyber 205. Some examples of performance improvements obtained by vectorizing a polymer refinement program are described in §4 and the impact of these machines is summarized in §5.

2. VECTOR COMPUTERS AND GENERAL CONCEPTS OF VECTORIZATION

In this section I describe some general principles of programming for supercomputers that apply to most machines. The Cyber 205 and Cray-1 are used as examples since these machines characterize two of the main architectures. Literature on the design of vector algorithms is scattered and the best source of information at present appears to be conference proceedings describing applications of supercomputers (Gary, 1984; Gentzsch, 1984; Kowalik, 1984; Devreese, 1985; Numrich, 1985; Feilmeir *et al*, 1986; Schonauer and Gentzsch, 1986). The book by Lazou (1986) is also good. Some specific references on vector algorithms useful in crystallographic calculations are: for linear algebra: Dongarra *et al* (1984), Duff (1984), Lambiotte (1984); for the FFT: Swarztrauber (1984), Temperton (1984) and sorting: Brock *et al* (1981).

Vector computers exploit several concepts to enhance their performance over conventional computers. These are *pipelining, chaining,* and *overlapping*. Pipelining involves splitting the function to be performed into smaller stages and allocating separate hardware to each stage. Consecutive operands (forming a *vector*) are streamed into a pipeline and after an initial *start up time* (the pipe length) results are produced every clock period provided the operands are supplied to the first stage every clock cycle. The rate is therefore independent of the length of the pipeline. For example, if two vectors of length m are added and the adder requires 3 clock periods to complete, 3+m clock periods are required rather than 3m in a conventional computer. Chaining is a technique in which the results of one vector operation are fed directly into a second functional unit for the next vector operation. After the start up time of both units, results of the combined operation are produced once per clock cycle. Operations involving a scalar and two vectors, called a *linked triad*, can be chained on many machines, doubling the peak performance. It is also possible to overlap two independent operations using different hardware with the second operation beginning one clock cycle after the first.

The key to utilizing these computers effectively is replace loops with vector instructions using long vectors (to minimize the effect of startup times) and to minimize memory references. Performance of an algorithm can be dominated by memory traffic rather than floating point operations. *Vector mode* refers to the use of vector instructions and *super-vector mode* refers to the use of chaining and minimization of data movement. Super-vector mode must be used to obtain the most out of a vector machine. Higher levels of parallelism are possible when multiple processors are available.

Code can be vectorized either automatically or explicitly (or both). Automatic vectorization refers to the use of compiler options, or pre-compilers such as VAST (provided by CDC) or KAP (Kuck and Associates, Savoy, Illinois 61874, USA), to automatically vectorize source code. Explicit vectorization refers to the programmer specifying where vector instructions should be executed, and reformulating algorithms to maximize the use of vector instructions. Explicit vectorization results in the most efficient code as it makes full use of the programmer's skill and knowledge of the algorithm. Execution times of different sections of the code must be measured to determine which will be most usefully vectorized. For example, vectorizing a section that consumes 10% of the execution time can produce a maximum speedup of 11% whereas vectorizing a section that uses 90% could produce an overall speedup of 1000%. Timing measurements should be made as vectorization proceeds as the proportion of time consumed by sections of the code changes as some are vectorized. Most systems provide program analysis tools that include a *profiler* to measure execution times in different sections and a *vectorization summary* that shows which loops the compiler vectorized, which it could not, any why.

The primary difference between the Cyber 205 and the Cray-1 is the way in which data is transferred between memory the functional units. The Cyber 205 is a memory-to-memory machine in that the pipes operate on data from main memory to main memory, but the Crays are register-to-register machines in which the pipes operate only with operands stored in special vector registers. On the Cyber 205, a vector is a series of contiguous memory locations. The Cray-1 has 8 64 word vector registers and it can fill these with values from a series of memory locations, the addresses of which are incremented by a constant *stride*. Also, the Cyber 205 is a virtual memory machine while the Cray-1 is not.

Because of the startup time of a pipeline, the actual performance approaches the peak performance only for long vectors (Fig. 1). The effect of vector length on performance can be described by the vector length $n_{\frac{1}{2}}$ required to obtain half peak performance (Hockney and Jesshope, 1981). On the Cyber 205 $n_{\frac{1}{2}}$ is about 50 for 64 bit arithmetic, and is about 10 on the Cray-1. Hence the use of long vectors is more important with the Cyber 205 than with the Cray-1. If a loop is short, it may be more efficient to process it in scaler, rather than vector, mode. For variable length loops, it is sometimes advantageous to incorporate both scalar and vector codes in the program; the former being executed for vectors shorter than the *scalar-vector crossover* and the later otherwise. On the Cyber-205 the optimum scalar-vector crossover is usually in the range 10 to 20 (Reynolds and Lester, 1984; Millane *et al*, 1985).

Explicit vectorization on the Cyber 205 makes use of an explicit vector notation. For example A(I,J;N) refers to the vector consisting of N elements starting with A(I,J). Cray-1 Fortran does not have an explicit vector notation but compiler

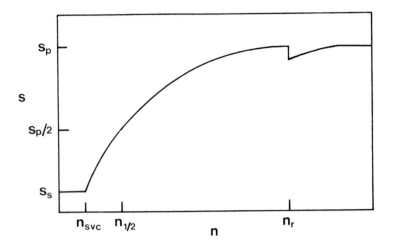

FIG. 1. Performance (speed s in MFLOPS) of a vector processor as a function of vector length n, where s_p is the peak speed and s_s is the scalar speed. n_{svc} is the scalar-vector crossover length, n_r is the vector register length and $n_{\frac{1}{2}}$ is described in the text. For the Cyber 205 n_r = 65535 and for the Cray-1 n_r = 64.

directives are used to control whether specific loops are executed using vector instructions. Vectors may be used to form expressions in the same way as ordinary (scalar) variables. Examples of vector expressions are

$$C(1;100) = A * B(1;100) \tag{1}$$

$$C(1;100) = A(1;100) + B(1;100) \tag{2}$$

In (1), each element in B is multiplied by the scalar A and the result stored, element by element, in C. In (2), each element of A is added to the same element of B and stored in the same element of C. The operands can often be grouped in such a way as to save execution time. For example, the expression

$$A * V(1;1000) * B \tag{3}$$

which involves two vector multiplies should be written as

$$(A * B) * V(1;1000) \tag{4}$$

which involves one scalar multiply and one vector multiply.

In many cases vectorization is made possible by simply interchanging the order of nested loops. This may change the order of memory references to make vectorization possible, remove dependencies or make the inner loop the longest. Consider for example the following typical scalar code to multiply the two NxN matrics A and B using the inner product method

```
      DO 10 I=1,N
      DO 10 J=1,N
      C(I,J) = 0.0
      DO 10 K=1,N
   10 C(I,J) = C(I,J) + A(I,K) * B(K,J)
```
(5)

On the Cyber 205, the inner loop cannot be vectorized since the array elements are not accessed sequentially in memory. However if the loops are reversed into the outer product form

```
      DO 10 J=1,N
      DO 10 I=1,N
   10 C(I,J) = 0.0
      DO 20 J=1,N
      DO 20 K=1,N
      DO 20 I=1,N
   20 C(I,J) = C(I,J) + A(I,K) * B(K,J)
```
(6)

then the inner loops of 10 and 20 can be vectorized, the later being a linked-triad which runs very fast on the Cyber 205. Using the Cyber 205 explicit vector notation this would be coded as

```
      DO 10 J=1,N
   10 C(1,J;N) = 0.0
      DO 20 J=1,N
      DO 20 K=1,N
   20 C(1,J;N) = C(1,J;N) + A(1,K;N) * B(K,J)
```
(7)

If the first dimension of C is N then the vector length in loop 10 can be increased from N to N*N by coding it as

```
   C(1,1;N*N) = 0.0                                       (8)
```

further increasing the speed. If the first dimension is not N, the elements are not contiguous and (4) would give an incorrect result.

For the Cray-1 however, the vector registers can be filled from memory locations separated by a constant stride so that the inner loops in both (5) and (6) can be vectorized. The inner loop of (1) is a vector multiplication with strides N for A and 1 for B. In fact, the fastest method on the Cray-1 is to use (5) and the Cray dot product function SDOT to perform the vector multiplication and the summation:

```
      DO 10 I=1,N
      10 10 J=1,N                                         (9)
   10 C(I,J) = SDOT(N,A(I,1),N,B(1,J),1)
```

The first argument in SDOT is the vector length and the third and fifth arguments are the strides with which the vectors A and B are read.

As a more complicated example, consider the conventional coding for forward substitution in solving a system of linear equations:

```
      Y(1) = B(1)
      DO 10 I=2,N
      Y(I) = B(I)                                         (10)
      DO 10 J=1,I-1
   10 Y(I) = Y(I) - A(I,J) * Y(J)
```

This doesn't seem to have a vector structure, even if the two loops are interchanged. However a closer examination of the algorithm shows that it can be rewritten in the vectorized form

```
      Y(1;N) = B(1;N)
      DO 10 I=2,N                                         (11)
   10 Y(I;N-I+1) = Y(I;N-I+1) - Y(I-1) * A(I,I-1;N-I+1)
```

which includes a linked triad.

Only loops that have a relatively simple structure can be vectorized directly. Loops that contain dependencies, recursion, conditional statements, branching, I/O statements, or subroutine calls cannot be vectorized. However, such loops can usually be spit into a number of loops, some of which can be vectorized. For others, special optimized scalar subroutines are often available. For example, the loop

```
    DO 10 I=2,N                                           (12)
 10 B(I) = B(I-1) + X
```

is recursive since the result of one pass through the loop is dependent on the result of a previous pass and so cannot be vectorized. The loop

```
    DO 10 I=1,N
    A(I) = A(I) * B(I)                                    (13)
    IF(A(I) .GT. X) A(I) = X
 10 CONTINUE
```

cannot be vectorized directly but can be split into two loops, one of which can be vectorized:

```
    A(1;N) = A(1;N) * B(1;N)
    DO 10 I=1,N                                           (14)
    IF(A(I) .GT. X) A(I) = X
 10 CONTINUE
```

Control Data have provided a useful vector library called MAGEV that contains a variety of optimized vector routines for linear algebra (matrix manipulation, matrix inversion, solution of linear equations, eigenvalues/vectors), sorting, FFT and a number of utilities. A similar library SCILIB is available for the Crays.

Now consider the memory traffic in example (7). If the arrays A, B and C remain in main memory during the computation there is no difficulty, but if they are too large, data transfers between memory and mass storage can consume large amounts of elapsed time. Also, each time data is fetched from mass storage the vector unit must be shut down and restarted. For each value of J, the K loop accesses every column of A, so that there are N complete passes over A. If A is too large to fit in main memory (together with some portion of B and C) then after memory is filled, new columns must be read from mass storage and other columns written to mass storage to make space for them. If Q columns of A can be held in main memory at one time and a block of Q columns is read in each time a column not in memory is needed (this would be the most efficient scheme), then the algorithm requires approximately N^2/Q read/write operations. Data transfers can be reduced by processing the matrix A in blocks of Q columns using

```
    C(1,1;N*N) = 0.0
    DO 30 L=1,M
    DO 20 J=1,N
    K1 = (L-1)*Q+1                                    (15)
    K2 = K1+Q-1
    DO 20 K=K1,K2
 20 C(1,J;N) = C(1,J;N) + A(1,K;N) * B(K,J)
 30 CONTINUE
```

where M=N/Q (assumed to be an integer for simplicity), which requires only N/Q read/write operations giving an N-fold reduction. Hence, for a 1000x1000 matrix, the second algorithm could reduce the I/O time from, say, 1 hour to 3.6 seconds. The Cyber 205 is a virtual memory machine so that arrays can be declared in a virtual address space that is much larger than the physical memory and the particular paging algorithm that the operating system uses must be considered when designing algorithms to minimize I/O activity. Matrix manipulation in a virtual memory environment is discussed by Du Croz *et al* (1981). On a non-virtual memory machine such as the Cray-1, the programmer is forced at the outset to consider how large arrays are to be accessed since they must be stored in random access files. The algorithm must then be designed to minimize I/O involving these files.

3. VECTOR PROGRAMMING ON THE CYBER 205

In this section some of the special vector programming facilities available on the Cyber 205 are described. Similar facilities are available on other machines. The Cyber 205 is accessed through a separate *front end* computer and runs under control of the Control Data Virtual Storage Operating System (VSOS). The executable file is called the *controllee file* on the Cyber 205. In order to minimize the size of the controllee file, large arrays are assigned to blank COMMON which is not mapped to the controllee. The *dynamic stack* is a region in virtual space in which the operating system allocates and deallocates space (blank common for example) as needed. The CDC Cyber VSOS, Fortan 200 and Hardware reference manuals are essential when programming for the Cyber 205. CDC provides a utility program called PROFILE that is used to measure program execution times. Putnam and Seaman (1987) is a good beginning tutorial on programming for the Cyber 205.

The vector length must not exceed 65535 except for COMPLEX and DOUBLE PRECISION vectors which must not exceed 32767 elements. To process a longer vector, it must be broken down into vectors of length less than or equal to these values which are processed sequentially in a (non-vectorized) loop. This process is known as *strip mining*. Operations using double precision are slow and should be avoided - they are usually unnecesary because of the long word length (64 bits) of single precision variables. Vectors of type BIT are also available which use one bit per element with the value 1 corresponding to .TRUE. and 0 to .FALSE. Half precision (32 bit) arithmetic is also available and runs at twice the peak performance of 64 bit arithmetic. Certain operations such as SCALAR*VECTOR+VECTOR form a linked triad that can be processed at twice the peak rate. The general linked triad is described in Cyber Hardware manual. Hence a 2-pipe machine has a peak speed of 400 MFLOPS for 32 bit linked triads. For 64 bit arithmetic $n_{\frac{1}{2}}$ is about 50, is 100 for 32 bit operations and 64 bit linked triads, and is 200 for 32 bit linked triads. Hence the use of long vectors is important to obtain maximum performance from the Cyber 205. Arrays can be declared ROWWISE in which case they are stored such that consecutive elements are obtained by varying the rightmost (rather than leftmost) subscript most rapidly.

Cyber Fortran provides a library of *vector intrinsic functions*. For example, the vector intrinsic VSIN in

$$S(1;N) = VSIN(X(1;N);S(1;N)) \tag{16}$$

or

$$S(1;N) = VSIN(X(1;N);N) \tag{17}$$

calculates the sines of each element of X and stores them in the same elements of S. The argument following the semicolon is an *output-argument* which must be provided for vector functions and provides temporary storage for computed results. The output-argument may be a vector or an integer specifying the amount of space (vector length) to be allocated on the dynamic stack. Although it appears superfluous in (17) it is essential in more complicated expressions such as

$$X(1;N) = VSIN(X(1;N);Y(1;N)) * VCOS(X(1;N);Z(1;N)) \tag{18}$$

Most of the scalar intrinsics have equivalent vector intrinsics.

Another class of functions available are *Q8 functions*. These perform more complicated operations on vectors and produce either scalar or vector results. There are about 40 Q8 functions that are described in Chapter 10 of the Cyber Fortran manual. For example the function Q8SDOT in

$$Y = Q8SDOT(X(1;N),Y(1,I;N)) \tag{19}$$

computes the dot product of X and the Ith row of Y. This

operation cannot be vectorized but is implemented efficiently using machine instructions. Many of these functions provide a means of efficiently processing data that are not contiguous in memory and some of the more useful ones are described below. Suppose we wish to add some of the elements of a vector A to the corresponding elements of a vector B and leave the remaining elements of B unchanged. This can be done using the *control store* function Q8VCTRL. A bit vector (or *control vector*) CV is created that has 1's in positions where the elements are to be added and 0's elsewhere. The code

```
T(1;N) = A(1;N) + B(1;N)                                    (20)
B(1;N) = Q8VCTRL(T(1;N),CV(1;N);B(1;N))
```

adds the two vectors into the temporary vector T and CV controls which elements of T are stored into B. Note that all the elements of A and B are summed. If the number of unused sums becomes large, then it is more efficient to use the *compress* (Q8VCMPRS) and *expand* (Q8VXPND) instructions as follows

```
TA(1;M) = Q8VCMPRS(A(1;N),CV(1;N);TA(1:M))
TB(1;M) = Q8VCMPRS(B(1;N),CV(1;N);TB(1;M))                  (21)
TB(1;M) = TA(1;M) + TB(1;M)
B(1;N) = B(1;N) + Q8VXPND(TB(1;M),CV(1;N);N)
```

where the required elements are compressed into temporary vectors (of length M), added, expanded back to their original positions and stored in B. Note that Q8VXPND puts zeros in the elements of the output vector for which the control vector element is zero so that the values of B in these positions would be changed if, for example, A and B were multiplied together rather than added. A more flexible (but slower) way of collecting elements into a contiguous vector is provided by the *gather* and *scatter* functions Q8VGATHR and Q8VSCATR. These use an *index vector* rather than a control vector which allows the order of the elements to be changed and, for the scatter function, an element of the input vector may be put into a number of locations in the output vector.   The statement

$$V(1;M) = Q8VGATHR(A(1;M),IN(1;N);V(1;M)) \tag{22}$$

gathers the elements from the vector A and puts them in V under
control of the index vector IN (Fig. 2).   The scatter function
does the reverse operation in the sense that the elements of the
index vector define position in the output, rather than input,
vector.

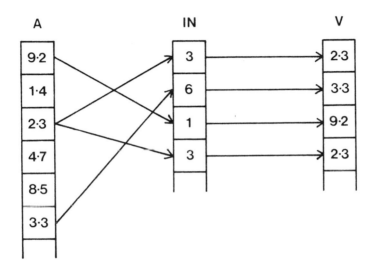

FIG. 2.   Illustration of the operation of the function Q8VGATHR.

A set of *special calls* are also provided that give the
Fortran   programmer   access   to   scalar   and   vector   machine
instructions.   There   are   some   instances   where   special   calls
provide operations not directly available in Fortran that are more
convenient or efficient.   They are more difficult to use than the
Q8 calls however and it is possible to make mistakes that the
compiler will not detect.   The programmer using special calls must
refer   to   the   Cyber   205   Hardware   Reference   Manual   to   find   an

adequate description of the vector machine instructions. The special call Q8VXTOV for example is a more general form of the gather instruction that allows transmission of groups of elements, rather single elements.

A further library of predefined subroutines are the STACKLIB subroutines. These are optimized scalar routines that can be used to replace some non-vectorizable loops, particularly those involving recursion. For example the STACKLIB call

```
CALL Q8A010(R(2),A(2),R(1),L-1)                          (23)
```

is equivalent to the recursive (non-vectorizable) loop

```
   DO 10 I=2,L                                           (24)
10 R(I) = A(I) + R(I-1)
```

The STACKLIB library is described in chapter 11 of the Fortran 200 reference manual.

The scalar block IF statement has a vector equivalent called the block WHERE statement. For example, the following code places the square roots of positive elements of X, and the magnitudes of negative elements of X, in the vector Y.

```
   WHERE (X(1;N) .GT. 0.0)
   Y(1;N) = VSQRT(X(1;N);Y(1;N))                         (25)
   OTHERWISE
   Y(1;N) = -X(1;N)
   ENDWHERE
```

The right hand side of the assignment statements are evaluated and assigned to the corresponding elements of the vector on the left hand side only if the corresponding element of the bit vector (or expression) following the "WHERE" contains a "1" in the case of the first block (before the "OTHERWISE"), or a "0" in the case of the second block (after the "OTHERWISE"). This provides one means of vectorizing some loops that contain conditional statements.

The Cyber 205 is a virtual memory machine, and program or data that is required during execution is transferred into memory by the operating system using a demand paging system. Program and data are stored in pages which are groups of contiguous addresses. There are two page sizes available; small pages of either 512, 2048 or 8192 words depending on the particular system configuration, and large pages of 65536 words. If a program needs to access data which is not in memory then a *page fault* exists and the operating system loads the required page into memory. If all the available memory is in use, a page must be removed to make room for the new page. Ideally the removed page should be one that will not be needed for a while so a least recently used (LRU) algorithm is used for page removal. The LRU algorithm is a generally good strategy but it can behave poorly in some circumstances. By careful programming, the data flow can be minimized under control of the LRU algorithm. By default, a program is assigned to small pages at load time. However, large arrays should be assigned to large pages to minimize the number of page faults and to get maximum efficiency from the vector unit. To use large pages, the large arrays should be placed in COMMON and selected common blocks can be assigned to large pages at load time using the GRLP option. The number of large pages that can be in memory simultaneously during program execution is specified using the RESOURCE statement. If a program places large temporary vectors on the dynamic stack it may be advantageous to map the stack to large pages.

As an example, consider the vectorized code (7) for matrix multiplication with $N = 2048$ so that each matrix fits on 64 large pages. Since the inner loop of (7) accesses each collumn of A sequentially, the algorithm references all pages of A in cyclic order. If all 64 pages do not fit in memory, then for each new page reference for A, the LRU algorithm generates a new page fault. Hence the algorithm requires 64×1024 page faults for A as well as 64 for each of B and C. If a page fault consumes 0.5 sec, say, of elapsed time then the algorithm would take about 18 hours. Consider now accessing A in blocks of columns as in

(15). If we assume that half of A, as well as one page of each of B and C fits into memory then we can choose M=2 in (15). Each page of A is then referenced only once, and each page of B and C twice. This gives a total of 320 page faults or about 3 minutes of elapsed time. This illustrates the importance of programming to minimize data transfers.

Program execution normally is suspended until an I/O operation is completed. Cyber Fortran provides some facilities for *concurrent I/O* which allows an I/O operation to be initiated and then execution of other statements in the program to continue. This reduces execution time by overlapping I/O and other computations. The greatest efficiency can be achieved by overlapping all I/O with other computations. The Q7BUFIN and Q7BUFOUT subroutines initiate concurrent I/O from or to mass storage respectively and Q7WAIT is used to check on the status of concurrent I/O. When using concurrent I/O, arrays must be aligned on page boundaries (using the load parameters GRSP or GRLP). The use of concurrent I/O is described in detail in chapter 11 of the Fortran 200 reference manual.

A Fortran program can interact with the operating system while it is executing using the *System Interface Language* (SIL). SIL subroutine names begin with "Q5" and are described in chapters 8 and 9 of the VSOS manual. There are about 100 SIL subroutines and I will briefly mention a few that are useful and related to some of the previous discussion. The SIL call Q5ADVISE is used to keep a specified region of the program's virtual space paged in (or out). The programmer can use this to minimize paging activity when the LRU algorithm would not efficiently handle paging of a particular array or part of an array. The call Q5DCDMSC retrieves information from the miscellaneous table which contains a variety of runtime information; in particular the current page usage. The calls Q5GETLP and Q5SETLP are used to get and set the current large page limit for the job. The call Q5MAPIN (or Q5MAPOUT) is used to associate (or disassociate) a mass storage file with a program's virtual address space. This allows a portion of the data to be "unplugged" from the executing program. The file is directly associated with an array and is accessed without read/write statements - i.e. the operating system pages the data in and out as required. This is referred to as *implicit I/O*. The array used must not be mapped to the controllee file - this is done by declaring it as blank (or named) common and specifying the common block with the GRSP or GRLP (or GROS or GROL) parameter on the LOAD statement. The file must be opened or created using the SIL routine Q5GETFIL.

4. VECTORIZATION OF THE LALS POLYMER REFINEMENT SYSTEM

Much of the linked atom least squares (LALS) refinement system for helical polymers (Smith and Arnott, 1978) has been vectorized (Millane *et al*, 1985). Some examples of increases in performance obtained are summarized here and the reader is referred to Millane *et al* (1985) for a more comprehensive description.

Strucure amplitude calculations usually consume a substantial amount of the time involved in crystallographic calculations. A comparison of the time required to calculate the structure amplitudes using the scalar and vectorized LALS codes as a function of the vector length (number of atoms) is shown in Fig. 3. This demonstrates the effect of vector length on performance. Since the calculation of structure amplitudes involves evaluation of triginometric functions, the speed increases are not as dramatic as in calculations involving only multiplications and additions. Some sections of the code involved in solution of the normal equations gave 20-fold speed increases for 500 x 500 matrices and even more for larger matrices. Table 1 shows the CPU times for one cycle of refinement of three different size structures using the CDC 6600 machine and the Cyber 205 with scalar and vector codes. Structure II would have been difficult to run on the 6600 and structure III impossible. For large structures, the vector code is at least 15 times faster than the

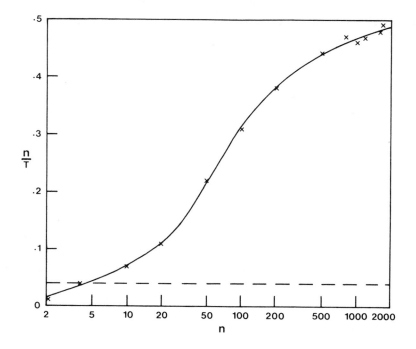

FIG. 3.   Comparison of the speeds (n/T) in μs^{-1} for calculation of structure amplitudes using scalar (---) and vector (——) codes of the LALS system, where T is the CPU time for n atoms (From Millane *et al*, 1985).

205 scalar code.   Further vectorization of the LALS code is possible which would further improve its performance.   Using the vectorized code on the Cyber 205 has allowed us to study nucleic structures that are more complex than those that could be examined previously (Millane *et al*, 1984; Chandrasekaran *et al*, 1985).

5.   CONCLUSIONS

Modern supercomuters provide high processing speeds and large memory capacities, allowing crystallographic analyses of molecules larger than was heretofore possible.   To achieve maximum

TABLE 1

Comparison of CPU times for one refinement cycle of structures of various sizes using vector and scalar codes of the LALS system (From Millane *et al*, 1985).

Struc-ture	Atoms	Para-meters	Data	Reflec-tions	CPU Time(s)		
					6600	205 scalar	205 vector
I	60	89	241	100	48.0	6.0	2.2
II	223	212	581	266	-	51.0	6.9
III	386	411	754	414	-	275.0	21.5

performance from these machines however, algorithms must be redesigned to make optimal use of the machine architecture. This involves maximizing the use of vector instructions and minimizing memory traffic. Multiple processor machines that allow concurrent processing that are being developed (Hillis, 1987) are potentially even more powerful but will provide greater challenges in the design of parallel algorithms.

In the future, high processing power may allow structures to be solved using fewer experimental data by, for example, performing exhaustive refinements starting with a variety of either phase sets or model structures. Finally, computationally more intensive calculations such as molecular dynamics, quantum mechanics and computer graphics may be used to complement crystallographic analysis and interpret crystallographic results.

This work was supported by the National Science Foundation.

REFERENCES

ARNOLD, E., VRIEND, G., LUO, M., GRIFFITH, J.P., KAMER, G., ERICKSON, J.W. JOHNSON, J.E. and ROSSMANN, M.G. (1987). *Acta Crystallographica*, **A43**, 346-361.

BROCK, H. K., BROOKS, B. J. and SULLIVAN, F. (1981). *BIT*, **21**, 142-152.

CHANDRASEKARAN, R., ARNOTT, S., HE, R. G., MILLANE, R. P., PARK, H. S., PUIGJANER, L. C. and WALKER, J. K. (1985). *J. Macromolecular Science - Physics*, **B24**, 1-20.

DEVREESE, J. T. and VAN CAMP, P. (eds.) (1985). *Supercomputers in Theoretical and Experimental Science*, Plenum, New York.

DONGARRA, J. J., GUSTAVSON, F. G. and KARP, A. (1984). *SIAM Review*, **26**, 91-112.

DU CROZ, J. J., NUGENT, S. M., REID, J. K. and TAYLOR, D. B. (1981). *ACM Trans. Mathematical Software*, **7**, 527-536.

DUFF, I. S. (1984). In Kowalik (1984), pp. 293-309.

FEILMEIER, M., JOUBERT, G. and SCHENDEL, U. (eds.) (1986). *Parallel Computing 85,* Elsevier, Amsterdam.

GARY, J. P. (ed.) (1984). *Cyber 200 Applications Seminar*, NASA Conference Publication 2295, NASA, Maryland.

GENTZSCH, W. (ed.) (1984). *Vectorization of Computer Programs with applications to Computational Fluid Dynamics*, Vieweg, Braunschweig.

HILLIS, W.D. (1987). *Scientific American*, **256** (6), 108-115.

HOCKNEY, R. W. and JESSHOPE, C. R. (1981). *Parallel Computers,* Arrowsmith, Bristol.

KOWALIK, J. S. (ed.) (1984). *High-speed Computation*, Springer-Verlag, New York.

LAMBIOTTE, J. J. (1984). In Gary (1984), pp. 243-256.

LAZOU, C. (1986). *Supercomputers and Their Use*, Clarendon Press, Oxford.

MILLANE, R. P., WALKER, J. K., ARNOTT, S., CHANDRASEKARAN, R. and RATLIFF, R. L. (1984). *Nucleic Acids Research*, **12**, 5475-5493.

MILLANE, R. P., BYLER, M. A. and ARNOTT, S. (1985). In Numrich

(1985), pp. 289-302.

NUMRICH, R. W. (ed.) (1985). *Suppercomputer Applications*, Plenum, New York.

PUTNAM, B. and SEAMAN, D. (1987). *Cyber 205 Performance Programming*, Purdue University Computer Center, West Lafayette, Indiana 47907, USA.

REYNOLDS, P. J. and LESTER, W. A. (1984). In Gary (1984), pp. 103-116.

SCHONAUER, W. and GENTZSCH, W. (eds.) (1986). *The Efficient Use of Vector Computers with Emphasis on Computational Fluid Dynamics*, Vieweg, Braunschweig.

SMITH, P. J. C. and ARNOTT, S. (1978). *Acta Crystallographica*, A34, 3-11.

SWARZTRAUBER, P. N. (1984). *Parallel Computing*, 1, 45-63.

TEMPERTON, C. (1984). In Kowalik (1984), pp. 403-416.

26
NETWORKING
FOR CRYSTALLOGRAPHERS

Philip E. Bourne

1   Introduction

Communication between crystallographers using global computer networks is expanding at a rapid rate. In March of 1987 the informal database of crystallographers accessible via computer network (Teeter, 1986) contained entries for 360 scientists in 13 countries. This number can be expected to continue to grow as national and international networks expand and more crystallographers gain access to this form of communication.

At the institutional level the trend towards more distributed data processing, either on individual workstations or small laboratory computers, requires an understanding of inter-computer communications.

This paper reviews the general features of computer networks with regard to both local institutional networks and remote international networks important to crystallographers. The paper concludes with a discussion of what can be expected in the near future and suggests how the crystallographic community should take advantage of this form of rapid and reliable communication.

2 General Features of Computer Networks

Simply stated, a network could be considered as a physical link between two or more computers permitting the exchange of information. Each computer on the network is referred to as a *node* (or *host*). Nodes on a network are characterized by a unique *node name* (or *host name*) which is usually from 1–8 alphanumeric characters. The computer to which a users terminal is connected is referred to as the *local node* (or *local host*), whereas all other computers on the network are referred to as *remote nodes* (or *remote hosts*).

The characteristics of the physical link between nodes is usually dependent on the distance which separates them. Nodes in the same, or adjoining buildings, are typically connected by a baseband coaxial cable which permits high speed data transmission rates up to 10^7 baud (bits per second). Such a connection is referred to as a *local area network* or *LAN*. Physical links which span larger distances, *long haul networks*, may be via satellite, microwave connections, leased telephone lines or dedicated wiring. The speed of transmission across these links ranges from 300 to 10^6 baud.

From the perspective of a user sitting at a terminal, there are four possible types of network access:

(i) *Remote terminal sessions* — permits a user to log in to a remote node and conduct a terminal session as if it were the local node.

(ii) *Remote device access* — permits a user to access a device on a remote node, for example a virtual disk, printer or tape drive.

(iii) *File transfer* — permits a user to send and receive files between a remote and local node.

(iv) *Electronic mail* — a special form of file transfer.

The availability of these four possible types of user access is dependent on the type of network. Generally, a *LAN* will permit all four types of access, whereas a long haul network will only permit file and electronic mail transfer.

3   Network Topology

In an ideal situation an organization would be composed of a number of *LAN*'s for each department or division, with each *LAN* connected to a central institutional backbone via a *bridge*. The function of the bridge is to restrict data flow between the institutional backbone and the *LAN* to that which is pertinent to the users of the *LAN*. This is particularly important in situations where the *LAN* is used as a *bus* for diskless workstations, where unwanted network data could hinder local performance.

A users terminal is either wired directly to the local node or attached to the network via a *terminal server*. The latter has the advantage of permitting direct connection to any node on the *LAN*. Hence, if one computer is unavailable an alternative can be selected. Terminal servers may also permit multiple terminal sessions to be conducted simultaneously on one or more nodes. A simple toggle mechanism can transfer control to an alternative session. As windowing software and larger screens become more common the ability to simultaneously view multiple terminal sessions will be possible.

The institutional backbone can be connected to a long haul network either by a bridge or by a *gateway* computer, thus linking geographically remote institutions.

4   Network Architecture

The most common way of describing a network uses the International Standards Organization Open System Interconnect (OSI) model. The adoption of this standard by most computer vendors has facilitated the connection of dissimilar computer systems. A detailed discussion of the OSI model is beyond the scope of this paper. However, for the purposes of discussion the model is considered in terms of:

- The physical connection between computers.

- The commonality required for data recognition — *the high level protocol.*

- The software interface between the network and the user.

4.1   The Physical Connection

Computers can be physically connected by a variety of different media, the characteristics of each are summarized in Table 1.

TABLE 1

Physical Media

Media	Speed (10^6 baud)	Distance (km)
Computer Bus	10	0.001
Twisted Pair	1	3
Baseband	10	10
Broadband	n/a	50
Fiber Optic	50	10

Computers which share the same bus blur the distinction between a network connection of distinct cpu's and a true multiprocessor which could be thought of as a single node. Multiprocessors will not be discussed further here. The majority of sl LAN's are constructed using baseband cable, although more use of fiber optic cable can be expected as the cost decreases. Broadband offers the opportunity to transmit both analog and digital signals. However, the cost of the interface equipment involved prohibits common use of the broadband medium at this time. Twisted pair is the least expensive option, and is used in some microcomputer based networks and for direct terminal to local node (or terminal server) connections.

4.2   Network Protocols

Regardless of the physical medium, data are moved across the network in segments called *packets*. Apart from the specific data being sent, each packet has encoded information describing the destination node, the local node name, a *high level protocol* identifier and a sequence identifier. The exact format of the packet is defined by the *low level protocol*. The Ethernet standard is an example of a lower level protocol. The receiving node checks the integrity of the data being sent and reconstructs the packets in the correct sequence. If an error occurs the data are retransmitted. Multiple high level protocols can simultaneously be transmitted on the same low level protocol and physical medium. Table 2 outlines some common high level protocols.

TABLE 2

Common High Level Network Protocols

ADCCP	Advanced Data Communications Control Procedure, used by the US Government in ARPANET
BSC	Binary Synchronous Communications (BISYNC) used by IBM in the Synchronous Network Architecture (SNA)
DDCMP	Digital Data Communications Protocol used by DEC in the Digital Network Architecture (DNA), which includes DECnet
RSCS	Remote Spooling and Communications Subsystem used by IBM
TCP/IP	Transmission Control Protocol/Internet Protocol used by Unix based processors
Token Ring	Used by IBM in the microcomputer based network
X.25	Used by Xerox

4.3   Network Software

The software available for a crystallographer to make effective use of a computer network is dependent on both the hardware and operating system of the local node. A variety of inexpensive or public domain software exists to support this interface. The name, function and distributor of a selection of these products is given in Table 3.

TABLE 3

Some Useful Network Software

Product	Purpose	Hardware/ Software	Distributor
Finger	View events on local and remote nodes	Various	Philip Bourne
Gmail	Facilitates inter-network communications (Gateway Mailer)	VAX/VMS	Ed Miller, Stanford Linear Accel. Center, Stanford, CA, USA.
Jnet	Emulates RSCS high level protocol for BITNET access	VAX/VMS	Joiner Associates, PO Box 5445, Madison, WI, USA.
Kermit	Inter-computer communications terminal emulation	Many	Computer Center, Columbia University, New York, NY, USA.
NET-MAILER	Similar to Gmail	VAX/VMS	Christoph Gatzka, Bredeneyerstrasse 141, 4300 Essen W. Germany.
RTP/1100	Emulates RSCS high level protocol	Sperry 1100's	Computer Science Center, Uni. of Maryland, College Park, MD, USA.
RNET	Emulates RSCS high level protocol	Prime	Education Marketing, PRIME Computer Inc., PRIME Park, Natick, MA, USA.
UREP	Emulates RSCS high level protocol	Unix BSD	Dept. of Computer Science, Pennsylvania State Uni., University Park, PA, USA.

The syntax used by the various software packages described in Table 3 differs significantly. The only consistency is found in the format used for the network address, which defines each network user uniquely. A network address consists of three segments:

> domain — defines the network
> nodename — defines the computer
> username — defines the user

Only the segments that make a user unique need be defined, for example, to address a crystallographer on the same network, but on a different node, the domain may be omitted. In using network addresses the three segments are separated by delimiters. Such delimiters are dependent on operating system type. The most common form is:

> username@nodename.domain

5   International Computer Networks

A number of national and international networks have developed over the past 18 years (Quarterman and Hoskins, 1986). These different networks are connected at strategic locations by *gateways*. Gateways are computers which perform the necessary high level protocol conversion (if necessary) and pass network traffic from one long haul network to another. From a users perspective, it is not essential to know the features of each network, only to understand the use of the software resident on the local node (if any) for accessing each network and passing network traffic through the gateway to a user at some remote node.

Table 4 summarizes the major networks available to the academic community. BITNET (Oberst and Smith, 1986), the network in most common use by crystallographers, is described in some detail. The other networks outlined in Table 4, which are accessible via BITNET using gateways, are summarized.

TABLE 4

Some National and International Research Networks

Name	Protocol	Description	For Further Information[1]
ACSNET	TCP/IP	Australian Computer Science Network	postmaster@munnari.oz.au
ARPANET	TCP/IP	US Department of Defense	institutional computer center
BITNET	RSCS	Sponsored by IBM for academic institutions	info@bitnic
CCNET	DDCMP	Connects several academic institutions in the NE of the US	Philip Bourne
COSAC	X.25	French research network	kintzig@france.csnet
CSNET	TCP/IP	Connects US Computer Science Depts.	info@sh.cs.net
DFN	X.25	West German research network	DFN–Verein, Pariser Strasse 44, 1000 Berlin 15, Bundesrepublik Deutschland
EARN	RSCS	European Academic Research Network, European counterpart of BITNET	info@bitnic
NSFnet	TCP/IP	US National Science Foundation	Director of Networking Nat. Sci. Found. Washington DC, USA
JANET	X.25	UK Joint Academic Network	Network Executive Rutherford Lab. Ox. UK
NETNORTH	RSCS	Canadian counterpart of BITNET	info@bitnic
UUCP	TCP/IP	Worldwide Unix to Unix copy	uucp-query@cbatt.UUCP

[1] Network addresses, where given, are for inquiries from BITNET.

5.1   BITNET

The 'Because It's Time' network was established in 1981 as a single link using a leased telephone circuit between the City University of New York (CUNY) and Yale University. At the time of writing these two nodes have grown to 1643 located at 548 different institutions worldwide, with gateways to 10 other networks. Currently the only requirement for joining the network is to install a communication link to a computer already part of BITNET. BITNET uses the Remote Spooling and Communication Subsystem (RSCS) high level protocol developed by IBM. However, at least one-third of the computers on BITNET are non-IBM machines. This is made possible by software that emulates the RSCS protocol and permits the standard features of file transfer, electronic mail and interactive message exchange (see Table 3).

BITNET is a *store and forward* network, that is, the computer sending the data keeps a complete copy until it has been forwarded successfully to the next computer along the chain leading to the final destination. Such *routing* is *explicit* — there is no intelligence built into the routing mechanism. Data will always be routed the same way towards a final destination unless there is some human intervention. Thus, if two alternative routes are available for data transmission the predefined route will be chosen, even if a computer further down the route is off-line. Although at first glance there would appear to be many points of failure, in practice this form of data exchange works very well. The practicality of BITNET is illustrated in Table 5, in which sample times for incoming data to reach New York City from remote destinations, are reported.

BITNET is administered jointly by EDUCOM, which developed a Network Information Center (BITNIC), and CUNY, which created a Development and Operations Center (BITDOC). Both will be funded by IBM until mid-1987, whereupon operating costs are to be met by the institutions using BITNET. BITNIC distributes *routing tables*, which define an explicit route to an adjacent node for every destination on the network, to all nodes on a monthly basis.

An interesting feature developed by BITDOC is NICSERVE, a database query system. Incoming data from remote network nodes are interpreted by NICSERVE as database queries. The result of the database retrieval are returned to the user making the query without any human intervention.

TABLE 5

BITNET Times To New York (hh:mm)

From	Shortest	Longest	Number	Mean
Alberta, Canada	0:11	4:04	3	1:34
Atlanta, US	6:06	6:06	1	6:06
Leeds, UK	0:05	0:39	3	0:18
Melbourne, Aus.	1:55	82:17	7	39:12
Rehovat, Israel	1:19	1:19	1	1:19
Sheffield, UK	1:17	1:17	1	1:17
Stanford, US	1:56	12:65	3	8:35
Sydney, Aus.	2:32	3:14	2	2:56
Tokyo, Japan	0:10	16:00	14	2:39
Uppsala, Sweden	74:59	74:59	1	74:59

5.2   ASCNET

ASCNET is the major Australian research network and was established in 1979 to support mail and file transfer between research, educational and industrial users. ASCNET connects to CSIRONET, which is a government research network. Connection between the approximately 300 nodes on this network is primarily by 1200 baud telephone lines.

5.3   ARPANET

The oldest US network, ARPANET was established in 1969, and currently has over 150 nodes. Many nodes are connected via 56,000 baud dedicated lines which permit, file transfer, mail transfer and remote terminal sessions. General access to ARPANET nodes is restricted to users with US Defense Contracts, although file transfer and mail transfer is possible to ARPANET nodes via gateways.

5.4 CCNET

CCNET connects nodes at several universities in the Northeast of the US, permitting file transfer, mail transfer, remote terminal sessions, and the addressing of remote devices. The data transfer rate between computers is at least 9600 baud.

5.5 COSAC

'COmmunications SAns Connections' is a French research network of approximately 30 nodes operating at 1200 baud or greater.

5.6 CSNET

CSNET comprises approximately 150 nodes, primarily in Computer Science Departments in US universities. Mail transfer is the only service supported over the whole network, although remote terminal sessions and file transfer are possible in some circumstances.

5.7 DFN

'Deutsche Forschungnetz' is the national research network of West Germany consisting of approximately 30 nodes connected at 9600 baud.

5.8 EARN

EARN is the European counterpart to BITNET and consists of over 150 nodes in 18 European countries. The routing tables for a node on BITNET also contain all the EARN remote node routing information.

5.9   NSFnet

Presently under development in the US, the first phase of NSFnet connects all the National Science Foundation (NSF) funded supercomputer sites with implicit links of at least 56 000 baud. Later phases will bring on-line many academic institutions which are presently part of ARPANET and CSNET. This will be a full function network, permitting file transfer, mail transfer, remote terminal sessions and remote device access.

5.10   JANET

The majority of university computer centers and research establishments in the UK are connected via JANET using leased telephone lines of at least 9600 baud. There are approximately 20 institutions on JANET with a total of 1500 nodes.

5.11   NETNORTH

NETNORTH is the Canadian counterpart of BITNET and has over 90 nodes which are included in the BITNET routing tables.

5.12   UUCP

'Unix to Unix Copy' is an extensive world-wide Unix based network which supports only mail transfer across standard telephone lines at speeds of 300, 1200 or 2400 baud depending on the modems in use. At predefined times a local node will dial up a remote node and transfer any mail. Initially this network required explicit routing, but with the availability of routing maps, software is now available which permits implicit routing.

6   Electronic Mail

Electronic mail, also known as *E-mail*, is a special form of data transfer, useful for sending messages, data and programs. Most operating systems support some form of electronic mail software for users to communicate on the local node. Communications to users on remote nodes may require special software for translation between different E-mail formats. If an E-mail message can be sent the message being received will typically be preceded by a header which contains the following:

- The senders network address.

- The receivers network address.

- The time the message was sent.

- The time the message was received.

- The subject of the message.

- Some routing information relating to gateways if applicable.

7   Public Networks

The potential of E-mail has not gone unnoticed by telephone companies and other public utilities. Some offer dial-in electronic mailbox facilities on their own computers, only requiring a terminal and telephone modem for access. Interface software for the most popular operating systems is available so that central institutional computers can connect directly to such dial-in services, permitting global communications, even when a long haul network link is unavailable.

8   The Future

Over the next several years the emphasis will be on increasing the speed and reliability of network transmissions. Crystallographers accessing BIT-NET should observe the speed of transmission increase to 56,000 baud, with a corresponding improvement in reliability achieved with the use of implicit routing. This level of service, which is currently only available on NSFnet (Jennings *et al.*, 1986), may be accompanied by a protocol change to the de facto TCP/IP standard. At higher transmission speeds interactive communication will be possible, although the lesson on ARPANET has been that the unexpected increase in the amount of network traffic nearly always precludes efficient interactive communications.

With global networks in place, the opportunity exists to develop an automated clearing-house for crystallographic information. The latest versions of software, protein coordinate data, employment opportunities and useful technology are just some examples of information that could be made available. Information could be requested in an manner similar to the NICSERVE database query system.

Crystallographers have always been alert to technology which might advance their field. It will be interesting to see what use we make of computer networks over the next few years.

REFERENCES

Jennings, D.M., Landweber, L.H., Fuchs, I.H., Farber, D.J. and Adrion, W.H. (1986). *Science*, **231**, 943–950.

Oberst, D.J. and Smith, S.B. (1986). *EDUCOM Bulletin*, **21**, 10–17.

Quarterman, J.S. and Hoskins, J.C. (1986). *Communications of the ACM*, **29**, 932–971.

Teeter, M.M. (1986). *ACA Newsletter*, **17**(2), 49–53.

Brief contributions

AN ANALYSIS OF THE ERRORS WHICH REMAIN AFTER PROTEIN REFINEMENT

D. E. Tronrud and B. W. Matthews

Institute of Molecular Biology, University of Oregon, Eugene, OR 97403 U.S.A.

Macromolecular refinement procedures have been applied to many protein structure problems in the last 15 years. However, in almost all cases the agreement between the structural models and the measured data is not as good as one would expect from the estimate of the amount of error in the data. An attempt is being made to determine the sources of the remaining discrepancies by analyzing these discrepancies by three different means.

The first method of attack is to determine how the difference density in final difference maps differs between several determinations of very similar structures. In our laboratory the structures of a number of mutants of T4 lysozyme have been solved, all of them in the same space group and at about 1.7Å resolution. Another project has been to compare different inhibitor complexes of Thermolysin. Both of these projects have resulted in a series of data sets and crystallographic refinements for very similar molecules. The examination of the similarities and differences in the residuals of these families structures will be discussed.

The second method is to examine how the residual difference density in (Fo – Fc) maps varies as a function of the distance from the nearest atom. This function can be displayed and the distribution of density can suggest reasons for the inadequacy of the model.

The final method is to consider the nature of the structural model and examine its limitations. Two cases will be considered. (1) The scale factors which are used to relate Fc to Fo. (2) The absence of hydrogens in the protein model. The importance of each of these souces of error will be assessed.

A PROGRAM WHICH CALCULATES SPACE–GROUP–SPECIFIC FFT'S FOR MANY SPACE GROUPS

D. E. Tronrud

Institute of Molecular Biology, University of Oregon, Eugene, OR 97403 U.S.A.

A program will be described which will calculate space–group–specific Fourier transforms for almost 50 different space groups. The program will handle all space groups without mirrors and centers of symmetry (all space groups in which chiral molecules can crystallize).

The program was written after a careful examination of the FFT method showed that only four special cases would be required to handle all of these groups. The program reads the symmetry operators, in *International Tables* format, and discards those operators which it cannot handle. It then determines which special case to use and procedes to read the atom list. It calculates the model electron density map, in memory, and performs the Fourier transform. The proper asymmetric volume in reciprocal space is written to disk.

INEXPENSIVE AUTOMATION OF A DIFFRACTOMETER USING A ZENITH Z–151 MICROCOMPUTER

J. F. Noonan and E. D. Stevens

Department of Chemistry, University of New Orleans, New Orleans, Louisiana 70148, U.S.A.

A system is described which can completely control X-ray diffraction data collection on a single crystal instrument. A General Electric XRD–490 and an Enraf–Nonius PAD–3 (both are quarter circle Eulerian geometry instruments) were rebuilt with modern stepper motors on each of the 4 axes. The motors are controlled by a stepper controller board in an IBM–compatible Zenith Z-151 microcomputer.

A C language program was written that will locate and center reflections, find the primitive reciprocal cell, calculate estimated standard deviations on the cell parameters, and carry out intensity data collection at user specified (hkl). Provision is also made for transforming the primitive cell to one of higher symmetry before data collection.

The price of the hardware including motors, interface board, and power supply was approximatelky $1400. Low cost generic brand IBM–compatible computers capable of serving as a host to the system can currently be purchased for as little as $800.

The authors note that for individuals at smaller institutions and those with limited equipment funds that such a system would be servicable alternative to a new and expensive automated diffractometer. We have found there are many usable old instruments gathering dust because their automation systems no longer function.

STRUCTURE ANALYSIS OF SERUM TRANSFERRIN AT 3.3Å

S. Bailey, H. Jhoti, R. Garratt, B. Gorinsky, P. Lindley and R. Sarra
Department of Crystallography, Birkbeck college, London, U.K.

Serum Transferrin is an iron-binding protein (M.W. = 80kD) whose function is central to iron metabolism. The structure has been solved using solvent flattening techniques. Initially a 3.3Å MIR map was calculated and the molecular boundary was observable, however, the map was uninterpretable. As the protein crystals contained high solvent (68%) solvent flattening techniques were employed. A manual designation of the molecular boundary was possible; all solvent density was set to zero, negative regions within the envelope were multiplied by 0.1 and the resulting map back-transformed into reciprocal space. The final combined phases were used to calculate a new map and the cycle repeated until convergence. After 5 cycles the resulting mean phase change was 31.3°. Wang's solvent flattening technique which designates the molecular boundary automatically was also used to calculate two solvent flattened maps. Averaging spheres of radius 8Å and 10Å respectively were used with a solvent content of 65% by volume. Convergence, after 4 cycles in both cases, produced mean phase changes of 33.7° and 34.5° respectively. The map produced using manual designation of the boundary resulted in a clearer definition of secondary structure that either of the Wang maps. However, in contrast, the Wang maps showed better connectivity in the loop regions. A complete chain trace was only possible using both techniques.

NUMERICAL ABSORPTION CORRECTION USING SHELX–76 COMPUTER PROGRAM

Babu Varghese[1] and K. Sivakumar[2]

[1]Regional Sophisticated Instrumentation Centre, Indian Institute of Technology, Madras 600 036, India. [2]Department of Physics, Anna University, Madras 600 025, India.

To compute numerical absorption correction using Shelx–76 computer program, the program requires as input, the normal distances to each morphological plane of the crystal used for intensity data collection from an interior point in the crystal. The authors present a computer program to calculate normal distance from an interior point to all morphological planes of a general polyhedra, knowing the lengths of edges and the Miller indices of the morphological planes of the crystal. The algebra will be presented. Also presented is another computer program to calculate the reverse incident beam direction cosines and diffracted beam direction cosines of each reflection using the reciprocal axis matrix (CAD–4 diffractometer geometry is assumed) to suit for HKLF 4 input.

A COMPUTER PROGRAM FOR THE INTERPRETATION OF X–RAY POWDER DIFFRACTION DATA

E. Wu

School of Physical Sciences, Flinders University of South Australia, Australia.

A program written in Fortran 77 is designed to interpret the S values (*mm*) for each reflection measured from a Guinier–Hagg X–ray film.

The input data is read in interactively. The program calculates the corrected values of 2θ and $\sin^2\theta$ using an internal standard (the program contains a set of 4θ values for silicon). The program then indexes the reflections, calculates the interplanar spacings and lattice parameters, and decides the Bravais lattice if the crystal system is supposed. The agreements of the calculations are shown by comparison of the calculated and the observed $\sin^2\theta$ values, and the minimal percentage of the found reflections to the predicted reflections (the systematic absences caused by glide planes and screw axes is not considered). The results then can be refined by a least squares method. If the agreements are unacceptable, the results can interactively be readjusted by changing the $\sin^2\theta$ values, the lattice parameters or the crystal systems. The formats of the output shown on the screen and saved in the file are sufficient, concise and easy to read and check.

An important advantage of this program is it is suitable to deal with the reflections which contain polyphases. The program would mark the reflections which could belong to a second phase. The program allows you to do above calculations eliminating these reflections. The phase of these reflections could then be identified by conversely doing calculations eliminating other reflections or checking their interplanar spacings from the Power Diffraction File.

THE MODULE IS THE METHOD: AN ALGEBRAIC ANALYSIS OF SYMBOLIC ADDITION.

J. S. Rutherford

Department of Chemistry, University of Transkei, Private Bag X 1001, Umtata, Transkei, Southern Africa.

In the space group $P\bar{1}$ the process of origin definition automatically produces a reference derivative reciprocal lattice, while in higher symmetries one may be chosen to suit the purpose of the analysis. In either case, any such choice redefines vector addition in reciprocal space in terms of a direct product of an infinite subgroup isomorphous with the three-dimensional lattice group and, linking the cosets in reciprocal space, a finite set of operations called the torsion subgroup. This latter is associated with a finite arithmetic, the overall system forming a 'module on a ring'. By identifying this underlying algebriac structure, and so classifying the reciprocal lattice points, it is possible to predict how the sign symbols will combine together in the sign determining process. With this knowledge symbols can be assigned most efficiently, and the reasons traced when subsets of the data are bypassed in the propagation process.

A program has been written which, starting from three arbitrary basis vectors for a derivative reciprocal lattice, converts the associated transformation matrix to the standard triangular form, and from it derives its prime power factors and their inverses, also in standard form. These inverse matrices then serve to classify the individual reciprocal lattice points.

Some simple applications will be given.

THE CRYSTAL AND MOLECULAR STRUCTURE OF HEXAKIS (PHENYLTHIO) CYCLOTRIPHOSPHAZATRIENE

M. Krishnaiah[1], L. Ramamurthy[1], M. Manohar[2], S. S. Krishnamurthy[2] and T. S. Cameron[3]

[1]Department of Physics, Sri Venkateswara University, Tirupati 517 502, India. [2]Department of Inorganic and Physical Chemistry, Indian Institute of Science, Bangalore-560 012, India. [3]Department of Chemistry, Dalhousie University, Halifax, N.S. Canada

Structural investigations of cyclophosphozene derivatives have become increasingly important in connection with the effect of the substituents on the conformation of the cyclophosphazene ring and the nature of bonds in the ring as well as those involving the exocyclic substituents. The crystal structure of the title compound has been elucidated. The hexakis (phenylthio) compound, $N_3P_3(SPh)_6$ crystallizes in the orthorhombic space group $P2_12_12$ with $a = 19.392(2)$, $b = 7.559(2)$ and $c = 12.685(2)$Å and $Z = 2$. The structure was solved by direct methods using 969 visually measured intensities from photographic data and subsequently refined by least squares with diffractometer data using anisotropic temperature factors to R-value of 0.029 for 1 540 observed reflections.

The molecule lies on a two-fold axis of symmetry. The phosphazene ring is planar. The structural features will be discussed and compared with those of $N_3P_3Cl_4(SPh)_2$ and $N_3P_3(OPh)_6$ solved earlier.

PHASE EXTENSION AND REFINEMENT AT LOW RESOLUTION IN MACROMOLECULAR CRYSTALLOGRAPHY

P. M. Alzari[1], J. Navaza[2], A. D. Podjarny[3], and R.J. Poljak[1]

[1]Institut Pasteur, Immunologie structurale, 25 rue du Dr Roux, 75724 Paris Cedex 15, France. [2]Faculté de Pharmacie, 92290 Chatenay-Malbry, France. [3] IBMC, 15 rue Descartes, 67084 Strasbourg, France.

The determination of the maximum entropy estimate of the electron density function using only the information of positivity and structure factor moduli is a formidable task and leads to very poor results. In fact, it was shown that the true phases are not even placed in a concave region of the entropy functional.

However, encouraging results were obtained in test cases when the upper bound of the acceptable maps was artificially reduced. The formalism used (Navaza, J. 1985, *Acta Cryst.*, **A41** pp. 232) which allows for solvent flattening and a variable upper bound in the protein region, gave rise to a procedure that was successfully applied in a macromolecular structure at low resolution (Navaza, J., Podjarny A. D. and Moras, D., 1986, ACA Meeting, Hamilton, Canada).

As the quality of the results is certainly resolution dependent, we analyze here applications of the method to some Fab structures for phase refinement and extension in different resolution ranges.

RECENT DEVELOPMENTS IN MITHRIL

C. J. Gilmore and S. R. Brown

Department of Chemistry, University of Glascow, Glascow G12 8QQ, Scotland.

We report here some recent extensions to the MITHRIL direct methods computer program:

(1) The use of weighting schemes to mimic the effect of errors in E-magnitudes measured at high $\sin\theta/\lambda$ values. The effect of such schemes is to downweight or remove triplets which involve more than one high-angle reflection. Such a process can be rationalised on the basis that such reflections have large standard deviations and hence can give rise to triplets which are seriously in error. We are currently exploring methods of exploiting the true standard deviations of the E's.

(2) The use of colour graphics to investigate E–maps which arise from phase sets that are seriously in error. Such maps are not readily accessible by simply selecting the peaks and interpreting them using chemical knowledge, but often useful information can be extracted by using the human eye to examine contoured map sections — a fact well-known to macromolecular crystallographers! A set of graphics routines have been written to allow easy manipulation of electron density maps.

(3) A symbolic addition program (based on LSAM) for centrosymmetric structures designed to speed up calculations in such situations.

(4) An experimental maximum–entropy interface in collaboration with Dr. G. Bricogne. In this option tangent formula expansion and refinement is replaced either wholly or in part by the application of the maximum extropy formalism using the exponential modelling method.

These developments have been very successful with both hitherto unsolved structures (up to 140 atoms in the asymmetric unit) and tests involving the Sheldrick data base of 'difficult' crystal structures.

MOLDRAW: A PROGRAM FOR THE GRAPHICAL MANIPULATION OF MOLECULES ON PERSONAL COMPUTERS

G. Borzani, P. Ugliengo and D. Viterbo

Istituto di Chimica Fisica, Via P. Giuria 7, 10125 Torino, Italy.

MOLDRAW is a new program for the graphic representation of molecules. It is written in BASIC (Microsoft Quick Basic Compiler) and runs on IBM and compatible PC's equipped with standard graphic features (CGA board). It can also use advanced graphic solutions with 16 colours at the same time and a resolution of 640 x 350 pixels as obtained with the EGA board.

The name of the molecule, the atom coordinates with cell parameters or the internal coordinates, using the standard z-matrix representation and other relevant data are input to the program through a free format pre-prepared file. All relevant results are also saved on a log file.

The program allows the following operations:

- selection of coordinate file (xxx.MOL);

- stick plot of the molecule;

- stick and ball plot of the molecule;

- full perspective view of the molecule as in PLUTO;

- labelling of selected or of all atoms;

- zoom by a selected factor;

- rotation around x,y,z axes either in steps or continuously;

- rotation around a given bond;

- translation of the molecule in the x-y plane;

- projection of the molecule on a plane defined by any 3 atoms;

- calculation of selected interatomic distances, bond angles and torsion angles;

- automatic display of short van der Waals contacts;

- calculation of the molecular moments of inertia;

- cutting of molecular fragments from a molecule;

- calculation of electrostatic and/or non-bonded energy;

- rotation of a molecular fragment around a given bond and plot of electrostatic and/or non-bonded energy as a function of the rotation angle;

- generation and graphic display of the unit cell content given the space group symbol;

- hardcopy facility on IBM standard graphic printer and fully compatible devices.

MOLDRAW is a very convenient tool not only for crystallographers but also for preparing the input and analysing the output from quantum mechanical and molecular mechanics calculations.

THEORETICAL CALCULATION OF RELATIVE AFFINITIES IN PROTEIN–SUBSTRATE BINDING

H. Luecke and F. A. Quiocho

Howard Hughes Medical Institute, Baylor College of Medicine, Houston, Texas, U.S.A.

Our laboratory has been engaged in the determination of the three–dimensional structures of several binding proteins from Gram–negative bacteria. These proteins, which are located in the periplasmic space, serve as essential components of osmotic shock–sensitive active transport systems for a large variety of carbohydrates, amino acids and ions. Several of the sugar-binding proteins also act as initial receptors for chemotaxis. With the highly refined structure of the L-arabinose-binding protein (ABP) and experimental values of free energy changes of binding for L-arabinose, D-fucose, D-galactose, and various deoxy-D-galactoses, we compute the difference in the relative free energy change for binding of different substrates using the thermodynamic cycle-perturbation method. This technique provides quantitative information about the contribution of individual hydroxyl-protein interactions to the overall binding energy of the sugar substrate, which can be compared with values obtained experimentally.

The thermodynamic cycle of interest is

$$
\begin{array}{ccc}
ABP : S_1 & \rightarrow & ABP : S_2 \\
\uparrow & & \uparrow \\
ABP + S_1 & \rightarrow & ABP + S_2
\end{array}
$$

with

$$ABP + S_1 \;\rightarrow\; ABP : S_1 \quad \Delta A_1 \tag{1}$$

$$ABP + S_2 \;\rightarrow\; ABP : S_2 \quad \Delta A_2 \tag{2}$$

$$ABP + S_1 \;\rightarrow\; ABP + S_2 \quad \Delta A_3 \tag{3}$$

$$ABP : S_1 \;\rightarrow\; ABP : S_2 \quad \Delta A_4 \tag{4}$$

where (3) and (4) denote hypothetical reactions. The relative free energy difference between binding S_1 and S_2 is

$$\Delta\Delta A = \Delta A_1 - \Delta A_2 \qquad (5)$$

or, hence A, the Helmholtz free energy, is a thermodynamical state function,

$$\Delta\Delta A = \Delta A_3 - \Delta A_4 \qquad (6)$$

where ΔA_3 refers to the relative free energy of hydration for molecules S_1 and S_2. The perturbation technique is used to compute ΔA_3 and ΔA_4. First, potential energy functions V_1 for the $ABP/S_1/$solvent system and V_2 for the $ABP/S_2/$solvent system are defined. Next, one defines a 'mixed' potential energy function V_λ such that

$$V_\lambda = \lambda V_2 + (1 - \lambda)V_1 \qquad (7)$$

The free energy is then computed by performing stepwise perturbation of the $ABP/S_1/$solvent system to the $ABP/S_2/$solvent system by changing the potential energy function parameters (atomic radii and partial charges) of S_1 in discrete increments to those of S_2. For each step, λ_i, a dynamical simulation yields the free energy for perturbation parameter values λ about simulation step values λ_i

$$A_3(\lambda) - A_3(\lambda_i) = -kT \ln < exp(-(V_\lambda - V_{\lambda_i})/kT >_{\lambda_i} \qquad (8)$$

where $<>_{\lambda_i}$ is a canonical ensemble simulation average for V_{λ_i}. The simulation stepsize, $\lambda_{i+1} - \lambda_i$, is chosen to assure adequate statistics in the region of perturbation parameter λ where the results $A_3(\lambda)$ from simulation i overlap those from $i + 1$, and $\Delta A_3 = A_3(\lambda = 1) - A_3(\lambda = 0)$ is obtained by piecing together the individual simulation steps. The potential energy $V_1(V_{\lambda=0})$ is computed for a large number of system configurations to obtain a statistically significant canonical simulation average (8). Then the 'mixed' potential energy, V_λ, is computed for each configuration saved from the simulation for discrete values of the perturbation parameter λ.

The computations were carried out on a Cray X-MP using molecular dynamic algorithms of the program system AMBER with a dynamic stepsize of 2 *fs*. The L-arabinose-binding protein, the primary ligand and 182 highly ordered waters from the X-ray structure consist of 3 406 atoms. In order to simulate substrate binding in solution, the protein-substrate complex has been immersed in a box of nearly 2 200 water molecules obtained from a previous liquid state simulation, raising the total number of atoms in the system to over 10 000.

The first calculation for the perturbation of D-galactose (S1) to 1-deoxy-D-galactose (S2) yielded 19.77 kcal/mol for ΔA_3 and 24.11 kcal/mol for ΔA_4, or a $\Delta\Delta A$ of -4.34 kcal/mol. This value differs only 3% from the experimental value of -4.21 kcal/mol.

THE COLLECTION, ANALYSIS AND REFINEMENT OF DATA FROM TWINNED CRYSTALS

A. David Rae

School of Chemistry, University of New South Wales, Kensington, N.S.W. 2033, Australia.

A systematic approach to the elucidation of twinned crystal structures includes subsections on the evaluation of 4–circle diffractometer parameters $\Delta\omega$, $\Delta\chi$, $\Delta\theta$ to describe relative positions of partially overlapped reflections and the use of a refineable Window function to describe this overlap; data collection strategies; programming considerations and the creation of difference maps on an absolute scale; pseudosymmetry, including the twinning operation as a pseudosymmetry element; implications for atom refinement and error estimation; solution of twinned pseudosymmetric crystals, including selection of spacegroup and origin; symmetry selected constraints for least squares refinement; diffraction symmetry enhancement without twinning; problems with twin overlapped systematic absences. A group theory approach is developed and is used to analyse the refinement of two 1:1 twin structures that crystallise in spacegroups that arise from the ordering of disordered structures in $P4_2/\text{mnm}$ obtained using only a subset of the observed data.

INDEX